# Fundamentals of Soil Science

# Fundamentals of Soil Science

**FIFTH EDITION**

**H. D. Foth**
Professor of Soil Science
Michigan State University

**L. M. Turk**
Late Profressor Emeritus of Soil Science and
Former Director of Agricultural Station
Michigan State University

**JOHN WILEY & SONS, INC.**

New York • London • Sydney • Toronto

***Library of Congress Cataloging in Publication Data***

Foth, H. D.
Fundamentals of soil science.

First-2d ed., by C. E. Millar and L. M. Turk, published in 1943 and 1951; 3d-4th ed., by C. E. Millar, L. M. Turk, and H. D. Foth, published in 1958 and 1965, respectively.
Includes bibliographical references.

1. Soil science. I. Turk, Lloyd Mildon, 1906-1971, joint author. II. Millar, Charles Ernest, 1885- Fundamentals of soil science. III. Title.

S591.F67 1972 631.4 78-38846
ISBN 0-471-26790-2

Printed in the United States of America.

10 9 8 7 6 5 4 3 2

# Preface

The rapidly growing recognition of the importance of Soil Science to environmental quality has created a greater awareness of and interest in soil science. This is reflected in the increasing enrollment of non-agriculture students in soil courses and the development of many environmentally related courses. This edition, the fifth, shows evidence of these changes by including discussions of nitrate pollution, and pesticide degradation, as well as inclusion of illustrations of the importance of soils in civil engineering, landscaping, geography, geology, biology, history, and social science. As I write this preface while on leave at the University of Hawaii, I am very aware of the great competing demands between agriculture and urbanization for the limited amount of land on an island. Only yesterday, I saw houses that had slid down a slope and cracked apart because they had been built on steep slopes where the soils contained expanding and contracting clays. By concerning itself with these recent important and growing events and interests, this edition should prove to be more interesting to the student.

Another important change taking place on college campuses today is the increasing recognition of the difference between learning and teaching. Learning focuses on students. Having a real concern for learning, I saw the need to spend many hours to develop complex concepts in a more logical and lucid manner. Furthermore many models were built and photographed to serve as illustrations. Many other photographs were taken specifically to serve the needs of this book. Therefore, the time you spend reading and studying this book should be worthwhile and rewarding.

Finally, I am indebted to many persons. It is difficult to name them all, but most important have been my colleagues in Soil Science at Mighican State University. Time after time their counsel was sought for knowledge and ideas. To my colleagues, I am very grateful, and although I take full responsibility for this book, the book is in a sense a collective product of all of us. Particular recognition for help with tropical soils goes to my colleagues at the University of Hawaii where I spent the summer of 1971. Thanks also to many persons for photographs and maps.

I'm sorry that my friend and coauthor, L. M. Turk, passed away shortly before this fifth edition was printed.

**Henry D. Foth**

# Contents

| | | |
|---|---|---|
| **Chapter 1** | Concepts of Soil | 1 |
| | General Definitions of Soil | 1 |
| | Soil as a Natural Body | 3 |
| | Soil as a Natural Resource | 8 |
| | Soil as the Interface Where We Live | 12 |
| **Chapter 2** | Soil as a Medium for Plant Growth | 13 |
| | Factors of Plant Growth | 13 |
| | Utilization of the Soil by Plants | 18 |
| | The Concept of Soil Productivity | 22 |
| **Chapter 3** | Physical Properties of Soils | 27 |
| | Soil Texture | 27 |
| | A Textural Grouping of Soils | 30 |
| | The Texture Profile | 35 |
| | Soil Structure | 38 |
| | Weight, Pore Space, and Air Relationships | 45 |

| | | |
|---|---|---|
| | Tillage and Soil Properties | 53 |
| | Soil Color | 58 |
| | Soil Temperature | 60 |
| **Chapter 4** | Soil Water | 63 |
| | Energy Concept of Soil Water | 64 |
| | Plant-Soil Water Relations | 76 |
| | Water Requirement of Crops | 84 |
| | Soil Drainage | 88 |
| **Chapter 5** | Soil Organisms | 97 |
| | Kinds and Nature of the Soil Flora | 98 |
| | Influence of Soil Conditions on the Microflora | 103 |
| | Important Roles of Soil Microflora | 105 |
| | Soil Microflora and Environmental Quality | 114 |
| | Soil Animals | 118 |
| **Chapter 6** | Soil Organic Matter | 127 |
| | Humus Formation and Characteristics | 127 |
| | Amount and Distribution of Organic Matter in Soils | 131 |
| | Some Organic Matter Management Considerations | 139 |
| **Chapter 7** | Chemical and Mineralogical Properties of Soils | 149 |
| | Chemical and Mineralogical Composition of the Earth's Crust | 150 |
| | Weathering and Mineralogical Composition of Soils | 152 |
| | Origin, Structure, and Properties of Clay Minerals | 158 |
| | Cation and Anion Exchange in Soils | 169 |

Chapter 8 Soil pH—Causes, Significance, and Alteration 179

Definition and Causes of Soil pH 179
Significance of Soil pH 184
Alteration of Soil pH 189

Chapter 9 Soil Genesis and the Soil Survey 203

Horizon Differentiation and Soil Genesis 203
Soil Development in Relation to Time 208
Soil Development in Relation to Climate 212
Soil Development in Relation to Vegetation 215
Soil Development in Relation to Parent Material 218
Soil Development in Relation to Topography 229
The Soil Survey 231

Chapter 10 Classification and Geography of the World's Soils 237

Soil Classification 237
Soil Orders—Properties, Distribution and Use 242
Soil Classification Categories Below the Order 268

Chapter 11 Nutrient Requirement and Mineral Nutrition of Plants 273

Elements Used by Plants 273
Nutrient Uptake 275
Factors Affecting Plant Composition 279
Functions and Effects of the Essential Elements 286
Nutrient Deficiency Symptoms 290
Situations Where Soil-Nutrient Deficiencies Are Likely to Exist 294

**Chapter 12** Composition, Manufacture, and Use of Fertilizers 299

Fertilizer Materials 300
Mixed Fertilizers 307
Evaluation of Fertilizer Needs 313
Use and Application of Fertilizers 317
Fertilizer Use and Environmental Quality 325

**Chapter 13** Animal Manures and Wastes 329

Production and Composition of Manure 329
Losses Occurring in Manure Handling 334
Management of Manure for Crop Production 336
Use of Soil for Animal Waste Disposal 341

**Chapter 14** Soil Erosion and Its Control 345

Soil Erosion Defined 345
Reasons for Employing Erosion Control Practices 346
Controlling Erosion Caused by Water 349
Controlling Erosion Caused by Wind 366

**Chapter 15** Soils and Agriculture of Arid Regions 375

Characteristics and Utilization of Soil in Arid Regions 376
Development and Management of Saline and Sodic Soils 383

**Chapter 16** Irrigation 395

Water Supply and Land for Irrigation 395
Irrigation Practices 400

**Chapter 17** Soil Resources and Population 413

Acreages of Producing and Potential Cropland in the United States 413
Acreages of Arable Land and Land Requirements 418
The Food and Population Problem of the World 420
**Glossary** 427
**Index** 443

# Fundamentals of Soil Science

# 1

# Concepts of Soil

The formulation of a brief definition of soil is made difficult by the great diversity of soils in the world. The goal of this book is not to arrive at a carefully developed definition of the word "soil" or to provide definitive answers to specific questions, but to enable the reader to develop the basis required for intelligent thinking about soils. This chapter is devoted to the development of several important general concepts which will be greatly expanded throughout the remaining chapters.

## GENERAL DEFINITIONS OF SOIL

Many words that we use every day have several meanings and may be used in various ways. The word *soil* is no exception. As a transitive verb it means "to make dirty" as in the case of soiled dishes or clothing. The noun soil is derived through Old French from the Latin *solum,* which means floor or ground.

### Commonplace Definitions of Soil

In general, soil refers to the loose surface of the earth as distinguished from solid rock. Many people when they think of the word soil, have in mind that material which nourishes and supports growing plants. This meaning is even more general, since it includes not only soil in the common sense, but also rocks, water, snow, and even air—all of which are capable of supporting plant life. The farmer, of course, has a more practical conception of soil; to him it is the medium in which crops grow. The civil engineer, on the other hand, looks upon soil as that material which supports foundations, roads, or airport runways (Fig. 1–1). In short, the word soil has many meanings and we must understand that the word can and will be used in various ways throughout the book.

## The Four Components of Soil

Soil may also be defined as a mixture of mineral matter, organic matter, water, and air. The volume occupied by each of these in a "desirable" lawn or garden surface soil that is in ideal condition for plant growth will be approximately as follows: mineral matter, 45 percent; organic matter, 5 percent; water, 25 percent; and air, 25 percent. It is interesting to note that about half the volume is pore space.

Proportions of the components vary from time to time and from place to place. The volume of water and air bear a direct reciprocal relationship with each other. Entrance of water into the soil excludes air. As water is removed, by drainage, evaporation, or plant growth, pore space that was occupied by water becomes filled with air once more. Subsoils are generally characterized by considerably less or-

**Fig. 1–1** The unstable roadbed of this city street resulted from a layer of muck (organic soil). The muck layer is being excavated and replaced by sand to provide support for heavy vehicles.

ganic matter than surface soils. An organic soil, like a muck or peat, has a greater volume occupied by organic matter than by mineral matter.

### Manufactured Soils

Soils used in greenhouses are manufactured in the sense that topsoil, sand, and organic matter are mixed together to provide a desirable proportion of the four components.

Soil as ordinarily found in the field is not suited for golf greens. When the soil surface is wet, traffic causes soil compaction, and the maintenance of favorable air and water relationships is difficult. In the construction of a golf course, the sites where the greens will be located are excavated and refilled with a base layer high in sand and gravel. This is overlain by a layer which is commonly 12 inches thick and composed of a mixture of sand, topsoil, and organic matter (peat). Figure 1–2 shows an experiment in which various materials and mixtures are being tested to gain more knowledge concerning the best kind of soil to use for a golf green.

## SOIL AS A NATURAL BODY

Each soil occupies space. It has areal extent and depth. Each soil is a natural body which is surrounded by other soils with different properties. The area of an individual soil may vary from less than an

**Fig. 1–2** An experimental area devoted to the testing of various mixtures for golf green use. Plots are 3 by 6 feet in size, are underlain with sand and gravel, and then covered with a layer containing various mixtures of sand, topsoil, and peat. Plots in the foreground are just being established, whereas those in the rear were established the previous year. Note the poor growth on some of these plots.

acre to more than a hundred acres. This concept was developed simultaneously in Russia by Dokuchaev and in America by Hilgard about 1870. Until this time, soil was thought of largely in terms of a mixture of organic and mineral matter as just discussed.

### Boundary Limits of a Soil

The surface of the earth is the soil's upper limit. The lower boundary is defined by the depth to which soil weathering has been effective or by the depth of root penetration or both. Generally the lower boundary is not sharply defined.

Lateral changes from one soil to another are commonly associated with changes in slope. Typically, a transition zone exists between two adjacent soils in which properties of both soils can be observed. A sharper than usual boundary exists between the two soils shown in Fig. 1–3 because the change in slope is abrupt. Differences in parent material or vegetation are other factors responsible for lateral changes in soils.

Each soil can be viewed as a section of a landscape in the sense of the two soils shown in Fig. 1–3. The landscape as a whole can be viewed as being comprised of many different soil bodies, each contributing to the whole as a piece of a jig-saw puzzle contributes to the overall pattern of the puzzle. The various soils on the landscape are the product of evolution—the topic to be considered next.

**Fig. 1–3** The lateral boundary between the organic soil in the foreground and the mineral soil on the slope is clearly visible.

**Fig. 1–4** An A/R soil. The A horizon is about one foot thick and has formed in parent material produced by the direct weathering of bedrock. Soils like this may develop from bedrock in as little as 100 years or as long as 100,000 years or more, depending on the hardness of the rock and the environmental conditions.

## Evolutionary Nature of Soil

The soils on the earth's surface are undergoing continual change, which escapes a casual study of the soil. Each soil has a life cycle in terms of geologic time. This dynamic and evolutionary nature is embodied in a definition of soil as a natural body of the earth's surface having "properties due to the integrated effect of *climate*, and *living matter* (plants and animals), acting upon *parent material*, as conditioned by *relief* (slope), over periods of *time*."[1]

Weathering of bedrock produces unconsolidated debris that serves as the *parent material* for the evolution of soil profiles that eventually reflect the integrated effect of climate, living matter, relief, and time. Exposure of parent material to the weather, under favorable conditions, will result in the establishment of photosynthesizing plants. Their growth results in the accumulation of some organic residues.

[1] *Soil*, USDA Yearbook, page 767, 1957.

Animals, bacteria, and fungi join the biological community and feed on these organic remains. Breakdown of organic residues sets free the nutrients contained therein for another plant growth cycle. The plants and animals feeding on the organic debris become a part of the total organic matter complex. When the surface layer of the soil attains a reasonable thickness and assumes a darkened color because of the accumulation of organic matter an *A horizon* comes into existence. Such a soil has only two horizons, as seen in Fig. 1–4.

The soil in Fig. 1–4 may eventually become many feet thick if the rate of soil removal by erosion is less than the rate of rock weathering. Erosion, however, is a fundamental process on the earth's surface and many soils form in sediments (a product of erosion) instead of from the direct weathering of bedrock. Where soil evolution occurs in sediments, horizon evolution may proceed rapidly by comparison. Pore spaces in sediments permit deep rooting by plants and facilitate the removal of soluble compounds by percolating water. Suspended

**Fig. 1–5** A soil profile with the three major horizons, A, B, and C. The soil profile has developed in a sediment consisting of material that weathered from bedrock at some other site and deposited here.

colloidal-sized particles may also be translocated by percolating water; however, the suspended colloidal particles tend to move only a few feet before they become lodged or precipitated. The result is the formation of a zone under the A horizon where colloidal particles accumulate. This zone is designated a *B horizon*. The most common colloids accumulating in B horizons are clay, organic matter, and oxides of iron and aluminum. A soil with a B horizon is shown in Fig. 1–5.

Soils that develop under grass typically have thick dark colored A horizons resulting from the profuse growth of roots throughout the A horizon. In the forest the addition of organic matter occurs largely from leaves and wood. The addition of leaves and wood on the top of the soil promotes the development of a thin, dark-colored layer enriched with organic matter. The leaching of material out of the A horizons of forest soils in humid regions causes the lower part of the A horizon to become "bleached" or light colored. In these cases the A horizon is subdivided into an A1 and A2, as illustrated in Fig. 1–6.

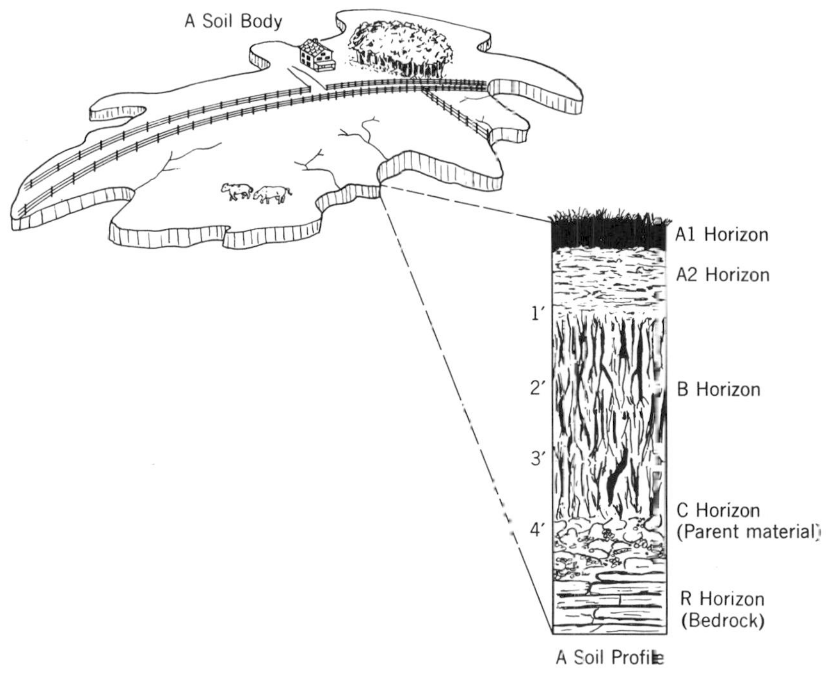


**Fig. 1–6** The A1, A2, B, and C horizon sequence is typical for soils developed under forest vegetation. Also shown is the relationship of soil horizons and profile to the soil body. (Courtesy of Dr. F. D Hole, Soil Survey Div., Wis. Geol. Nat. Hist. Survey, University of Wisconsin.)

The relationship of the soil body to the soil profile is also shown in Fig. 1–6.

The concept that present-day soils are evolving into soils which will have different properties is well documented and is one of the most useful concepts in soil science today. To a limited extent, however, some soils appear to have reached an end point in their development and any additional significant change is hard to imagine. Theoretically, soil development proceeds to the point where the soil becomes incapable of supporting vegetation if left undisturbed for a sufficiently long period of time in a humid environment. The release of nutrients from minerals by weathering and their removal by water may be nearly completed. The soil in Fig. 1–7, Nipe clay, is representative of soils that are essentially "weathered out." The Nipe soil contains about 60 percent iron oxide and this accounts for the ability of the soil to retain the vertical exposed faces. The original rock that served as parent material contained 6 percent magnesium oxide whereas the soil contains only a trace. Extensive areas of Nipe-like soils on the earth's surface are not common, relatively speaking, because erosion, deposition, volcanism, and other geologic forces are so active.

## SOIL AS A NATURAL RESOURCE

Changes in the kinds of use and intensity of soil utilization are a part of the story of civilization. A brief review of some of these changes will help to clarify the present and future emphasis that is likely to arise regarding the use of soil in our society.

### Land Use in a Food-Gathering Society

In a food-gathering society the land supports plants and animals that man gathers for his food, clothing, and shelter. This frequently results in a nomadic life because man follows animal migrations or migrates with the seasons to obtain sufficient food. The land is used as it is, usually having low productivity and capable of supporting a small human population. It has been estimated that two square miles or 1280 acres of "good" land is required to support one person in a food gathering economy.[2]

### Land Use in an Agricultural Society

Man domesticated plants and animals and ushered in the agricultural revolution in the hills surrounding the fertile crescent in south-

[2] *Challenge of Man's Future*, H. Brown, The Viking Press, 1954.

**Fig. 1–7** A very old soil, the Nipe clay (Puerto Rico). It is largely an accumulation of iron oxide.

western Asia about 10,000 years ago. The change from food gathering to food production at first gave more certainty to the source of food than to the abundance of supply. It was not until about 4000 B.C. when irrigation had developed sufficiently on the floodplains of southern Mesopotamia that productivity of the soil was great enough to give rise to cities.[3]

[3] R. J. Braidwood, "The Agricultural Revolution," *Sci. Am.*, September 1960.

In the absence of the knowledge required to maintain soil fertility with cropping systems, animal manures, or chemical fertilizers, "productive" agriculture was necessarily restricted to the fertile soils of the arid regions that could be irrigated or to the alluvial soils subject to renewal of nutrients by periodic flooding. Later, the ingenuity of the oriental farmers enabled man to maintain soil fertility over extensive areas in the humid regions for a long period of time, as shown in the book entitled, *"Farmers of Forty Centuries."*[4] The Chinese farmers carefully utilized plant and animal wastes, including human wastes, grew legumes to fix atmospheric nitrogen, and returned canal mud to the land. When F. H. King visited several Chinese farms shortly after 1900, he found that in some instances a square mile (640 acres) supported 1783 people, 212 cattle or donkeys, and 399 swine. It is not uncommon in China today to find large areas of farmland supporting over 2000 persons per square mile or 3 persons per acre. Stated in another way, an area equal to many residental city lots (about $100' \times 150'$) supports one person. How adequate would you feel if you had to produce your food, and perhaps some fiber for clothing, on such a small parcel of land?

By contrast, in the humid tropics on very infertile soils, about 200 million farmers practice shifting cultivation and use about 50 acres or more to support one person. Under shifting cultivation some land is cleared of brush each year and some land is returned to brush each year. In brush the land is rejuvenated by the accumulation of nutrients in the top soil through the addition of organic matter in the form of tree roots, wood, and leaves. After about 10 to 20 years, the land is again cleared and used for a couple more years of cropping. Even though both the Chinese and shifting cutivation systems use very little technology, there is an enormous difference in the intensity of land use. Most agricultural societies today use the land with an intensity somewhere between these two extremes.

## Land Use in a Technological Society

Certainly the earliest of farmers recognized that soils had different capacities or abilities to produce a particular crop. With limited knowledge and limited ability to alter the soil, however, they had to use the soil in large part as it was. Today, in an industrial society, many additional alternatives exist. Crops may be grown in a given location because the climate is ideal, markets are close, or water is available for irrigation instead of because "good soil" exists. For

[4] Farmers of Forty Centuries, F. H. King, Madison, Wisconsin, 1911.

**Fig. 1–8** The apparently uninterrupted glass roof illustrates one of the most intensive uses of land for market gardening in the world. One-tenth of the world's market gardening under glass occurs in this district located between the Hague and Rotterdam in the Netherlands. (Photo courtesy Ministry of Agr. and Fisheries, Netherlands.)

example, in Florida, naturally infertile sand soils are used for citrus, and organic soils for vegetable production. The year round growing season and the location near centers of population make it feasible to greatly alter the existing soil so these crops can be grown successfully. Many acres of land have been encased with glass (greenhouses) in the Netherlands to intensify land use (see Fig. 1–8). Thus, some of the most productive soils in society owe their usefulness to man's ability to alter their characteristics instead of to their original fertility or nature.

The emphasis of modern soil science is on changing soil to better serve the needs of man. The question we commonly ask is, "How much input must be used on a given soil to produce a given quantity of product?"[5] This is comparable to the question, "How much will it cost to produce a million square feet of office space on an acre?"

When the questions like the one just raised are pursued deeply in a technological society, one finds fantastic production possibilities.

[5] "The Changing Model of Soil," M. G. Cline, *Soil Sci. Soc. Am. Proc.*, **25**:445, 1962.

In fact, there is evidence that with current technology, a person's food can be produced on an area smaller than that needed to "live" in. However, many persons are becoming quite concerned about the impact of technology on the quality of the environment.

## SOIL AS THE INTERFACE WHERE WE LIVE

Another concept relevant to soil utilization is that of the soil as the interface between the atmosphere and lithosphere. We live on the soil, this is our home. Besides being a basic resource for food production, the soil collects and purifies water, and disposes of wastes. Soil itself can be a pollutant as dust in the air and as sediment in waters. For the future a new concept of soil utilization will need to be forged —a concept that considers the impact of soil utilization on all aspects of man's life, including the quality of the environment. We may find new meanings in the words of one of America's greatest conservationists, Aldo Leopold,

"A thing is right only when it tends to preserve the integrity, stability, and beauty of the community, and the community includes the soil, water, fauna, and flora, as well as the people."

The chapters ahead contain information and ideas that, hopefully, will enable you to formulate some productive concepts about the nature of soil and its importance in your life.

# 2

# Soil as a Medium for Plant Growth

Soil influences our livelihood in many ways. It is important for lawns, as a foundation material for engineering structures, for sewage disposal, and for recreation. Some highly weathered soils of the humid tropics are rich in iron or aluminum and are mined as ore. Since the agricultural revolution began, however, man's primary interest in soil has centered on its potential to support plants which provide food, fiber and forest products. The emphasis in this chapter, is on soil as a medium for plant growth.

## FACTORS OF PLANT GROWTH

Basically the growth of continental plants depends on the soil for water and nutrient elements. Beyond this, the soil must provide an environment in which roots can function. This requires pore spaces for root extension. Oxygen must be available for root respiration and the carbon dioxide produced must diffuse out of the soil, rather than accumulate. An absence of inhibitory factors, such as a toxic concentration of soluble salts, is essential. Roots anchored in the soil also hold the plant erect. A summary of plant growth factors is given in Fig. 2–1 and a discussion of these factors follows

### Support

One of the most obvious functions of the soil is to provide support for the plant. Roots anchored in soil enable growing plants to remain

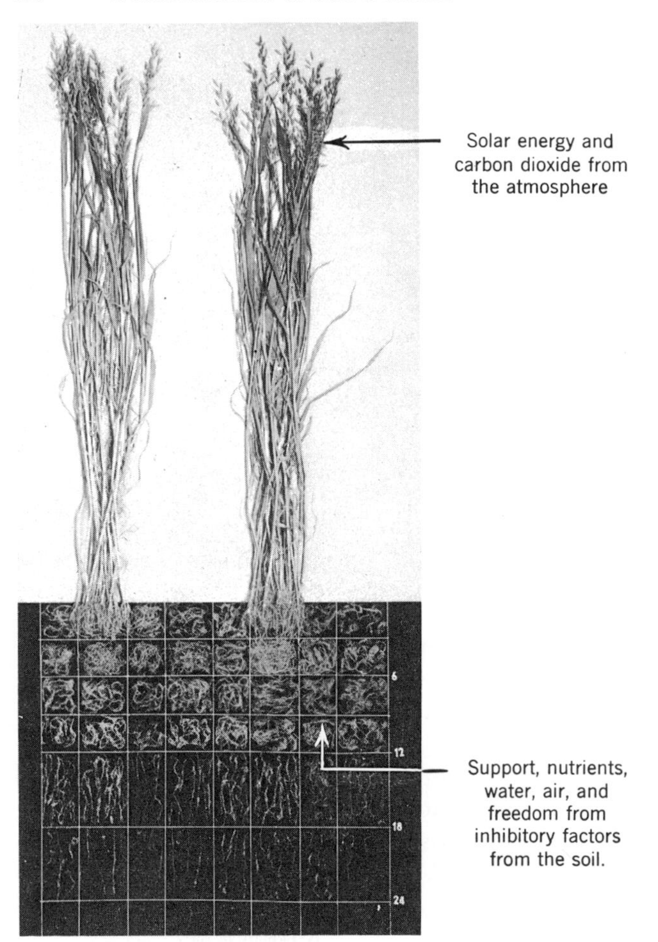


**Fig. 2–1** A summary of plant growth factors. (The background plants are oats.)

upright. Plants grown by hydroponics are commonly supported by a wire network. There are soils in which the impermeability of the subsoil or B horizon or the presence of a water table close to the surface of the soil often induces shallow rooting. Shallow rooted trees are easily blown over by the wind in a phenomenon called *windthrow.*

## Essential Nutrient Elements

At least 16 elements are currently considered necessary for plant growth. The carbon, hydrogen, and oxygen, combined in the photosynthetic reactions, are obtained from air and water. They comprise 90 percent or more of the dry matter. The remaining 13 elements are obtained largely from the soil. Nitrogen, phosphorus, potassium, calcium, magnesium, and sulfur are required in rather large quantities and are referred to as the *macro or major* elements. Nutrients required in considerably smaller quantities are called the *micro or trace* elements and include manganese, iron, boron, zinc, copper, molybdenum, and chlorine.

More than 40 additional elements have been found in plants. Some plants accumulate elements that are not essential but have a beneficial effect. The absorption of sodium by celery is an example, and, results, in this case, in an improvement in flavor.

Most of the nutrients exist in mineral and organic matter and as such are insoluble or *unavailable* to plants. Nutrients become *available* through mineral weathering and organic matter decomposition. It is a rare soil, indeed, which is capable of supplying all of the essential elements for long periods of time in quantities needed to produce high yields.

The ratio of nutrients is also important. An excess of one nutrient in available form may cause a deficiency of another element. This is illustrated by the fact that potassium deficiencies are sometimes caused by a high level of soluble calcium or magnesium which interferes with the absorption of the potassium. (For another example see Fig. 12–9.) It is interesting that all plants do not react in the same manner under the same conditions and while a deficiency of potassium will occur in one crop, there will be no harmful effect on some other crop.

The nutrients are absorbed from the soil solution or from colloid surfaces as cations and anions. Cations are positively charged; anions are negatively charged. Thirteen essential elements, their chemical symbols, and the forms in which they are commonly absorbed by plant roots are given in Table 2–1.

## Water Needs of Plants

It requires about 500 pounds of water to produce one pound of dry plant material. About 5 pounds or 1 percent of this water becomes an integral part of the plant. The remainder is lost through the stoma of the leaves during the course of carbon dioxide absorption. Atmospheric conditions such as relative humidity and temperature play a major role in determining how quickly water is lost and consequently in determining the amount of water plants require.

**Table 2–1** Chemical Symbols and Common Ionic Forms of the Essential Elements Absorbed by Plant Roots from Soils

| Nutrient | Chemical Symbol | Ionic Forms Commonly Absorbed by Plants |
|---|---|---|
| Macronutrients | | |
| Nitrogen | N | $NO_3^-$, $NH_4^+$ |
| Phosphorus | P | $H_2PO_4^-$, $HPO_4^{2-}$ |
| Potassium | K | $K^+$ |
| Calcium | Ca | $Ca^{2+}$ |
| Magnesium | Mg | $Mg^{2+}$ |
| Sulfur | S | $SO_4^{2-}$ |
| Micronutrients | | |
| Manganese | Mn | $Mn^{2+}$ |
| Iron | Fe | $Fe^{2+}$ |
| Boron | B | $BO_3^{3-}$ |
| Zinc | Zn | $Zn^{2+}$ |
| Copper | Cu | $Cu^{2+}$ |
| Molybdenum | Mo | $MoO_4^{2-}$ |
| Chlorine | Cl | $Cl^-$ |

Since the growth of virtually all economic crop plants will be curtailed when a shortage of water occurs, even though it may be temporary and the plants are in no danger of dying, the ability of the soil to hold water against the force of gravity becomes very important unless rainfall or irrigation is frequent. The need for removal of excess water from soils is related to the need for oxygen and is discussed in the following paragraphs.

### Oxygen Requirements

Roots have openings called *lenticels* which permit gas exchange. Oxygen diffuses into the roots cells and is used for respiration whereas the carbon dioxide diffuses into the soil. Respiration releases energy which the plant needs for synthesis and translocation of organic compounds and the active accumulation of nutrient ions against a concentration gradient.

Some plants (rice, for example) grow in standing water because they have morphological structures that permit the internal diffusion of atmospheric oxygen down into the root tissues. Successful production of most plants in water culture requires aeration of the solution. Great differences exist between plants in their ability to tolerate low oxygen levels. Sensitive plants may be wilted or killed by saturating the soil

**Fig. 2–2** The soil in which these tomato plants were growing was saturated with water. The stopper at the bottom of the crock on the left was immediately removed and the excess water quickly drained away. The plant on the right became severely wilted within 24 hours by the saturation treatment.

with water for a day as shown in Fig. 2–2. The wilting is believed to result from a decrease in the permeability of the root cells to water, a result of a disturbance of metabolic processes due to an oxygen deficiency.

Aerobic micro-organisms, bacteria, actinomyces, and fungi utilize oxygen from the soil atmosphere and are primarily responsible for the conversion of nutrients in organic matter into soluble forms that plants can reuse.

## Freedom from Inhibitory Factors

A soil should provide an environment free of inhibiting factors such as extreme acidity or alkalinity, disease organisms, toxic substances, excess salts, or impenetrable layers (see Fig. 2–3).

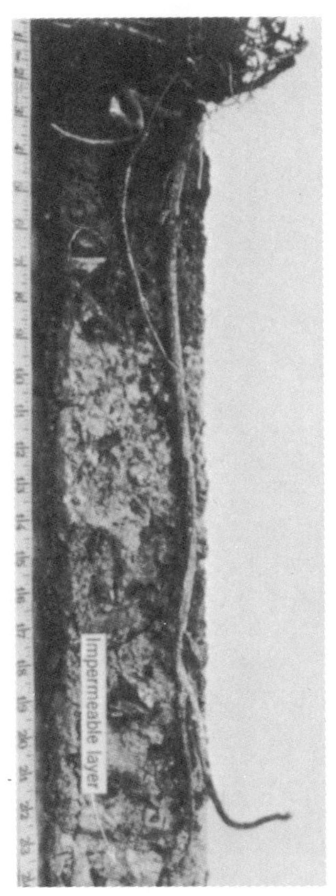


**Fig. 2–3** The marked change in the direction of the alfalfa taproot at a depth of 23 inches resulted from the impermeability of the underlying horizon.

## UTILIZATION OF THE SOIL BY PLANTS

The density and distribution of roots affect the plants efficiency in utilizing a soil. Perennials like oak or alfalfa do not reestablish a completely new root system each year; this gives them a distinct advantage over annuals like corn or cotton. This partly explains the difference in the water and nutrient needs of plants. Vegetable and cereal crops often respond to added phosphorus. Established apple trees retain from year to year, most of the phosophorus necessary for their life processes and rarely respond to phosphorus fertilizer. Let's take a look at the extensiveness of root systems and the extent to which the soil is in direct contact with root surfaces.

## Extensiveness of Root Systems

It is only reasonable to assume that there is as much difference in root systems as in the tops of plants. Root growth is influenced by environment; therefore root distribution and density are a function of both the kind of plant and nature of the root environment.

Root extension occurs by cell division and cell elongation directly behind the root cap. Actively growing roots have been known to increase several inches in length in 24 hours. It is not surprising that the lateral and vertical extension of the roots of some plants is great. By time time corn is "knee high," roots may have ramified soil midway between adjacent rows spaced 42 inches apart. By late summer they may extend more than 6 feet deep in well-drained, permeable soil. Alfalfa tap-roots commonly penetrate 6 to 10 feet and have been known to reach a depth of 20 feet or more. Cereal crops like oats have a moderately deep root system and effectively utilize the soil to a depth of 24 to 30 inches. Tree roots commonly extend 50 to 80 feet from the base of the tree.

The soybean root system shown in Fig. 2–4 was removed from a Conover loam on the University Farm at East Lansing, Michigan. Metal frames were used to collect the soil samples and running water was used to separate roots from soil. The roots appear to have been effective to a depth of about 3 feet. The roots extended 14 inches laterally to each side of the row, which indicates that the soil between the rows was effectively used.

Considerable uniformity in lateral distribution of roots through much of the soil (see Fig. 2–4) is explained on the basis of two factors. First, there is a random lateral distribution of large pore spaces, resulting from cracks or channels formed by roots of previous crops or earthworm activity. Second, as roots permeate soil they remove water and nutrients which makes such soil a less favorable location for additional root growth. Roots then preferentially grow in areas of the soil where roots have not yet grown. In this way the soil is permeated rather uniformly by soybean roots and by the roots of many other crops. (See Figs. 2–1 and 3–7 for the root distribution of oat plants.)

## Extent of Root and Soil Contact

A rye plant was grown in a cubic foot of soil for four months at the University of Iowa by Dittmer.[1] The root system was carefully re-

[1] "A Quantitative Study of the Roots and Root Hairs of a Winter Rye Plant," H. J. Dittmer, *Am. J. Botany*, **24**:417–420, 1937.

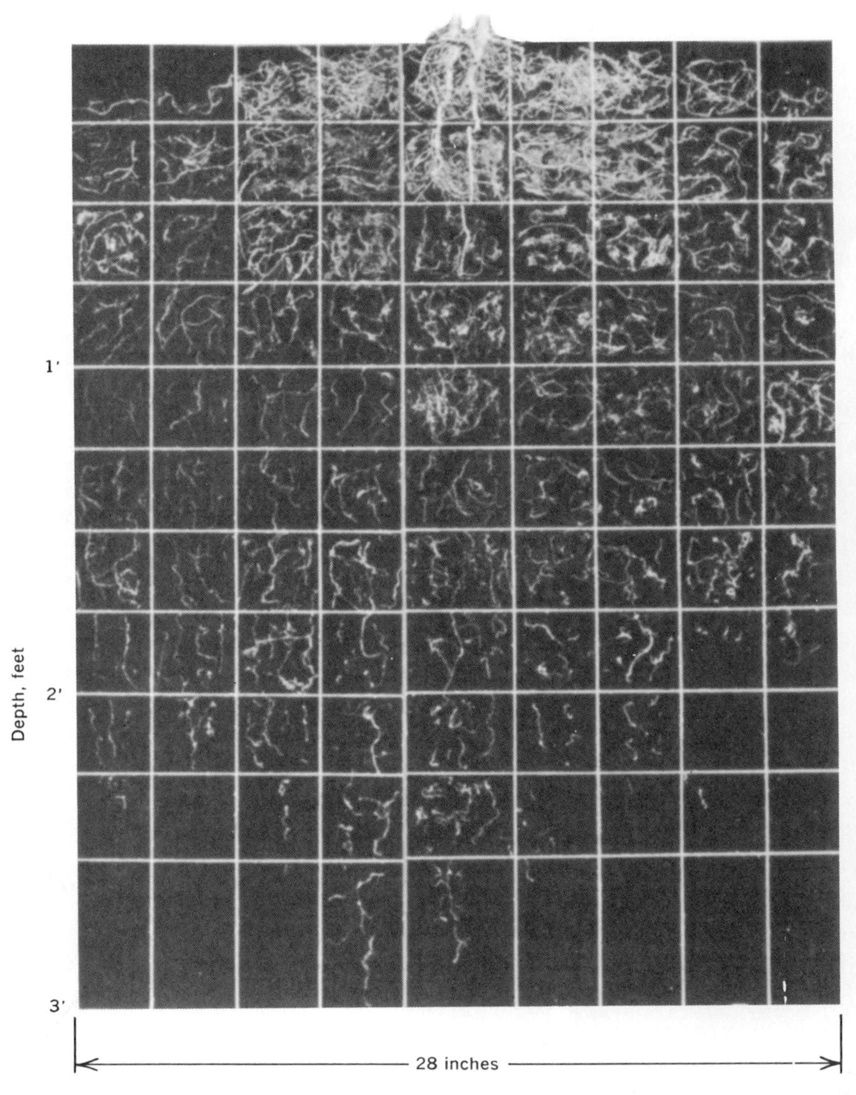


**Fig. 2–4** Root distribution of mature soybean plants grown in rows spaced 28 inches apart. They penetrated at least 3 feet and easily grew laterally to the centers of adjacent rows.

moved from the soil by using a stream of running water and the roots were counted and measured for size and length. It was determined that the plant had 385 miles of roots which had 2550 square feet of surface. The root hairs had a length of 6600 miles and 4320 square feet of surface. Because a loamy soil was used, at most only 1 percent of the total soil surface was estimated to be in direct contact with root surfaces. The average distance between roots, however, would be in the order of thousands of an inch (A cubic foot of soil would be cut almost 7000 times if root surfaces were spread in a plane. This would give $^{12}/_{7000} \approx 0.002$ inch between root surfaces). The movement of capillary water and the nutrients dissolved in the water, even over very small distances, enables plants to utilize the soil water and nutrients to a remarkable extent.

### Pattern of Soil Utilization by Plants

A seed is a dormant plant. Placed in moist soil where the temperature is favorable, it may absorb water by osmosis and swell. Enzymes then become activated and food[2] reserves in the endosperm move to the embryo and are utilized for growth. With the development of green leaves and the initiation of photosynthesis, the plant becomes independent of the seed for food. With extension of roots through the soil, the plant becomes independent of the seed for nutrient elements. The plant becomes totally dependent on the atmosphere and soil for its sustenance.

In a sense this is a critical period in the life of the plant because the root system is small. It may not be able to contact a sufficiently large amount of soil surfaces or volume of water to supply the phosphorus needs because phosphorus is very immobile in soils. Plants frequently show phosphorus deficiency symptoms at this time although the symptoms gradually disappear as root extension through the soil enables the plant to utilize an ever-increasing supply of phosphorus. The placement of soluble phosphorus fertilizer in the pathway of young roots has become a common practice in the production of many annual crops. It becomes apparent that the more mobile a nutrient is in the soil water, the more readily it will move to the root and be taken up by the plant. The limited distance that capillary water and dissolved nutrients can readily move requires continual ramification of the soil by the roots in order to sustain the growth of an annual plant.

[2] Plant food consists of compounds like carbohydrate, protein, or fat which contain reduced carbon, while a plant nutrient is an inorganic element.

As the plant continues to grow, extension of roots into the subsoil will probably occur. The subsoil environment will be different in terms of the supply of water, nutrients, and oxygen and in other growth factors. This causes roots in different horizons to perform different functions or the same functions to a varying degree. For example, most of the nitrogen will probably be absorbed by roots from the A horizon because most of the organic matter is concentrated there and nitrogen is made available by organic matter decomposition. In contrast, deeply penetrating roots might penetrate less weathered and leached horizons where little available nitrogen is found but where there is an abundance of calcium. Roots here will tend to absorb a disproportionately large amount of calcium.

The plow-layer frequently becomes depleted of moisture in dry periods while an abundance of moisture still exists in underlying horizons. This results in relatively greater dependence on nutrient and water absorption from the subsoil. Subsequent rains which remoisten the upper layer of soil cause a shift to greater dependence again on the surface soil. This is probably due, in part, to better aeration nearer the surface of the soil. Thus we see that the manner in which an annual plant utilizes the soil is complex and changes continually through the season. In this regard the plant may be defined as "an integrator of a complex and ever-changing set of environmental conditions."

## THE CONCEPT OF SOIL PRODUCTIVITY

From the foregoing section it is apparent that the utilization of the soil by plants is complex. Add to this the fact that plant requirements are diverse; it is readily seen that it is impossible that a given soil be productive for the growth of all plant species. A brief discussion of some of the differences in plant requirements will aid our understanding of the soil productivity concept.

### Diversity of Plant Requirements

The requirement of many economic plants will be satisfactorily met if the soil is well aerated, near neutral in reaction, without layers that inhibit root penetration, without excess salt, and has sufficient water and an ample nutrient supply. Corn and sugar cane are profitably grown under a very wide range of soil conditions, but blueberries and azaleas are capable of survival only under a narrow range of conditions.

Blueberries and pineapples thrive very well on extremely acid soils.

Alfalfa, red clover, and table beets are only slightly tolerant of acidity and require nearly neutral soils for highest yields.

Tobacco to be used for cigarette manufacturing is planted on soils low in organic matter to insure that the plant will become nitrogen deficient several weeks before harvest. This results in a lower nicotine content and in a more desirable taste and aroma. Well-drained, sandy soils low in organic matter are well suited for this purpose.

Trees like willow, black spruce, white cedar, and tamarack can tolerate wet soil conditions for long periods of time whereas maple, red pine, and fruit trees require well-drained and well-aerated soil environments. In addition, red pine cannot tolerate alkaline soil whereas white cedar thrives very well as shown in Fig. 2–5.

## Soil Productivity Defined

Soil productivity is defined as "the capability of a soil to produce a specified plant (or sequence of plants) under a specified system of management." For example, the productivity of a soil for cotton is

**Fig. 2–5** Through the center of the photograph (in the foreground and background) are small white cedar trees. Many years ago a limestone road existed in this area. Red pine was originally planted in rows to serve as a windbreak, but was unable to survive on the alkaline roadway which was later planted to the white cedar. (Photograph taken on the Conservation Nursery at Boscobel, Wisconsin.)

commonly expressed as pounds of lint cotton per acre when using a particular management system in which are specified such things as planting date, fertilization, irrigation schedule, tillage, and pesticide control. Soil scientists determine soil productivity ratings of soils for various crops by measuring yields (including tree growth or timber production) over a period of time in a "reasonable" number of management systems that are currently relevant. (For an example see Table 3–2.) Included in the measurement of productivity are the influence of climate and the nature and aspect of slope. Thus, soil productivity is an expression of all the factors, soil and nonsoil, that influence crop yields.

Soil productivity is basically an economic concept and not a soil property. Three things are involved: (1) inputs (a specified management system), (2) outputs (yields of particular crops), and (3) soil type. By assigning costs and prices, net profit can be calculated and used as a basis for determining land value, which is important in loan appraising and tax assessment. For planning management programs, two important aspects of soil productivity are presented in Fig. 2–6. First, different soils have different capacities to absorb inputs for profit maximization. Second, different crops have different capacities to absorb management inputs for profit maximization on a given soil type.

## Soil Fertility Versus Soil Productivity

Soil *fertility* is defined as "*the quality* that enables a soil to provide the proper compounds, in the proper amounts, and in the proper bal-

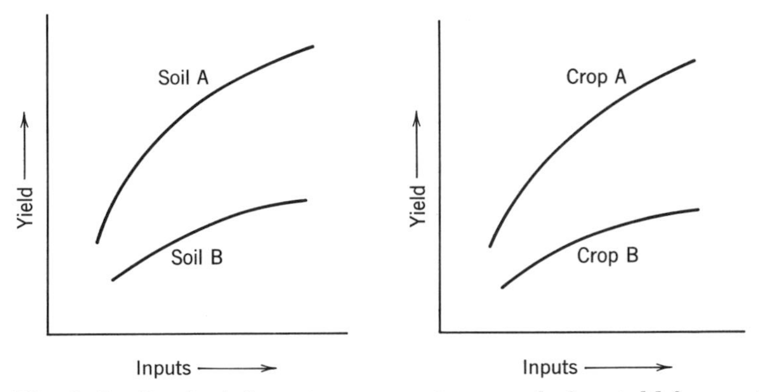


**Fig. 2–6** On the left, as imputs are increased, the yield from soil *A* increases more rapidly than from soil *B*. Soil *A* is a more productive soil than soil *B*. On the right, the soil represented is more productive for crop *A* than crop *B*; Crop *A* has a greater profit potential than crop *B*.

ance, for the growth of specified plants when temperature and other factors are favorable." Soil *productivity*, on the other hand, is defined as "*the capability* of a soil for producing a specified plant or sequence of plants under a specified system of management." For a soil to be productive it must of necessity be fertile. It does not follow, however, that a fertile soil is productive. For instance, many fertile soils exist in arid regions, but under systems of management which do not include irrigation they cannot be productive for corn or rice.

### Soil Productivity and Land Use

Sometimes the cost involved in making a soil capable of returning a reasonable profit for a certain crop is prohibitive. In these cases more attention is given to the selection of those crops or uses which allow successful utilization of the soil with a minimum of investment. Sandy soils are frequently used for recreation or forestry. For timber production emphasis is placed on selecting a species that will profitably grow on the soil as it is. It is apparent that a thorough knowledge of soil properties is a requisite to intelligent soil utilization. The role of physical properties of the soil will be discussed in Chapter 3.

# 3

# Physical Properties of Soils

The physical properties of a soil have much to do with its suitability for the many uses to which man puts it. The rigidity and supporting power, both wet and dry drainage and moisture-storage capacity, plasticity, ease of penetration by roots, aeration, and retention of plant nutrients are all intimately connected with the physical condition of the soil It is pertinent, therefore, that persons dealing with soils know to what extent and by what means man can alter those properties. This is true whether the soil is to be used as a medium for plant growth or as a structural material in the making of highways, dams, and foundations for buildings or for other engineering purposes, in manufacturing brick and tile, or in building golf courses and athletic fields, or for waste disposal systems. Texture is perhaps the most permanent and important characteristic of the soil and will be discussed first.

## SOIL TEXTURE

The relative size of the soil particles is expressed by the term *texture,* which refers to the fineness or coarseness of the soil. More specifically, texture is determined by *the relative proportions of sand, silt, and clay.* The rate and extent of many important physical and chemical reactions in soils are governed by texture because it determines the amount of surface on which the reactions can occur. The determination of the amount of the various separates present in the soil is called a *mechanical analysis* or a *particle size analysis*.

### Soil Separates—The Particle Size Categories

An outstanding fact in the examination of soils is their composition by particles of different sizes. These particles have been divided into groups entirely on the basis of size, that is, without regard to their chemical composition, color, weight, or other properties. The groups of particles are termed soil *separates*.

In Table 3–1 the names of the separates are given, together with their diameters, the number of particles in 1 gram, and the surface area in square centimeters exposed by the particles in 1 gram of the separate. The importance and magnitude of the surface area are presented in Fig. 3–1.

Sand particles are of comparatively large size and hence expose little surface compared to that exposed by an equal weight of silt or clay particles. Because of the small surface of the sand separates, the part they play in the chemical and physical activities of a soil is almost negligible. Because sand is relatively large but low in surface area, its chief function in soil is to serve as a framework around which the more active part of the soil is associated. Unless present in too small proportion, the sands increase the size of spaces between particles, thus facilitating movement of air and drainage water.

Since sand and silt consist mainly of particles resulting from the breakdown of rocks and minerals, they differ primarily in size in a given soil. As a consequence, silt has more surface area per gram (Table 3–1), and a faster weathering rate and release of soluble nutrients for plant growth than sand. Silt particles feel smooth like a powder and have little tendency to stick together or adhere to other particles. Soils with the greatest capacity to retain water against the pull of gravity are characterized as being high in clay. Silty soils have great capacity to hold *available* water for plant growth but at the same time, silt in a roadbed may present a frost heaving hazard for highway construction because of the expansion of water on freezing.

The great increase in surface area per gram of clay, as compared to silt and very fine sand, suggests that the difference between silt and sand and the clay cannot be accounted for on the basis of size alone. Chemical weathering on the surfaces of rocks, sand, and silt particles results in the solution of ions that regroup or recombine to form fine-sized particles of clay size. The clay fraction in most soils is composed of minerals that differ greatly in composition and properties as compared to sand and silt. The data for the clay separate in Table

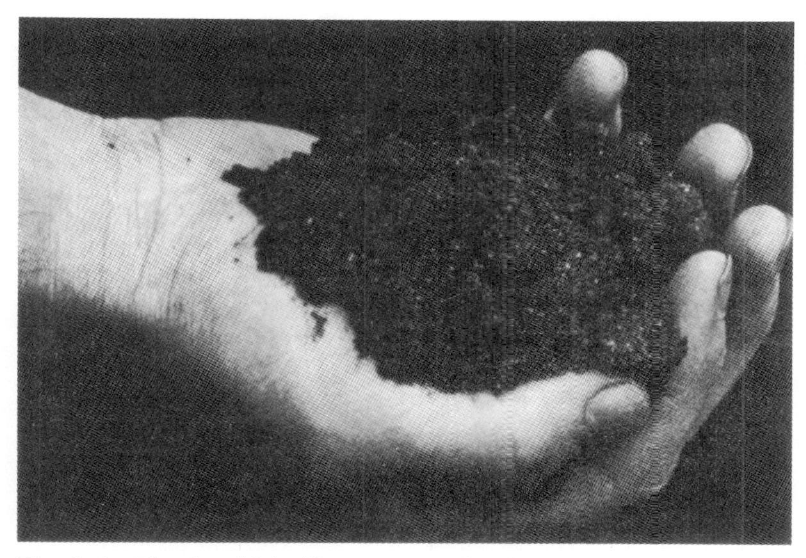

**Fig. 3–1** This handful of loam-textured soil has an estimated 5 to 10 acres of surface area. Surface area is important in its relation to plant growth, since most of the available nutrients and water are adsorbed on the surfaces of soil particles.

**Table 3–1** Some Characteristics of Soil Separates

| Separate | Diameter, mm[a] | Diameter, mm[b] | Number of Particles per Gram | Surface Area in 1 Gram, sq cm |
|---|---|---|---|---|
| Very coarse sand | 2.00–1.00 | . . . | 90 | 11 |
| Coarse sand | 1.00–0.50 | 2.00–0.20 | 720 | 23 |
| Medium sand | 0.50–0.25 | . . . | 5700 | 45 |
| Fine sand | 0.25–0.10 | 0.20–0.02 | 46,000 | 91 |
| Very fine sand | 0.10–0.05 | . . . | 722,000 | 227 |
| Silt | 0.05–0.002 | 0.02–0.002 | 5 776,000 | 454 |
| Clay | Below 0.002 | Below 0.002 | 90,260 853,000 | 8,000,000[c] |

[a] United States Department of Agriculture System.

[b] International Soil Science Society System.

[c] The surface area of platy shaped montmorillonite clay particles determined by the glycol retention method by Sor and Kemper. (See SSSAP, Vol. 23, p. 106, 1959.) Other separates assumed to be spheres and calculation is based on largest size permissible.

3–1 are for an unusually fine-sized, high surface area clay. Since a large part of the water in the soil is held as a film on the surface of the clay particles, the amount of clay in the soil has a great influence on its *total* water-holding capacity. In addition, certain available nutrients are held on the surface of clay particles. Therefore, clay acts as a storage reservoir for both water and nutrients. Clay may have thousands of times more surface area per gram than silt and nearly a million times more surface area than very coarse sand (Table 3–1).

### Particle Size Analysis by the Hydrometer Method

Bouyoucos devised the hydrometer method for determining the sand, silt, and clay content of the soil without separating them. The soil sample is first soaked overnight in a 5 percent Calgon solution in order to facilitate dispersion. Then it is placed in a metal cup with baffles on the inside and is dispersed for 2 minutes by the soil mixer running at a speed of 16,000 rpm (Fig. 3–2). The soil mixture is poured into the cylinder and water is added to bring the contents to the 1130 milliliter mark. With the help of a stirrer, the soil suspension is thoroughly resuspended and the time is immediately noted. Two hydrometer readings are taken of the soil suspension using a special soil hydrometer. A reading taken at 40 seconds determines the total percentage of sand (2.0 to 0.05) mm) and the 2-hour reading determines the total percentage of clay (below 0.002 mm) in the sample. The silt (0.05 to 0.002 mm) is calculated by difference—add the percent sand and the percent clay and subtract from 100.[1]

After the hydrometer readings have been obtained, the soil suspension can be poured over a screen to recover the entire sand fraction. After drying, the sands are sieved to obtain the various sand separates listed in Table 3–1.

## A TEXTURAL GROUPING OF SOILS

Suppose the results of a mechanical analysis showed that a soil contained 15 percent clay, 65 percent sand, and 20 percent silt. The logical question is, "What is the texture of the soil?"

### Soil Classes Used to Designate Texture

A soil is never composed of only one separate. Usually at least small quantities of the majority of the separates are present. The first step

[1] "Hydrometer Method Improved for Making Particle Size Analyses of Soils," G. J. Bouyoucos, *Agron. J.*, **54**:464–465, 1962.

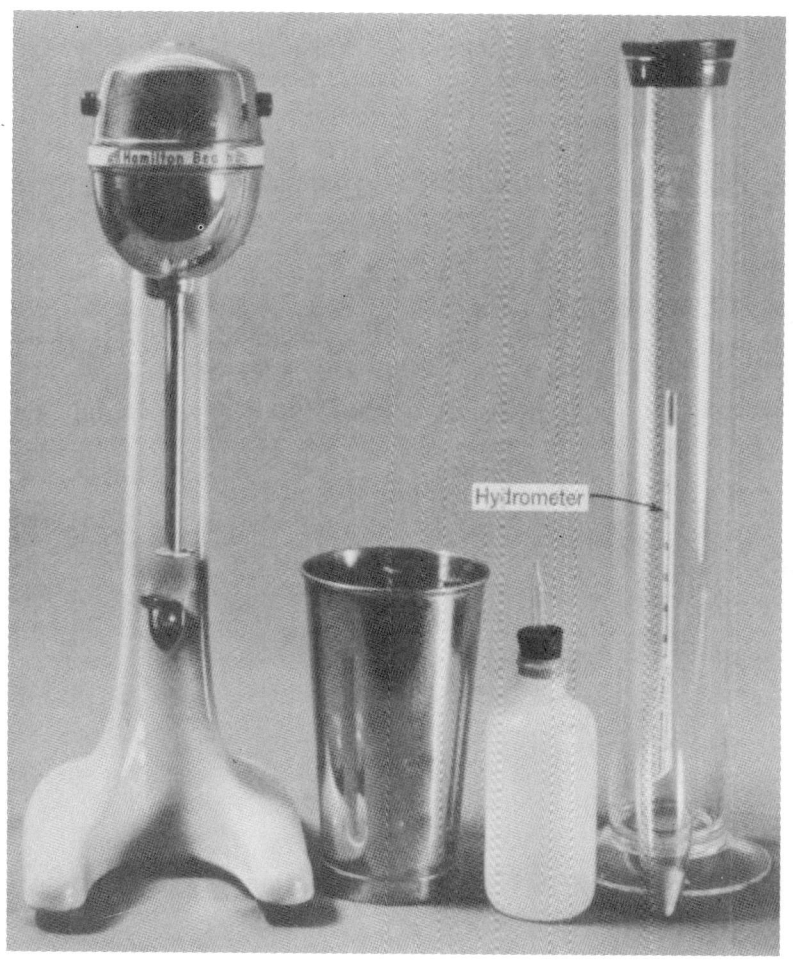


**Fig. 3–2** Equipment used in making a mechanical analysis by the hydrometer or Bouyoucos method. The cup in which the soil is stirred has baffles, which are essential if complete dispersion is to be obtained.

in a textural classification, therefore, is to group them on the basis of the proportion of the different separates present. These groups are designated as soil *classes,* which are named according to the separate or separates which contribute most to their characteristics. This does not mean that a class is necessarily named after the separate present in largest quantity. It usually takes a very large quantity of coarse particles to exert as much influence on soil properties as a comparatively small quantity of the finest particles—clay. Clay is the most potent separate in imparting its properties to a mixture of separates, and hence the adjective *clay* is found in the class name of many soils which contain a higher percentage of the other separates than they do of clay.

The proportions of the separates in classes commonly used in describing soils are given in the textural triangle shown in Fig. 3–3. The sum of the percentages of sand, silt, and clay at any point in the triangle is 100. Point *A* represents 15 percent clay, 65 percent sand, and 20 percent silt; the textural class name for this sample is *sandy loam.* A soil containing equal amounts of the three separates is a *clay loam* (point *B* in Fig. 3–3). The various soil classes are separated from one another by definite lines of division in Fig. 3–3. Their properties do not change abruptly at these boundary lines, however, but one class grades into the adjoining classes of coarser or finer texture.

## Class Names and Soil Properties

Strictly speaking, the class name describes only the particle-size distribution. Plasticity, rigidity, permeability, ease of tillage, droughtiness, fertility, and productivity may be closely related to the textural

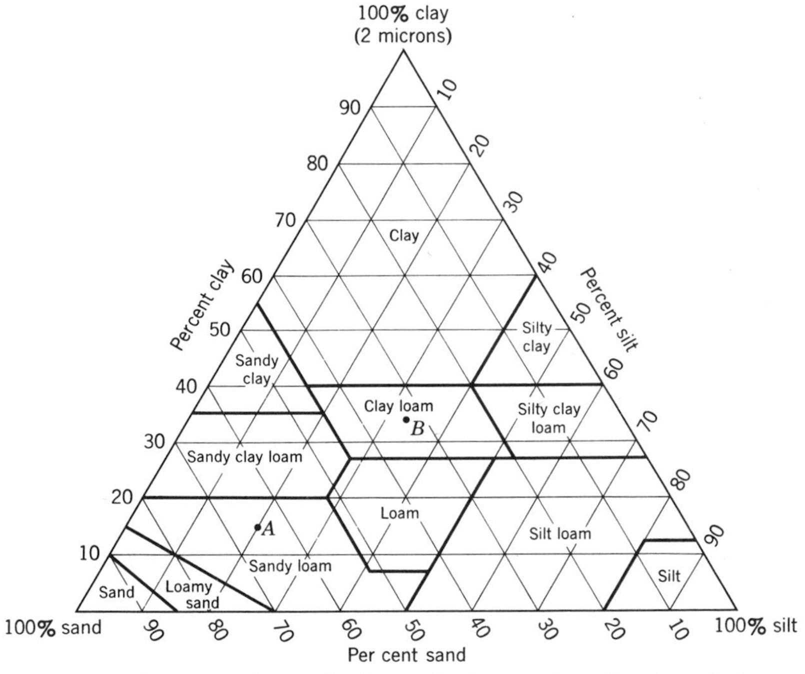


**Fig. 3–3** The textural triangle shows the limits of sand, silt, and clay contents of the various texture classes.

classes in a given geographical region; but, due to the great variation that may exist in the mineralogical nature of the separates, no broad generalizations of the world's soils can be made. Thus, a soil with 25 percent clay, which is montmorillonitic, will be much more plastic than a soil with over 70 percent clay which is composed mainly of hydrous oxides of iron and aluminum. The sand and silt in some soils are composed mainly of minerals rich in essential nutrients, whereas in others they are dominated by quartz ($SiO_2$). The infertility of sandy soils is commonly related to their high content of quartz. The discussion of minerals in Chapter 7 will aid in clarifying this point.

Many useful crop-yield and soil-texture relationships have been established for certain geographic areas. The dominant texture in the upper 3 feet of the profile and the yield of four crops in Michigan are given in Table 3–2. Corn and oak yields were highest on loam and yields of potatoes and pine were highest on sandy loam. Data such as these are useful in appraising land for loan or taxation purposes and in planning cropping systems or reforestation programs.

**Table 3–2** Relationship of Texture and Average Crop Yields or Soil Productivity

| Dominant Texture of the Upper 3 Feet of Soil Profile | Yield, per Acre[a] | | | |
|---|---|---|---|---|
| | Corn, bushels | Potatoes, cwt | Pine, cords | Oak, cords |
| Clay | 80 | . . . | 0.3 | 1.3 |
| Loam | 100 | 250 | 0.6 | 1.5 |
| Sandy loam | 80 | 300 | 1.3 | 1.3 |
| Loamy sand | 60 | 250 | 0.8 | 0.8 |
| Sand | 120 (irrigated) | 300 (irrigated) | 0.3 | . . . |

[a] Corn and potato yields with good management; yield of cord wood is annual sustained yield. Based on data from various state and federal agencies operating in Michigan.

Many engineering properties of soils are related to texture and important relationships between texture and engineering properties have been recognized. A high shrink-swell potential, degree of volume change with wetting and drying, causes some high clay content soils to cave in basement walls, disrupt pipe lines, and crack walls (see Fig. 3–4). The American Association of State Highway Officials has classified soil materials into seven groups for suitability as a road base. Group A-1 consists of gravelly soils with high load bearing potential when wet. Group A-7 consists of high clay soils with low load bearing potential when wet.

**Fig. 3–4** This house was built on soil with a high shrink-swell potential. The soil's swelling and shrinking with changes in water content made the wall crack.

## Influence of Coarse Fragments on the Class Name

Some soils contain significant amounts of gravel, stones, or other coarse fragments that are larger than the size of sand grains. An appropriate adjective is added to the class name in these cases. For example, a sandy loam with 20 to 50 percent of the volume made up of gravel is a *gravelly* sandy loam. If 50 to 90 percent of the volume was gravel, it would be a *very gravelly* sandy loam. *Rockiness* is used to express the amount of the land surface composed of exposed bedrock.[2]

## Organic Soils and Texture Designation

Coarse-textured soils with 20 percent or more organic matter and fine-textured soils with 30 percent or more organic matter by weight

[2] For a more detailed discussion see, *Soil Survey Manual*, U.S. Department of Agriculture Handbook 18, pp. 205–223, 1951.

have properties dominated by the organic fraction rather than the mineral fraction. Where such soils are over 1 foot thick, they are called organic soils.[3]

Muck, peat, mucky peat, and peaty muck are used in place of the textural class names for organic soils. *Peat* soils contain organic matter which still retains some of the original structure of the plant tissues that can be identified. In *muck* soils the organic matter is in an advanced state of decomposition and in a fine state of subdivision. Peat soils, therefore, can become muck soils upon further decomposition and this happens when they are drained and used for crop production. Mucky peat and peaty muck are intergrade classifications.

## THE TEXTURE PROFILE

The textures of the various horizons in a soil profile are usually different. When this occurs, the soil has a *texture profile.* Since the texture has a great influence on the properties of soil horizons, the development of the texture profile must be considered.

### The Texture Profile and the Bt Horizon

You will recall that clay particles are slowly moved from A horizons and deposited in B horizons by percolating water. The majority of the world's soils have clay-enriched B horizons, which are indicated by the subscript t as in Bt. The symbol t comes from the German word "Ton" meaning clay. When the Bt horizon has at least 20 percent more clay than the horizons above, it qualifies as an *argillic horizon.* For our purposes we can use Bt and argillic horizon interchangeably.

A texture profile exists in the Miami loam as shown by the clay-depth distribution curve of Fig. 3–5. Similar percentages of clay exist in both the A1 and A2 horizons. The maximum clay content in the Bt horizon is about twice as great as that in the A horizon. The presence of *clay skins* (cutans) or coatings of clay on the aggregates surfaces and in the pore spaces of the Bt horizon is evidence that translocation of clay has played a role in the development of the texture profile. This clay-distribution curve is typical of many mature soils.

### Influence of the Texture Profile on Plant Growth

The presence of the texture profile can be beneficial or detrimental, depending on the degree to which it has developed. Up to a certain point, an increase in the amount of clay in the subsoil is desirable. It

[3] "Organic Soils," J. F. Davis and R. E. Lucas, *Mich. Agr. Exp. Sta. Spec. Bull.* **425**, p. 5, 1959.

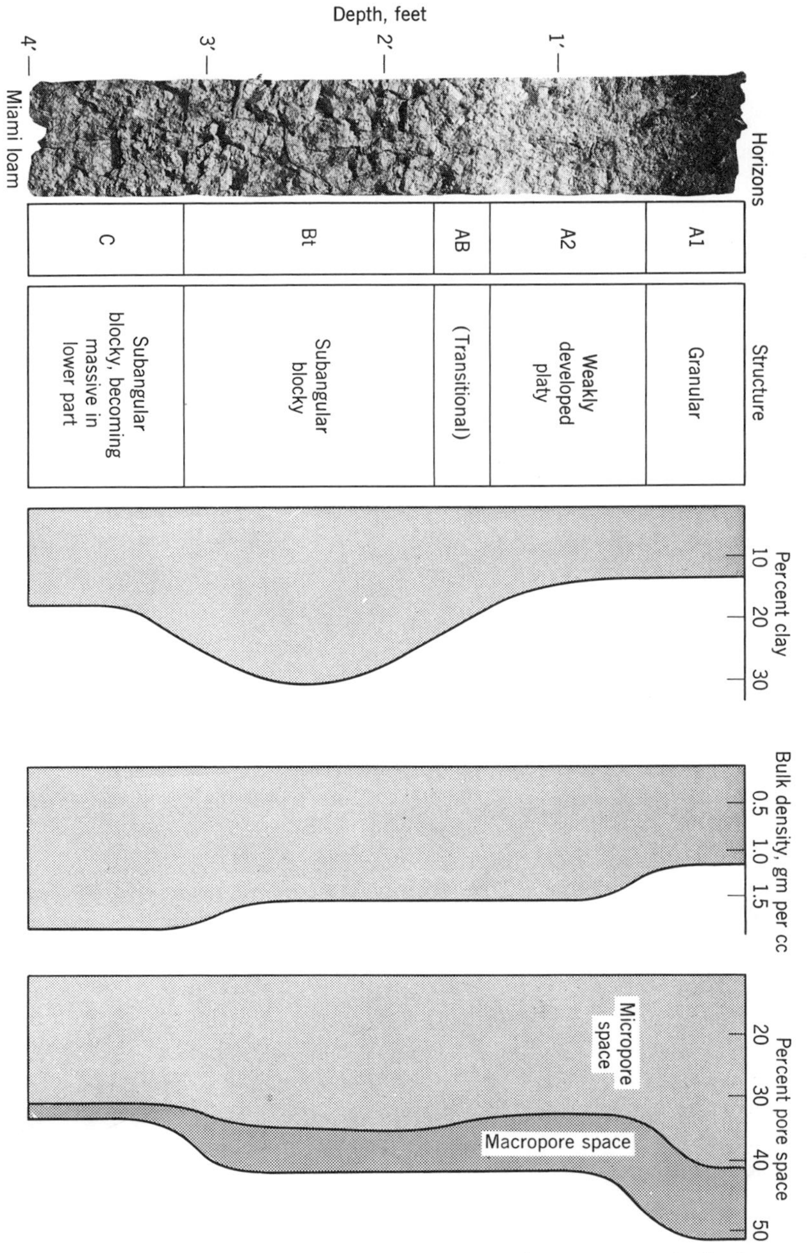


**Fig. 3–5** Horizon designations, structure, clay content, bulk density, and percentage pore space of the horizons of the Miami loam (Alfisol). (University of Illinois Agriculture Experimental Station Bulletin 665, 1960.)

can increase the amount of water and nutrients stored in that zone. By slightly reducing the rate of water movement through the soil, it will reduce the rate of nutrient loss through leaching. If, however, the accumulation of clay is great, as in the case of a clay-pan soil, it will severely restrict the movement of air and water and the penetration of roots in the Bt horizon. It will also tend to increase the amount of the rainfall that will occur as runoff on sloping land.

### The Texture Profile and Deep Plowing for Wind Erosion Control

Deep plowing has been used to control wind erosion where sandy-surface soils are underlain by Bt horizons containing 20 to 40 percent clay. Scientists of the United States Department of Agriculture and the Agricultural Experiment Stations of Kansas and Texas observed that at least one inch of subsoil should be plowed up for every two inches of surface soil thickness in order to control wind erosion. In many cases, the soil must be plowed to a depth of 20 to 24 inches. This kind of plowing increased the clay content of the surface soil 5 to 12 percent in some cases. Since it was found that about 27 percent clay in the surface soil was required to halt soil blowing, deep plowing by itself resulted in only partial and temporary control of wind erosion. When deep plowing was not accompanied by other erosion control practices, wind erosion removed the clay from the surface soil and the effects of deep plowing were short-lived.

### Alteration of Soil Texture

Fabrication of soils to meet certain textural specifications is commonplace for golf greens, tree nurseries, and greenhouse use. Alteration of texture in the field is only occasionally attempted because of the enormous weight of such large quantities of soil. Deep plowing is used in some cases to break up root-inhibiting layers as well as to control wind erosion.

An interesting case of texture alteration occurs in the Netherlands. The low load-bearing capacity of peat lands limits the number of cattle that can be grazed in pastures. Cattle hoofs make deep holes in the peat soil, destroying the sod. Where these peat lands have sand within three feet of the surface, giant plows are used to bring up the sand to form a surface with greater load-bearing potential and thus increase the number of cattle that can be pastured per acre. The high value of these land for pasture also makes it profitable to use a machine that augers sand to the surface from depths as great as 10 or more feet.

## SOIL STRUCTURE

The term *texture* is used in reference to the size of soil particles, but, when the arrangement of the particles is being considered, the term *structure* is used. Structure refers to the *aggregation of primary soil particles (sand, silt, clay) into compound particles,* or clusters of primary particles, which are separated from the adjoining aggregates by surfaces of weakness. The structure of the different horizons of a soil profile is an essential characteristic of the soil just as are color, texture, or chemical composition.

Structure *modifies* the influence of texture in regard to moisture and air relationships, availability of plant nutrients, action of microorganisms, and root growth. A good example of this occurs in the "black-lands" of Alabama and Texas, where the content of highly plastic clay is as high as 60 percent. These soils would be of limited value for crop production if they did not have a well-developed granular structure, which facilitates aeration and water movement.

### Types of Soil Structure and Importance

Soil aggregates or peds are classified on the basis of shape as spheroidal, platelike, blocklike, or prismlike. These four basic shapes give rise to seven commonly recognized types as listed in Table 3–3. A brief description and the horizon in which they are commonly found is also given in the table. The types of structure in the Miami loam soil are shown in Figure 3–5.

The macroscopic size of most peds results in the existence of interped spaces that are much larger than those that can exist between adjacent sand, silt, and clay particles. It is this effect of structure on the pore space relationships that makes structure so important. Move-

**Table 3–3** Diagrammatic Definition and Location of Various Types of Soil Structure

| Structure Type | Aggregate Description | Diagrammatic Aggregate | Common Horizon Location |
|---|---|---|---|
| Spheroidal granular | Relatively nonporous, small and spheroidal peds; not fitted to adjoining aggregates. | | A horizon |
| Crumb | Relatively porous, small and spheroidal peds; not fitted to adjoining aggregates. | | A horizon |

**Table 3–3** Diagrammatic Definition and Location of Various Types of Soil Structure (Cont.)

| Structure Type | Aggregate Description | Diagrammatic Aggregate | Common Horizon Location |
|---|---|---|---|
| Platelike | Aggregates are platelike. Plates often overlap and impair permeability. |  | A2 horizon in forest and claypan soils |
| Blocklike | Blocklike peds bounded by other aggregates whose sharp angular faces form the cast for the ped. The aggregates often break into smaller blocky peds. |  | B horizon |
| Subangular blocky | Blocklike peds bounded by other aggregates whose rounded subangular faces form the cast for the ped. |  | B horizon |
| Prismlike | Columnlike peds without rounded caps. Other prismatic aggregates form the cast for the ped. Some prismatic aggregates break into smaller blocky peds. |  | B horizon |
| Columnar | Columnlike peds with rounded caps bounded laterally by other columnar aggregates which form the cast for the peds. |  | B horizon in solonetz soils |

From *Soils Laboratory Exercise Source Book*. Am. Soc. of Agron., 1964.

ment of air and water is facilitated. The interped spaces also serve as corridors for root extension. The effect of structure on the gross morphology of oat roots is presented in Figure 3–6.

Structure is also described in terms of size and grade (distinctness) of peds. Those in the Bt horizon of the Miami loam shown in Fig. 3–5 are very distinct and the grade is strong. In the A2 horizon the structure is indistinct, and the grade is called weak. Careful observation is required to see the peds in the A2 horizon, and furthermore only part of the soil mass is aggregated. Where no observable aggregation or definite orderly arrangement of peds is present, the grade is *structureless.*

Two structureless grades worthy of mention are the *single-grained* and the *massive.* If each particle in a soil functions as an individual, that is, if it is not attached to other particles, the condition is called single-grained. Such a structural condition may exist in very sandy soils. Some of the particles in many soils exist in the single-grained condition, but to find all the particles so functioning is unusual. If soils containing considerable amounts of clay are tilled or are tramped by grazing animals when too wet, a portion of the colloidal material, including the clay, tends to fill the pore spaces. This makes the soil more dense and upon drying it forms large clods. The condition is called massive and is commonly referred to as *puddled.*

## Formation of Soil Aggregates

An understanding of the causes for the development of structure in soils is of practical importance because structure has a great influence on plant growth and is also greatly altered by tillage operations and traffic.

Structure develops from either a single-grained or massive condition. In order to produce aggregates there must be some mechanism which groups particles together into clusters and also some means by which they are bound rather firmly so that the structural forms may persist. Evidence points to the colloidal fraction as the active constituent since, without its presence, single-grained structure prevails. Three groups of colloidal matter important as cementing material in aggregate formation are: (1) clay minerals, (2) colloidal oxides of iron and manganese, and (3) colloidal organic matter, including microbial gums.

There are several theories concerning the process by which aggregation is brought about, but the following is probably as widely accepted as any. Because colloidal particles are charged bodies, dipolar water molecules attach themselves readily and firmly to them. As

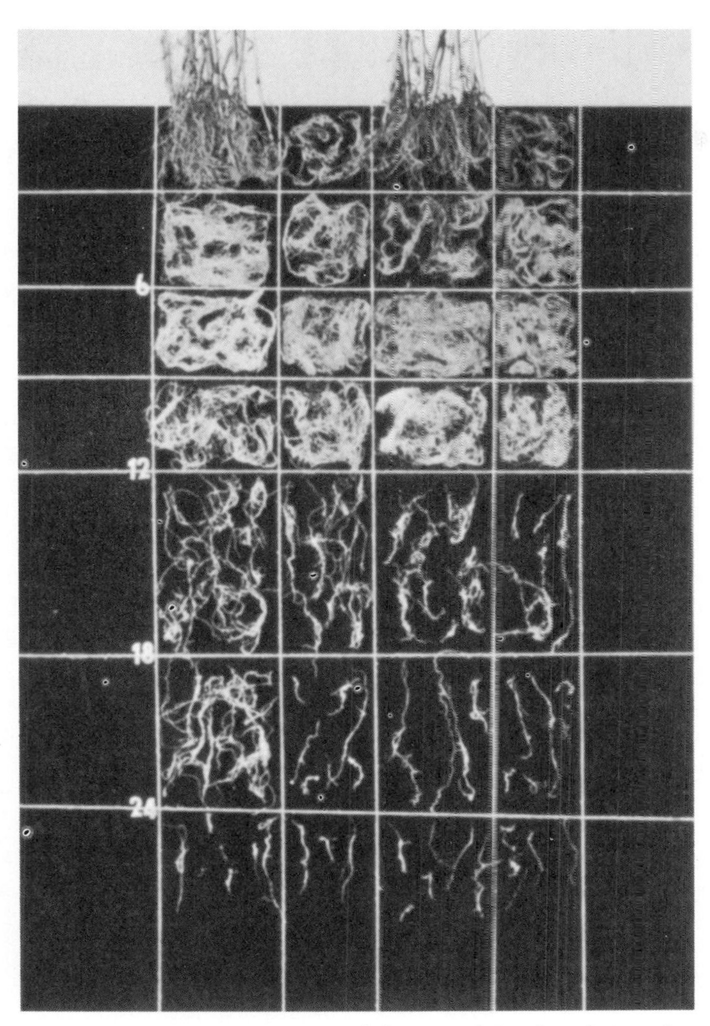


**Fig. 3–6** Oat roots recovered from a slab of soil 4 inches thick. Note the marked change in amount and kind of roots at a depth of 12 inches. This is the depth where the granular structure of the plow layer changes to the blocky structure of the Bt horizon. Roots in the B horizon were fewer in number, larger in diameter, and were growing primarily in the vertical spaces between structural units, old root channels, and in earthworm channels.

water molecules and colloidal nuclei carry both positive and negative charges, it is conceivable that oriented water molecules may connect colloidal particles. The linkage may include cations because aggregation seems to be affected by the cations adsorbed (see Fig. 3–7).

Such water molecules are held very tightly, and as water evaporates from the soil, the length of each linkage becomes shorter and stronger and so pulls the colloidal particles close together. Larger soil particles to which the colloids are attached may be drawn into the cluster, and some may be surrounded as the cluster or floccule develops. As more water is lost and the colloidal material becomes further dehydrated, the colloids stick or cement the particles into an aggregate. Thus it is seen that water, acting in conjunction with the colloids, constitutes the major force inducing granulation of soil and that colloidal material is the final binding agent.

Repeated cycles of wetting and drying occur in soils and the permanence of aggregates depends on two conditions. First, the soil along the ped faces must not disperse during rewetting or rehydration. Second, the colloids must be able to hold the particles within the peds together when the soil becomes wet. Let's consider next the importance of the various colloids in aggregate formation.

### Importance of Colloids in Aggregation

Wet-sieving is commonly used to measure aggregation. Dry aggregates are placed on a sieve that is gently lowered and raised in water. When the dry aggregates are immersed in water, the water moves into the aggregates from all directions and compresses the air in the pore spaces. Aggregates unable to withstand the pressure exerted by the entrapped air are disrupted and fall through the sieve along with the unaggregated soil. The aggregates remaining on the sieve after a standard period of time are considered water stable.

Researchers at the University of Wisconsin determined the aggregation of four soils in which they also determined the content of

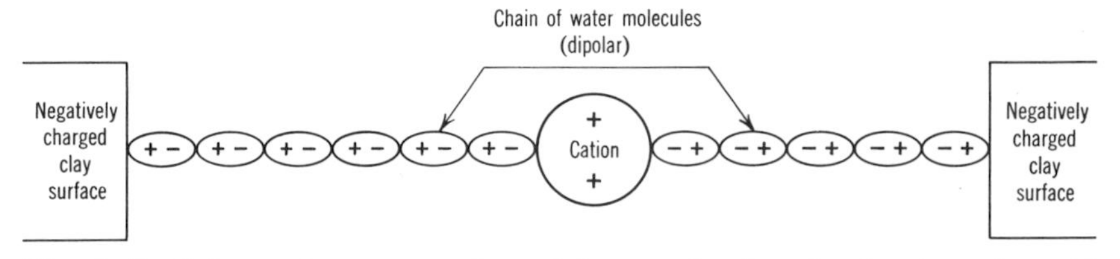


**Fig. 3–7** Schematic presentation of the mechanism in structure formation. As the water evaporates, the clay plates are pulled closer together, and with cementing agents, structure develops.

**Table 3–4** Order of Importance of Soil Constituents in Formation of Aggregates over 0.5 mm in Diameter in Four Wisconsin Soils

| Soil Type | Order of Importance of Soil Constituents in Aggregate Formation |
|---|---|
| Parr silt loam | Microbial gum > clay > iron oxide > organic carbon |
| Almena silt loam | Microbial gum > iron oxide |
| Miami silt loam | Microbial gum > iron oxide > organic carbon |
| Kewaunee silt loam | Iron oxide > clay > microbial gum |
| All soils | Microbial gum > iron oxide > organic carbon > clay |

From "Soil Aggregation in Relation to Various Soil Constituents," G. Chesters, O. J. Ottoe, and O. N. Allen, *Soil Sci. Soc. Am. Proc.*, **21**:276, 1957.

microbial gum, iron oxide, organic carbon (organic matter), and clay. The results are presented in Table 3–4. Microbial gum was the most important agent in producing aggregation in three of the four soils. Iron oxide was most important in one soil and second most important in three soils. For the four soils grouped together, the order of importance was microbial gum, iron oxide, organic carbon (organic matter), and clay. The great importance of microbial gum warrants consideration of the importance of microbial activity in aggregate formation.

### Microbial Activity and Aggregate Formation

Various experimenters have found that the addition of plant residues to soil without microbial activity has little if any effect on aggregation. It is the microbial activity supported by the organic matter that is closely related to aggregate formation. There appears to be several factors involved. First, the organisms like fungal mycelium serve to bind soil particles together. Various substances synthesized by microorganisms also play an active role. The gums, along with other substances, act as cements that exhibit some resistance to rehydration. Fats and waxes also tend to give aggregates the ability to resist wetting and the subsequent compression force of entrapped air that tends to disrupt aggregates. Lastly, some microbially produced substances have negative and positive charged groups that link soil particles in a manner analogous to that shown in Fig. 3–7.

### Aggregation Induced by Freezing and Thawing

As soil water freezes in a small crevice of the soil, it draws moisture from the surrounding soil as the ice crystals grow. Thus, under slow freezing the ice crystals tend to break clods and at the same time exert a pressure on the surrounding soil particles which presses them together and stimulates the development of new clusters or granules. Furthermore, as moisture is drawn from the adjoining soil, the colloidal material is partially dehydrated and shrinks and tends to cement the particles into an aggregate. Repetitions of freezing and thawing from night to day accentuate these processes until they culminate in a thoroughly granulated soil.

## Soil Aggregation and Crop Yields

From the previous paragraphs it is obvious that the aggregates in the soil are in a stage of continual change. Wetting and drying, freezing and thawing, tillage, and biological activities all contribute to the continual breaking down and building up of aggregates. The structure of the plow layer is affected by management practices and where aeration and drainage limit plant growth, cropping systems that maintain greater aggregation are likely to result in higher crop yields. Such was the case on the Paulding clay soils in Ohio where the percentage of the soil that was aggregated was positively correlated with the yield of corn (Fig. 3–8). When corn was grown in cropping systems that included a legume, the legume undoubtedly contributed to the higher yields because of an increase in the amount of available nitrogen. However, the fact that the use of fertilizer on the poorly aggregated plots resulted in yields of about 30 bushels, whereas yields from well aggregated nonfertilized plots were about 70, indicates that the physical conditions of the poorly aggregated plots were limiting the yield of corn.

Fig. 3–8 shows that as cropping systems went from continuous corn to a corn-oats system (corn and oats grown in alternate years) to a corn-oats-legume system (each crop grown every third year) there was a marked increase in aggregation. Greatest aggregation existed where cultivation was the least frequent and where the land was most often supporting an actively growing plant canopy. A plant canopy absorbs the raindrop impact and lets the water fall gently on the soil without breaking the aggregates apart. Further, an actively growing crop is associated with more microbial activity because there is a continual addition of organic matter to the soil which serves as food for microorganisms. In addition, several means by which plant roots cause aggregation may be summarized as follows: (1) Roots and root hairs penetrating the soil produce lines of weakness along which the clod or soil mass may break into granules. (2) The pressure exerted by developing roots may induce aggregation. (3) Root secretions may flocculate colloids, stabilize, or cement aggregates. (4) Use of moisture

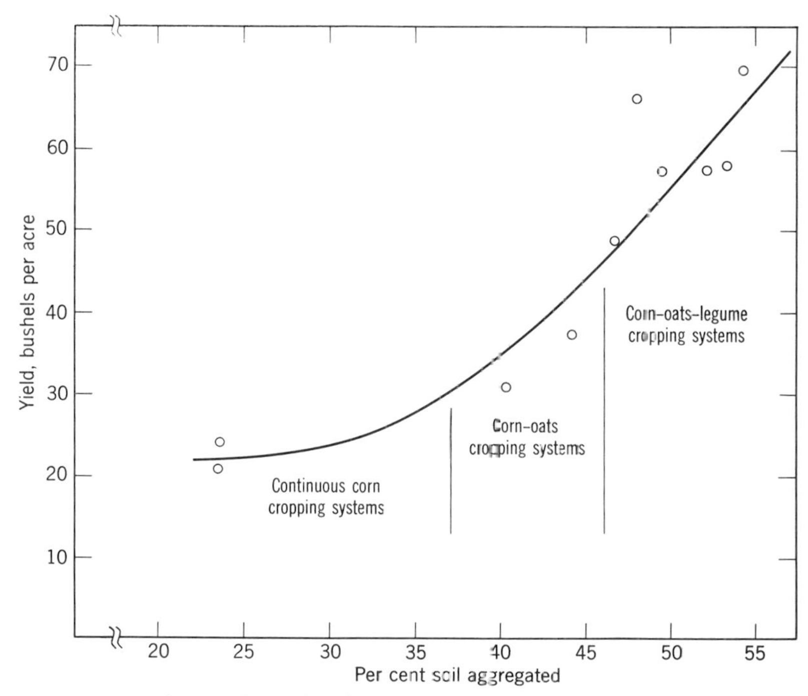


**Fig. 3–8** Relationship of soil aggregation and the yield of corn grown on the Paulding clay as affected by cropping system. Aggregation of soil was determined at the time of corn harvest and is the average of three years. (Data from "Cropping Systems and Soil Properties," J. B. Page and C. J. Willard, *Soil Sci. Soc. Am. Proc.*, Vol. 11, p. 87, 1946.)

by roots may cause dehydration of colloids, thus resulting in shrinkage and finally in cementation.

Before we leave the topic of structure it should be noted that iron oxide is commonly the most effective colloid producing aggregation in subsoils where the content of organic matter and the level of microbial activity is low. Iron oxide is also very important in the formation of aggregates in red-colored tropical soils where the content of iron oxide is unusually great. In fact, the aggregates in some of these red-colored tropical soils may be so hard and stable that the land can be plowed or cultivated when it is raining without serious consequences.

## WEIGHT, PORE SPACE, AND AIR RELATIONSHIPS

Pore space is important because soil pores are largely filled with water and air. Water and air (gases) also move through pore spaces.

Thus, the supply of water and oxygen for plant growth and the rate of water movement through the soil are related to the amount and size of soil pores.

Because the weight of soil is related to the amount of pore space, the pore space and weight relationships of soils are discussed together. Both weight and pore space vary from horizon to horizon as do the other soil properties, and both are affected by soil texture and structure.

**Fig. 3–9** Technique for obtaining bulk density cores. The light-colored metal core fits into the core sampler just to the left. This unit is then driven into the soil in pile-driver fashion using the handle and weight unit.

## Bulk Density of Soil

The bulk density is the weight per unit volume of *oven dry* soil, commonly expressed as grams per cubic centimeter. Core samples used to determine bulk density are obtained with the equipment shown in Fig. 3–9. Care is exercised in the collection of cores so that the natural structure of the soil is preserved. Any change in the structure of the soil is likely to alter the amount of pore space and likewise the weight per unit volume. Four or more cores are usually obtained from each soil horizon to obtain a reliable average value.

Bulk density cores obtained in the field are brought to the laboratory for oven drying and weighing. The weighing and calculation are shown in Fig. 3–10. In this case the bulk density is 1.5 grams per cubic centimeter. Bulk density can be expressed in other units, such as pounds per cubic foot; in this case it would be 93.6 (62.4 × 1.5). When expressed in grams per cubic centimeter, the bulk density of fine-textured *surface* soils will commonly be in the range 1.0 to 1.3. Coarse-textured surface soils will usually be in the range 1.3 to 1.8. The greater development of structure in the fine-textured surface soils accounts for their lower bulk density as compared to more sandy soils.

**Fig. 3–10** The core obtained in Fig. 3–9 was dried in an oven for 24 hours and weighed. The oven-dried soil core weighed 261 grams and since the volume was 174 cc, the bulk density is 1.5 grams per cc.

The bulk densities of the various horizons of the Miami loam given in Fig. 3–5 show that the parent material is the densest layer. It has a bulk density of 1.8 grams per cubic centimeter. Formation of structure during soil development caused the overlying horizons to have lower bulk densities than the original parent material.

The Bt horizon in the Miami has a greater clay content than the A1 horizon. Its bulk density is greater than the A1 horizon and thus has a lower percentage of pore space. Clay deposition in the Bt horizon tended to fill existing pore spaces and made the horizon more dense as the clay content increased. The general rule that fine-textured soils have more pore space and lower bulk densities than coarse-textured soils may hold when comparable structural conditions exist as is the case when samples from plow layers are compared. This point should provide a basis for understanding that the greatest amount of available water may not necessarily be in the horizon with the highest clay content, since water is stored in the pore space.

Organic soils have very low bulk density compared to mineral soils. Considerable variation exists depending on the nature of the organic matter and the moisture content at the time of sampling to determine bulk density. Values ranging from 0.2 to 0.6 gram per cubic centimeter are common.

## Weight per Acre Furrow Slice

Weight per acre furrow slice is the oven dry weight of the soil over one acre to a depth of 6 to 7 inches. A soil with a bulk density of 1.5 grams per cubic centimeter would have a density 1.5 times greater than water and would have an acre furrow slice weight equal to:

$$\underset{\text{(pounds per cubic foot)}}{(1.5)\,(62.4)} \times \underset{\text{(cubic feet in acre furrow slice)}}{(43{,}560)\,(7/12)} = 2{,}478{,}376 \text{ pounds}$$

It is customary to consider that an average acre furrow slice has a bulk density of 1.3 grams per cubic centimeter and weighs about 2,000,000 pounds or 1000 tons. On this basis an acre furrow slice that contains 1 percent organic matter on a weight basis would contain 20,000 pounds of organic matter. Soil losses are usually expressed as tons per acre. An average annual soil loss of 10 tons per acre would result in the removal of soil equal to the furrow slice about every 100 years. We also use the weight per acre furrow slice as the basis for calculating the amount of fertilizer and lime to apply per acre. From these examples it should be clear that the weight per acre furrow slice is a very useful soil property.

## Particle Density of Soil

In determining the particle density of soil, consideration is given to the solid particles only. Thus, the particle density of any soil is a constant and does not vary with the amount of space between the particles. It is defined as the mass (weight) per unit volume of soil particles (soil solids) and is frequently expressed as grams per cubic centimeter. For many mineral soils the particle density will average about 2.65 grams per cubic centimeter. It does not vary a great deal for different soils unless there is considerable variation in content of organic matter or mineralogical composition.

## Calculation of Total Pore Space

The percentage of pore space in a soil may be calculated from the bulk density and particle density if both are expressed in the same units of measurement. The following formula will give the percentage of the soil which is solid particles:

$$\left(\frac{\text{Bulk density}}{\text{Particle density}}\right) \times 100 = \%\ \text{solids}$$

This percentage, taken from the total volume (100 percent), will give the percentage of pore space, hence the formula

$$100\% - \left(\frac{\text{bulk density}}{\text{particle density}}\right) \times 100 = \%\ \text{pore space}$$

Substitution in the formula of 1.3 for bulk density and 2.65 for particle density gives 50.3 percent total pore space, which is considered rather typical for medium-textured plow layers. Using the same formula, the soil sample in Fig. 3–10 has 43 percent pore space and the C horizon of the Miami (Fig. 3–5) has only 32 percent total pore space. Soils research often requires an experimental determination of total pore space as discussed in the next section.

## Determination of Total Pore Space

The soil cores used for the determination of bulk density can also be used to determine total pore space. To determine the pore space the cores are placed in a pan of water until completely saturated and then the cores are weighed. The difference in weight between saturated and oven dry cores represents a volume of water equal to the volume of the pore space in the soil. For a 174 cubic centimeter core that contained 87 grams (87 cc) of water at saturation, the total pore space of the soil would be 50 percent.

## Pore Size and Its Importance

In the discussion of bulk density, it was pointed out that sandy surface soils usually have a greater bulk density than do clayey soils. This means that in a dry condition sandy soil has less volume occupied by pore space. Yet, our everyday experiences tell us that water usually moves much faster through a sandy than through a clayey soil. The explanation for this seeming paradox lies in the size of the pores that are found in each soil.

The total pore space in a sandy soil may be low, but a large proportion of it is composed of large pores which are very efficient in the movement of water and air. The percentage of the volume occupied by small pores in sandy soils is low, which accounts for their low water-holding capacity. In contrast the fine-textured surface soils have more total pore space and a relatively large proportion of it is composed of small pores. The result is a soil with a higher water-holding capacity. Water and air move through the soil with difficulty because there are few large pores. Thus we see that the size of the pore spaces in the soil may be as important as the total amount of pore space.

In a moist, well-drained soil the large pore spaces are usually filled with air, and consequently they have been called *aeration pores* or *macropores*. The smaller pores usually tend to be filled with water and are commonly called *capillary* or *micropores*.

## Distribution of Pore Space in the Profile

The pore space distribution in the profile of a mature soil is shown in Fig. 3–5. Development of granular structure in the A1 horizon results in high total porosity as well as favorable amounts of both micro- and macropore space. As pointed out earlier, clay deposition in the pores of the Bt horizon has been responsible for decreasing porosity. The Bt horizon of this soil has less of both types of pore space than does the A1 horizon.

## Soil Permeability and the Percolation Rate

Water in a capillary tube does not move or drain out because the attraction between water and glass offers greater resistance than can be overcome by gravity. As a result, a substance can be very porous and yet be slowly permeable to water. In large (noncapillary size) tubes or pipes, water movement varies as the fourth power of the radius. Thus, as the diameter of a pipe is doubled, the rate of water flow increases 16 times ($2^4$). Since water molecules are strongly adsorbed onto soil surfaces, as is the case with glass, *pore size* is of great importance in regard to the flow or movement of water into (infiltration) and through (percolation) soil. By contrast, the insignificant attraction between soil particles and air results in the air movement being primarily related to the *volume* of the vacant soil pores, not *size* of pores.

Permeability is the ability of the soil to transmit water or air. Permeability is commonly measured in terms of the rate of water flow through the soil in a given period of time and is commonly expressed as inches per hour. Soil permeability rates and classes are given in Table 3–5. The percolation rate is given as minutes for one inch of water to percolate through the soil. The percolation rate is a common expression used by sanitary engineers when referring to the permeability of the soil in regard to the discharge of septic tank effluent through soil.

**Table 3–5** Soil Permeability and Percolation Classes and Rates

| Classes | Permeability inches per hour | Percolation minutes per inch |
|---|---|---|
| Slow: | | |
| 1. Very slow | less than 0.05 | more than 1200 |
| 2. Slow | 0.05–0.20 | 300–1200 |
| Moderate: | | |
| 3. Moderately slow | 0.20 to 0.80 | 75–300 |
| 4. Moderate | 0.80 to 2.50 | 24–75 |
| 5. Moderately rapid | 2.50 to 5.00 | 12–24 |
| Rapid: | | |
| 6. Rapid | 5.00 to 10.00 | 6–12 |
| 7. Very rapid | more than 10.00 | less than 6 |

Important decisions that are based on a knowledge of soil permeability include (1) determination of the distance between lines of drainage tile, (2) size of area of seepage beds for septic tank systems, and (3) size of terrace ridges and the slope of terrace channels for erosion control. For example, note the characteristics of a claypan soil (Planosol) as presented in Figure 3–11. The high clay content of the Bt horizon is associated with little aeration pore space and very little permeability. It would take hours for an inch of water to move through the least permeable layer, in this case the Bt horizon. The claypan soil is unsuited for drainage by tile. If the soil existed on a slope, the claypan soil would be very susceptible to erosion resulting from a high water-runoff rate. During periods of rainy weather the lower

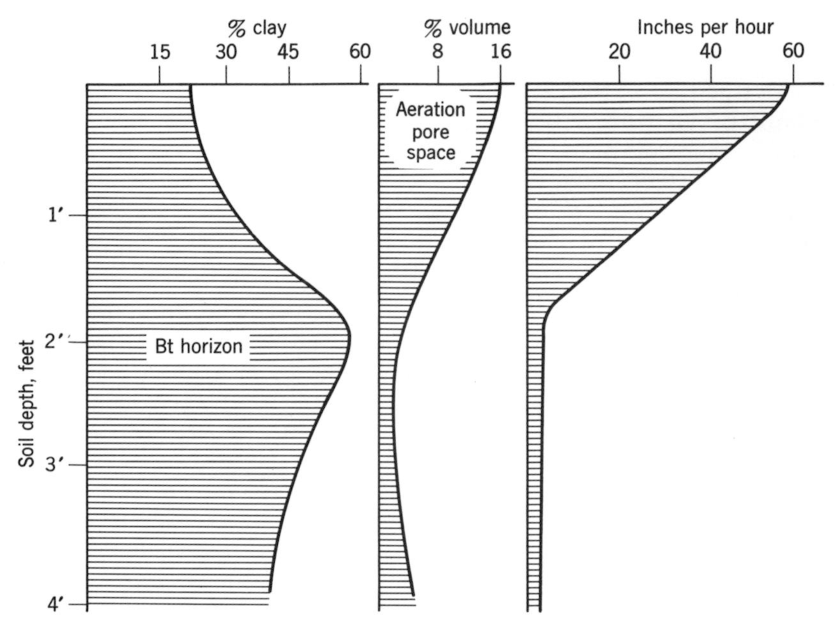


**Fig. 3–11** Distribution with depth of clay, aeration pore space, and permeability of a claypan soil (Planosol). (Based on data of Ulrich for Edina silt loam from *Soil Science Society of America Proceedings*, Vol. 14, 1950.)

part of the A horizon would become saturated with water. The lack of oxygen in the saturated soil would inhibit root growth. Let's consider next the relationship of soil aeration and root growth.

## Soil Aeration and Plant Growth

After a rain the smallest (capillary) pores in moist soil will be filled with water and the largest (aeration) pores will be air filled. Obviously, the amount and size of pores will affect both the water and air content of the moistened soil. If the rain storm is prolonged, a slow rate of water movement through less permeable underlying soil horizons may result in a "build up" of water in the surface soil layer and cause an increase in the water content and a decrease in the air content. Throughout the root zone, however, roots and organisms consume oxygen and liberate carbon dioxide. The soil air commonly contains 10 to 100 times more carbon dioxide and slightly less oxygen than normal air. The atmosphere contains 21 percent oxygen and 0.03 percent carbon dioxide. Unless oxygen can diffuse down into the root zone and carbon dioxide can diffuse out of the soil, an oxygen deficiency or carbon dioxide toxicity will eventually develop.

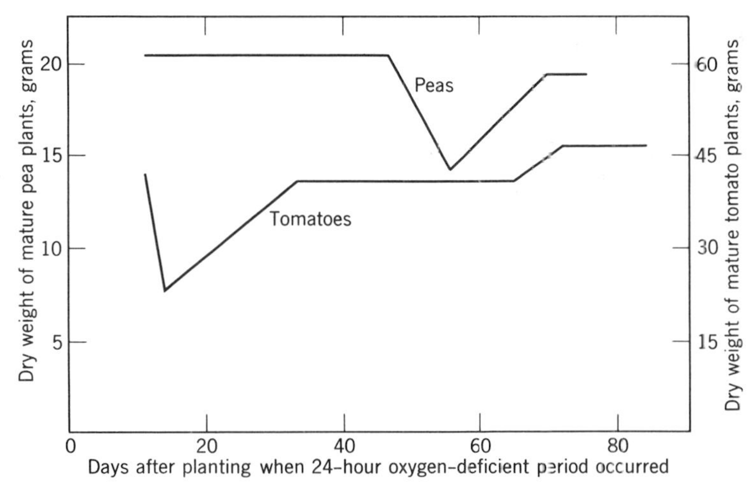


**Fig. 3–12** The effect of a 24-hour oxygen-deficient period on the growth of peas and tomatoes. Note that tomatoes were most sensitive in the early part of the season whereas peas were injured the most near flowering time. (From data of A. E. Erickson and D. M. Van Doren, *7th Intern. Cong. Soil Sci. Trans.*, Vol. 3, p. 431, 1960.)

The data plotted in Fig. 3–12 show that the growth of peas and tomatoes was reduced by an oxygen deficiency that lasted only 24 hours. It is interesting to note that a deficiency early in the season was the most detrimental for tomatoes and for peas the greatest reduction in growth resulted from a deficiency occurring later in the season (near blossoming). Thus, two important aspects of oxygen deficiency are the length of the oxygen-deficient period and the stage of plant growth when the oxygen deficiency occurs.

## TILLAGE AND SOIL PROPERTIES

The beginning of agriculture marks the beginning of soil tillage. Crude sticks were probably the first tillage tools used to establish crops. Paintings on the walls of ancient Egyptian tombs, dating back 5000 years, depict oxen yoked together by the horns, drawing a plow made from a forked tree. The Greeks improved the Egyptian plow by adding a metal point. By Roman times tillage tools and techniques had advanced to the point where thorough tillage was a recommended practice for crop production.

There was little improvement in the plow until American Colonial days, when a wooden-framed plow, having a moldboard covered with metal strips and carrying a curved point or share, was constructed. The "modern" metal moldboard plow was cast in about 1800. Development and improvement of tillage tools and methods contributed to greater food production and an increase in the human population. Childe[4] considered the discovery and development of the plow one of the nineteen most important discoveries or applications of science in the development of civilization.

There are three commonly accepted purposes of tillage: (1) to kill weeds, (2) to manage crop residues, and (3) to alter soil structure. Let us consider next the reasons for tillage and some modern tillage techniques and problems.

## Tillage and Weed Control

Weeds compete with crop plants for nutrients, water and light. If weeds are eliminated without tillage, can cultivation of row crops be eliminated? An answer to this question can be developed by considering the data in Table 3–6. In three different six-year-long experiments the average yield of corn where weeds were controlled by cultivation was 60 bushels per acre. Cutting the weeds off at the surface of the ground with a scraper produced an average yield of 61 bushels per acre. Yields were very low when weeds were allowed to grow. Data from these experiments support the conclusion that the major benefit of cultivating the corn was weed control. On many soils herbicides have been used for weed control with good success. There are, however, some instances where cultivation during the growing season may be justified to improve soil aeration or to increase the infiltration of water.

**Table 3–6** Effect of Weed Control Methods on the Yield of Corn

| | Weed Control Method | | |
|---|---|---|---|
| Soil Type and State | None, Weeds Allowed to Grow | Cultivation | Scraping of the Soil Surface |
| Brown silt loam, Illinois | 7 | 53 | 53 |
| Miami silty clay loam, Ohio | – | 56 | 60 |
| Brookston silty clay loam, Ohio | – | 72 | 70 |
| Average | – | 60 | 61 |

[4] *Man Makes Himself,* V. E. Childe, The New American Library of World Literature, Inc., 1951.

## Effect of Tillage on Soil Structure

All tillage operations change the structure of the soil. The lifting, twisting, and turning action of the moldboard plow leaves the soil in an aggregated and loose condition. Cultivators, discs, and packers tend to crush some of the soil aggregates and leave the soil in a more compact state.

Cultivation of a field to kill weeds may have the immediate effect of loosening the soil, increasing soil aeration and infiltration of water. The long-time (few weeks or months) effect of cultivation resulting from crushing of soil aggregates may be a less well-aggregated and more compact soil. Exposed cultivated land also suffers from disruption of aggregates by raindrop impact in the absence of a vegetative cover. We have already observed in Fig. 3–8 that cropping systems with the least frequent tillage (corn-oats-legume) are associated with highest percentage of aggregated soil. The most frequent tillage occurred in continuous cropping systems (continuous corn) where soil aggregation was the least.

It is not uncommon to find a compact subsurface layer that has been caused by the pressure exerted by the tires of tractors or other heavy equipment. These compact layers are called *traffic pans*. Most of the tire pressure is applied downward to soil directly underneath the tire tread and the *depth* of compaction increases with increasing soil moisture content. On wet soil compaction may occur to a depth of 24 inches or more. However, the greatest pressure is exerted on the soil immediately underneath the tire. Repeated trips over a field with heavy equipment tend to produce a traffic pan. In plowed fields the pan occurs underneath the plow layer because subsequent tillage operations may not break up the soil underneath the plow layer. Such pans may inhibit root penetration and percolation of water. It can be seen that tillage operations may improve soil structure or contribute to an undesirable soil structure.

## Tillage and Management of Crop Residues

Crops are generally grown on land that contains the plant residues of a previous crop. The moldboard plow is widely used for burying such crop residues in the humid region. Fields free of trash permit precision placement of seed and fertilizer at planting and easy cultivation of the crop during the growing season. In the subhumid and semiarid regions, by contrast, the need for wind erosion control and conservation of moisture have led to the development of machines that can successfully establish crops without plowing as shown in

**Fig. 3–13** A chisel planter designed to plant corn in the residues of the previous crop without plowing the land. In one operation the land is prepared, planted, fertilized, and sprayed with herbicide. (Photo USDA Soil Conservation.)

Fig. 3–13. The plant tops remain on the surface and provide some protection from water and wind erosion. The plant residues left on the land over the winter may also cause snow to become lodged, which later melts and increases the water content of the soil.

### The Minimum Tillage Concept

It is obvious that plants grow without tillage of the soil. Sooner or later it was inevitable that the extent to which tillage was necessary would be questioned in the search for ways to maintain the soil in good physical condition and to produce high yields at minimum cost. Even though plants may grow very well in experiments without tillage, production of most crops will generally require at least some tillage. For the production of sugar beets it is interesting to observe in Table 3–7 that on an experimental field the yields were lowest when the land was not worked at all between plowing and planting and when the land was worked 4 or more times. The highest yields occurred where the land was worked only 1 or two times between plowing and planting. Clearly, this experiment shows that it is easy to overtill the soil. Some tillage is necessary for the practical production of crops, but the concept of "thorough tillage" is declining. The result has been considerable progress in the development of minimum tillage systems.

**Table 3–7** Beet Yields as Affected by the Number of Times a Field Was Worked Prior to Planting

| Times Worked | Tons per Acre |
|---|---|
| None | 14.0 |
| One | 16.8 |
| Two | 16.7 |
| Three | 15.2 |
| Four | 14.8 |
| Five | 14.2 |
| Six | 14.3 |

From "1958 Production Practices of Michigan Sugar Beet Farmers," R. L. Cook, J. F. Davis, and M. G. Frakes, *Quart. Bull. Mich. Agr. Exp. Sta.*, November 1959.

Minimum tillage systems employ fewer operations to produce crops. Machinery like that in Fig. 3–13 prepares the land, plants the seed, and applies fertilizer and herbicide in one trip over the field. Under favorable conditions, no further tillage will be required during the growing season. Press wheels behind the planter shoes pack soil only where seed is placed; leaving most of the soil surface in a state very conducive to water infiltration. Experimental evidence shows that minimum tillage reduces water erosion on sloping land. It also makes weed control easier because planting immediately follows plowing. Weed seeds left in the loose soil are at a maximum disadvantage. Crops yields are similar to those where more tillage is used, but the use of minimum tillage in crop production reduces costs.

## Tillage Operations of the Future

The minimum tillage concept represents a revolution in tillage practices. Rapidly expanding human populations and food requirements will probably intensify the use of the best agricultural lands and encourage another revolution in tillage techniques. Some futuristic planners foresee the day when many fields will be leveled, fields will be ten miles long, and all production operations will be carried out with machines that run on tracks. Obviously, there would be no soil compaction. Such systems or machines would permit precise tillage operations and precise placement of seeds, fertilizers, and pesticides, and exact application of irrigation water and harvesting. Although these changes are being anticipated in the world where advanced technology exists, 70 percent of the world's farmers have only a hoe or wooden plow as their sole tillage implement.

## SOIL COLOR

The color of soil serves both farmer and soil scientist, provided that they understand the causes of the various colors and are able to interpret them in terms of soil properties. Organic-matter content, drainage condition, and aeration are soil properties related to color which are of interest to farmers. The investigator uses color as an aid in soil classification and draws from the color of the different horizons information about conditions pertaining to and forces active during soil formation.

### Factors Affecting Soil Color

The minerals occurring in appreciable quantities in most soils are light in color. As a result, soils would be of a light gray color if composed of crushed minerals which had undergone little chemical change. Accordingly, for an explanation of the dark gray, brown, red, and yellow colors in soils we must look to chemical changes in the constituents (especially iron) of the minerals and to the addition of organic matter. There are a few instances, however, in which the proportion of colored minerals is sufficient to give the soil a decided color.

The dark color of soils is generally due to the highly decayed organic matter they contain; in fact, with some practice the percentage of organic matter in many soils may be judged with reasonable accuracy from their colors. Organic matter imparts a gray, dark gray, or dark brown color to soils unless some other constituent such as iron oxide or an accumulation of salts modifies the color.

If soils are poorly drained there is usually a greater accumulation of organic matter in the surface layers, thus giving a very dark color. The lower soil layers, which contain very little organic matter, on the other hand, are of a light gray color, indicating the poorly drained condition. If drainage is intermediate, the gray of the subsoil is likely to be broken by flecks of yellow.

When drainage permits aeration, and moisture and temperature conditions are favorable for chemical activity, the iron in soil minerals is oxidized and hydrated into red and yellow compounds. Highly hydrated iron oxides are yellow, but as hydration diminishes reds replace the yellows. Accordingly, we find shades of red in soils extending from the southern deserts across the semiarid and subhumid states of the Southeast. The red and yellow colors of the subsoils in the southeastern states immediately catch the eye of the traveler. The low organic matter content of many soils in this area leaves undulled the brilliant colors of the iron oxides. With an appreciable humus content the red colors are converted into mahogany colors.

## Measurement of Soil Color

The color of light can be accurately described by measuring its three principal properties, hue, value, and chroma. Hue refers to the dominant wavelength or color of the light. Value, sometimes called brilliance, refers to the total quantity of light. It increases from dark to light colors. Chroma is the relative purity of the dominant wavelength of light. It increases with decreasing proportions of white light.

The Munsell notation of color is a systematic numerical and letter designation of each of the three variable properties of color. The relationships of the colors to one another can be shown by use of a solid, for example a cube, in which hue, value, and chroma are plotted along the three edges. Each possible color represents a point in this cube and is completely defined by the three coordinates of that point, which is its Munsell notation. The three properties are always given in the order hue, value, and chroma. For example, in the Munsell notation 10YR 6/4, 10YR is the hue, 6 is the value, and 4 is the chroma.

The Munsell notation for a given soil sample can be quickly determined by comparison of the sample with a standard set of color chips (Fig. 3–14). The chips are mounted in a notebook with all the colors of a given hue on one page. Each page then corresponds to a slice through the color cube parallel to its front. The pages are arranged in the order of increasing or decreasing wavelength of the dominant color in order to facilitate the matching of the unknown soil color with the color of the standards.

Many soil horizons have a single dominant color. Horizons that are dry part of the year and wet part of the year tend to exhibit a mixture of two or more colors. These colors are intermediate between those of well-drained and poorly drained soils. When several colors are

**Fig. 3–14** Color is determined by comparing a sample of soil with the color chips in the Munsell color book as shown on the right. These two soil scientists are writing a soil profile description. Other properties that will be recorded for each horizon besides color include texture, structure, thickness, and pH. The latter is being determined by the soil scientist on the left.

present in a spotted or variegated pattern the word *mottled* is used to describe the condition. In these cases several of the dominant colors may be recorded.

## SOIL TEMPERATURE

As temperatures decrease, the life processes of both plants and animals are slowed down until finally they cease altogether. Growth processes of most agriculturally important plants are very sluggish at temperatures about 40°F and increase until temperatures ranging from 70° to 90°F are reached. The chemical processes and activities of micro-organisms which convert plant nutrients into available forms are also materially influenced by temperature. Plant growth is thus seen to be importantly influenced by both air and soil temperature.

### Heat Balance of a Soil

The heat balance of a soil consists of the gains and losses of heat energy. Solar radiation received at the soil surface is partly reflected back into the atmosphere and partly absorbed by the surface of the soil. A dark-colored soil and a light-colored quartz sand may absorb, respectively, about 80 and 30 percent of the incoming solar radiation. Of the total solar radiation available for the earth, about 34 percent is reflected back into space, 19 percent is absorbed by the atmosphere, and 47 percent is absorbed by the earth.

Absorbed heat is lost from the soil by (1) evaporation of water, (2) reradiation back into the atmosphere as longwave radiation, (3) heating of air above the soil, and (4) heating of the soil. In the long run the gains and losses balance each other. For the short run considered as daytime or summer, heat gains exceed heat losses and soil temperatures increase. During the night and winter the reverse is true.

## Heat Capacity of Soil

Mineral particles require a comparatively small amount of heat to raise their temperature. The quantity of heat required to raise the temperature of a gram of soil particles 1°C is only about one-fifth as much as is required to warm a gram of water the same amount. In other words, the specific heat of dry soil particles is 0.2, and it is evident from this fact that moisture content is an important factor in determining soil temperature. Soil high in water content will warm up slowly in the spring and will cool down slowly in the fall. Drainage therefore exerts a major influence on soil temperature.

## Control of Soil Temperature

As has just been pointed out, the removal of excess water from a soil will facilitate changes in soil temperature. By providing drainage, man may exert some influence on the temperature relations of soils which are so situated that they hold excessive quantities of water. By use of mulches and various shading devices, the amount of solar radiation absorbed by the soil, loss of heat energy from the soil by radiation, infiltration of water, and loss of water by evaporation can be altered.

Light-colored organic matter mulches: (1) reflect a large part of solar radiation, (2) retard heat loss by radiation, (3) increase infiltration of water, and (4) reduce evaporation of water from the surface of the soil. The net effect of a light-colored organic matter mulch is to reduce soil temperature. In regions where summers are cool the reduced soil temperature has been found to reduce crop yields. Dark-colored plastic mulches: (1) absorb most of the solar radiation, (2) reduce heat loss from the soil by radiation, and (3) reduce the evaporation of water from the surface of the soil. The net effect of black plastic mulches is to increase soil temperature in the soil under the

**Fig. 3–15** A black plastic mulch increases soil temperature and hastens the beginning of the harvest period for muskmelons. Weeds are also controlled.

mulch when used as shown in Fig. 3–15. The higher soil temperature increases crop yields in regions with cool summers. In Michigan plastic mulches are used to increase the growth rate of muskmelons which results in an earlier harvest. The earlier-harvested melons have a sufficient price advantage to make the mulching practice profitable. A further advantage of a plastic mulch that is opaque is the control of weeds where the mulch has been used.

## Location and Temperature

In the northern hemisphere, soils located on southern and southeastern slopes warm up more rapidly in the morning than those located on the level or on the northern slopes. The reason is that they are more nearly perpendicular to the sun's rays, and hence a maximum amount of radiant energy strikes a given area. Soils with a southern or southeastern exposure often are selected for the growing of early vegetables and fruits.

# 4

# Soil Water

Water is the most common substance on the earth and is necessary for all life. The supply of fresh water on a long-time sustained basis is equal to the annual precipitation, which averages 26 inches for the world's land surface. The soil, located at the atmosphere-lithosphere interface, plays an important role in determining the amount of precipitation that runs off the land and the amount of precipitation that enters the soil for storage and future use. Approximately 70 percent of the precipitation in the United States is evapotranspired and returned to the atmosphere as vapor with the soil playing a key role in water retention and storage. The remaining 30 percent of the precipitation represents the long-time annual supply of fresh water for use in homes, industry, and irrigated agriculture. About one-fifth of this 30 percent is currently being used in the United States.

Fortunately, water is not easily destroyed. The earth has as much water now as it did thousands of years ago. However, the water is unevenly distributed by rainfall, changes form, moves from place to place, and can be polluted. For students interested in pollution of the environment, resource use, and plant science this chapter contains important concepts and principles that are essential for gaining an understanding of the soil's role in the intelligent management of water resources.

## ENERGY CONCEPT OF SOIL WATER

As water cascades over a dam, the energy content (ability to do work) of the water decreases. If water that has gone over a dam is returned to the reservoir, work will be required to lift the water back up into the reservoir, and the energy level of the water will be restored. Water movement in soils and from soils into plant roots, like water cascading over a dam, is from regions of higher energy water to regions of lower energy water. Thus, "water runs down hill." For this reason it is necessary to consider the forces that determine the physical state or energy content of water in order to understand the behavior of water in soils.

### Surface Forces Control the Physical State of Soil Water

The change of water from a vapor to a liquid (condensation) is accompanied by a great reduction in the movement of the molecules and their energy content. Energy is released as heat when water changes from vapor to liquid. The liberation of heat from the formation of raindrops is a major source of energy for storm systems. When raindrops fall on dry soil and are adsorbed on the surface of soil particles, a further reduction occurs in the motion and energy content of the water molecules. The adsorbed water may still be in a liquid state, but the tendency of the water molecules to move has been further reduced. This change in energy content can be explained by considering the forces operating between soil particles and water molecules.

Water molecules are dipolar and attract each other through hydrogen (H) bonds (see Fig. 4–1). Soil particles are also charged and have

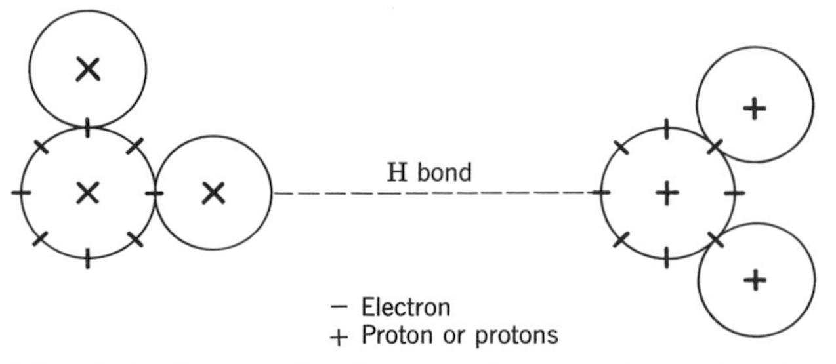


**Fig. 4–1** Schematic diagram of two water molecules. The sharing of electrons by the oxygen and hydrogen produces water molecules that are negatively charged on the oxygen side and positively charged on the side where the proton of the hydrogen sticks out. Molecules attract each other by H bonding; the attraction of the proton of one water molecule for the negatively charged oxygen side of an adjacent water molecule.

negative and positive charged sites. However, the bonding energy (strength of attraction) of the electrical charged sites of soil particles is much greater than the bonding energy of the H bonds of water molecules. The strong attraction of the soil for the water molecules (adhesion) results in a spreading of the water over the surface of the soil particles as a film when liquid water comes in contact with dry soil particles. The adsorption of water on the surface of soil particles produces (1) a reduction in the motion of the water molecules, (2) a reduction in the energy content of the water, and (3) the release of heat associated with the transformation of water to a lower energy level. You can observe the release of heat, called *heat of wetting*, by adding water to oven-dry soil in your hand and noting the increase in soil temperature.

Several layers of water molecules are strongly adsorbed to the soil particles basically because of these strong adhesive forces (see Fig. 4–2). This water is called *adhesion water*. Adhesion water moves little, if at all, and some scientists believe that the innermost layers of water molecules exist in a crystalline state similar to the structure of ice. Adhesion water is not available to plants and is always present in the normal soil (even in the dust of the air), but the adhesion water can be removed by drying the soil in an oven.

Beyond the sphere of strong attraction of the soil particles, water molecules are held in the water film by cohesion (H bonding between water molecules). This outer film water is called *cohesion water*.

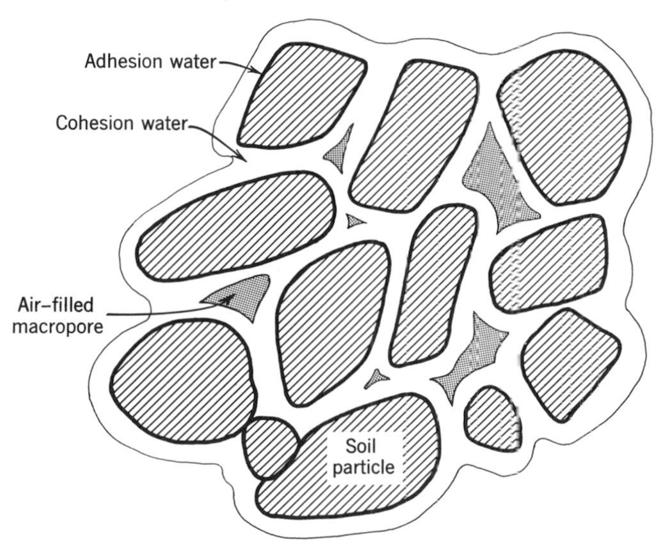


**Fig. 4–2** Schematic drawing showing the relationship of adhesion and cohesion water with respect to soil particles and air-filled macropores.

Molecules of cohesion water, compared to adhesion water, are in greater motion, have a higher energy level, and move more readily. The water film (including adhesion and cohesion water) in soils may be as much as 15 to 20 molecular layers thick. The outer approximately two-thirds of the film can be considered available to plants and constitutes the major source of water for plant growth (see Fig. 4–2).

You can demonstrate the existence of the forces important in holding water in soils by placing two clean microscope slides in water and bringing them together so that their flat sides are flush against each other. Then, try to pull them apart. Your failure to pull the slides apart demonstrates the existence of attractive forces between glass and water molecules. You can also observe that the forces are effective only over a very short distance, since the glass slides must be very close together before a strong attraction develops between the slides.

## Energy and Pressure Relationships

We have just noted that water exists in the soil over a range of energy contents. This energy content of water can be expressed in terms of water pressure. Because it is much easier to determine the pressure of water, as contrasted to the energy content or level, we usually categorize water on the basis of pressure. For this reason we must understand the relationship between the energy content and the water pressure.

The hull of a submarine must be sturdily built to withstand the great pressure encountered in a dive far below the ocean's surface. The head of water above the submarine exerts pressure on the hull of the submarine. If the water pressure exerted against the submarine was directed against the blades of a turbine, the gravitational energy in the water could be used to generate electricity. The greater the water pressure, the greater is the tendency of water to move and do work and the greater is the energy content of the water. Thus, the existence of a relationship between the pressure of water and the energy of water is established. Our next consideration will be that of the water-pressure relationships in saturated soils, which is analogous to the water-pressure relationships in a beaker of water or any body of water.

## Water Pressure in a Beaker or in Saturated Soil

The beaker in Fig. 4–3 has a bottom with an area of 100 square centimeters. The height of the water in the beaker is 20 centimeters.

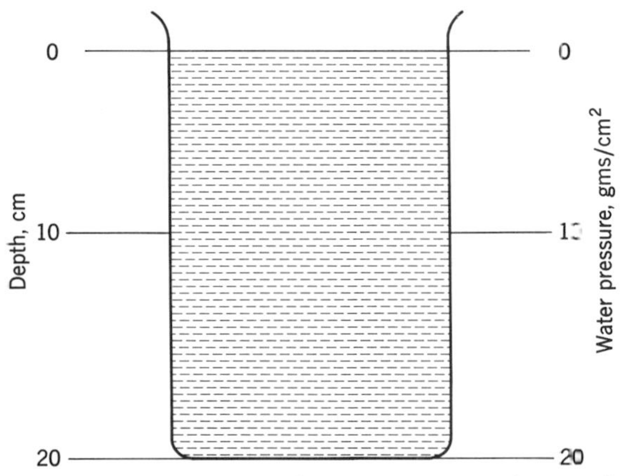


**Fig. 4–3** A beaker with a cross-sectional area of 100 square centimeters contains 2000 grams of water when the water is 20 centimeters deep. At the water surface the water pressure is 0 and the water pressure increases with depth to 20 grams per square centimeter at the bottom.

The water has a volume of 2000 cubic centimeters, weighs 2000 grams, and exerts a total force on the bottom of the beaker equal to 2000 grams. The pressure of the water at the bottom of the beaker is equal to:

$$P = \frac{\text{force}}{\text{area}} = \frac{2000 \text{ grams}}{100 \text{ cm}^2} = 20 \text{ grams per square centimeter}$$

The water pressure at the bottom of the beaker could also be expressed simply as equivalent to a column of water 20 centimeters high.

At the 10-centimeter depth the water pressure is half of that at the 20-centimeter depth and is, therefore, 10 grams per square centimeter. The *water pressure* decreases with distance toward the surface and becomes zero at the free water surface (see Fig. 4–3).

To apply the considerations of water in a beaker to soil, we must consider soil that is *saturated* with water. At the top of the water table or at the top of the soil saturated with water, the water pressure is zero. The water pressure increases with increasing depth below a water table as it does in a beaker or any body of water. The water existing in the macro or aeration pores of a *saturated* soil has a positive pressure determined by the distance below the surface of the saturated zone. Water in the macropores of saturated soil is under pressure and will freely flow through macropores from regions of higher pressure to regions of lower pressure (basically from higher elevations to lower elevations). The water that "freely" flows or drains out of the soil is called *gravitational* water. Where gravitational water exists in soil adjacent to the walls of the basement of a building, water pressure may force water through the cracks of walls and floor. If a soil remains saturated with water, one can assume that an impermeable layer is inhibiting the flow of gravitational water from the soil. Drainage ditches and tile drains are used to remove gravitational water from saturated soils.

After the gravitational water has drained out of the soil, the soil becomes unsaturated and the remaining water is held by the attractive forces between soil particle surfaces and water molecules (adhesion) and between water molecules (cohesion). A clear differentiation between saturated and unsaturated soils is needed to gain a meaningful understanding of the behavior of soil water. We shall consider next the water-pressure (energy) relationships in unsaturated soil.

### Water Pressure in a Capillary Tube or in Unsaturated Soil

If the tip of a capillary tube is inserted in a beaker of water, the attraction between glass and water molecules (adhesion) causes water molecules to migrate up the interior wall of the capillary. The cohesive force between water molecules causes other water molecules to be drawn up the capillary. Now we need to ask, "What are the pressure relationships in the capillary tube?" and "Why is this knowledge important?"

We have already seen that the water pressure decreases from the bottom to the top of a beaker and at the top of the water surface the water pressure is zero. Beginning at the water surface of the beaker and moving upward into the capillary tube, the water pressure continues to decrease. Thus, the water pressure in a capillary tube is less than zero or is *negative.* From Fig. 4–4 one can observe that the water pressure decreases in a capillary tube with height above the water in a beaker. At a height of 20 centimeters above the water surface in a beaker, the water pressure in a capillary tube is equal to −20 grams per square centimeter. Figure 4–4 also shows that at this same height above the water surface the water pressure in *unsaturated* soil is the same; −20 grams per square centimeter at the 20 centimeter height. The two pressures are the same at any one height and there is no net flow of water from the soil or capillary after an equilibrium condition is established.

Applying the considerations of water in a capillary tube to an unsaturated soil, we can make the following statements; (1) water in unsaturated soil has a *negative* pressure or is under *tension,* (2) the

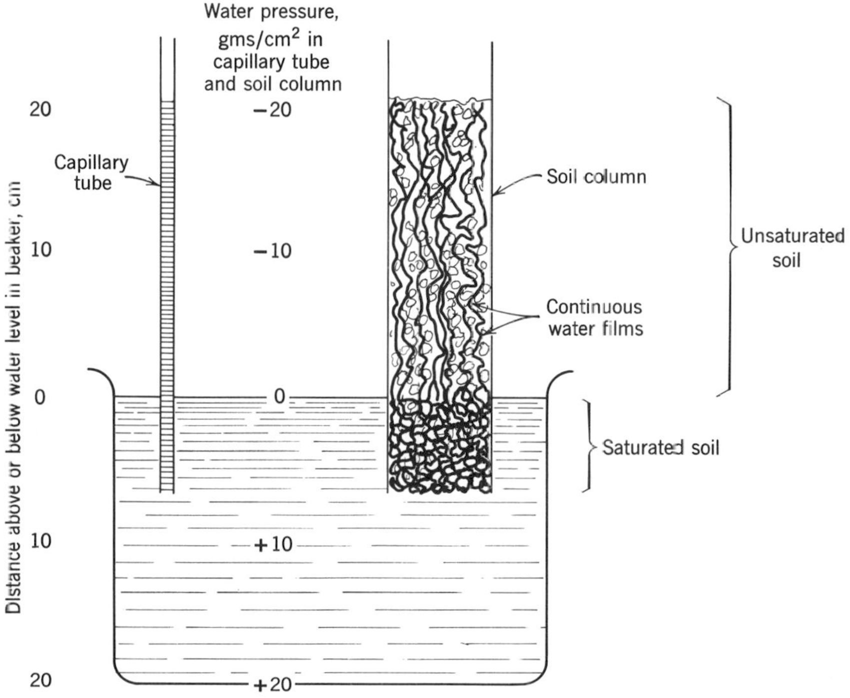


**Fig. 4–4** Water pressure in a capillary tube decreases with increasing distance above the surface of the water in the beaker and is −20 grams per square centimeter at a height 20 centimeters above the water surface. Since the water column of the capillary tube is continuous through the beaker and up into the soil column, the water pressure in the soil 20 centimeters above the water level of the beaker is also a −20 grams per square centimeter.

water pressure in unsaturated soil decreases with increasing distance above the surface of a water table, and (3) water in unsaturated soil, compared to saturated soil, has a lower pressure and lower energy level. Let's consider next the importance of the nature of water in unsaturated soil in relation to water movement.

## Water Movement in Unsaturated Soil

Several important consequences follow from the nature of water in unsaturated soils. Water in unsaturated soil exhibits very little tendency to move. Movement is very slow and mainly by adjustment of the thickness of water films on soil particles. We can visualize the water as occurring as surface films and as wedges in the angles of adjoining soil particles as shown in Fig. 4–5. There is a tendency for

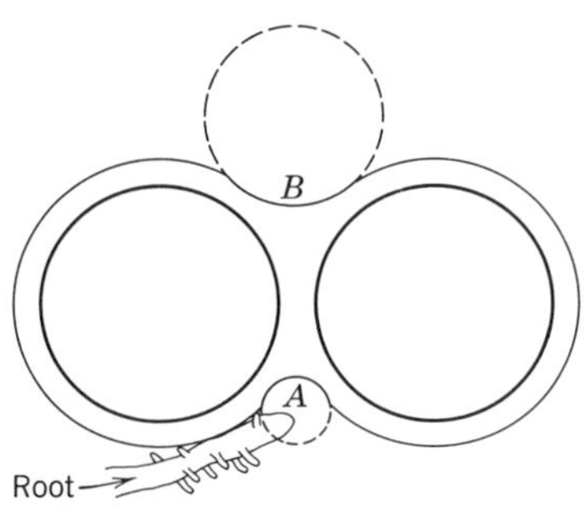


**Fig. 4–5** As the root absorbs moisture from the accumulation between two soil particles, the film curvature increases, as is shown by the projected circles. As the force tending to draw water into a given portion of the film varies inversely with the radius of the curvature ($p = 2T/r$), it follows that moisture will move to the feeding point of the root.

spherical droplets to form at points $A$ and $B$ in Fig. 4–5. However, droplet formation is prevented because the two soil particle surfaces are acting against each other. Now, if the curvature of one film is greatly increased through the removal of moisture by a root, as is shown at $A$, its pulling power will be proportionately increased since the pull exerted by such a curved film is inversely proportional to the radius of curvature. As a result, water will be drawn toward $A$ until the curvature of films $A$ and $B$ is equal. The equation for capillary movement may be written: $p$ (pull) $= 2T/r$, in which $T$ is surface tension and $r$ the radius of curvature. If it is assumed, for example, that the $r$ of $A$ is 1 and of $B$ is 2, the pull exerted at the two points will be $p = 2T/1$ compared to $p = 2T/2$. As the liquid in each case is water, $T$ will be equal, and so the pull exerted by the film $A$ will be twice as great as that exerted by film $B$. By film adjustment we can explain the movement of water in unsaturated soils in the direction of drier soil as a result of absorption of water by plant roots. Significant water movement occurs over distances of the order of less than an inch. As stressed in Chapter 2, the limited mobility of water in well-aerated soils (unsaturated soils) requires extension of plant roots into all soil horizons from which water is absorbed.

When moist unsaturated soil is adjacent to the basement walls of a building, water may slowly move by capillarity through small pores or cracks in the walls or floor. The edges of cracks may appear moist but pools of water will not collect on a basement floor, if water is moving by capillarity. Can you explain this on the basis of energy and pressure relationships?

### Water Pressure Versus Soil Moisture Tension (SMT)

A few crops, like paddy rice, are grown on saturated soils. In most situations plants are growing on unsaturated soils where the water is under tension analogous to water in a capillary tube. The water pressure is a negative quantity and expression of the pressure of soil water is more conveniently done with positive tension values. Thus, a negative pressure of 20 grams per square centimeter becomes a soil moisture tension (SMT) of 20 grams per square centimeter or:

$$(\text{water pressure}) \times (-1) = \text{soil moisture tension (SMT)}$$

Since it is the downward pull of a 20-centimeter column of water that produces 20 grams per square centimeter of tension (see Fig. 4–4), the SMT as grams per square centimeter can be directly converted into equivalent lengths of water columns. That is, a tension of 20 grams per square centimeter is equivalent to the tension produced by a 20-centimeter-long water column. It is analogous to saying a pressure of 14.7 pounds per square inch is equal to one atmosphere.

Soil moisture tension is commonly expressed in atmospheres. As a first step to express SMT in atmospheres consider the following equivalents:

1 atmosphere = 33 foot long water column = 14.7 pounds per $ft^2$

1 atmosphere = 1036 centimeter column of water = 1036 grams/$cm^2$

To convert the SMT at the top of the soil column in Fig. 4–4 to atmospheres, the 20 must be divided by 1036:

$$\frac{20 \text{ grams/cm}^2}{1036 \text{ grams/cm}^2} = 0.019 \text{ atmosphere}$$

When the SMT is 0.019 atmosphere, an opposing tension or drawing force (suction) greater than 0.019 atmosphere must be exerted by a plant root before water will move from the soil into the root. Shortly, in the consideration of the biological aspects of soil water, it will become clear that this tension (0.019 atmosphere) is very low and water at this tension is very available to plants.

To help clarify this concept we shall consider a practical example that shows the utility of knowing what SMT is and how it is expressed. Suppose that a flower pot has recently been watered and the excess water has just ceased draining from the large hole (noncapillary sized) in the bottom of the pot as shown in Fig. 4–6. The soil in our illustration is 20 centimeters thick. The large hole in the bottom of the pot

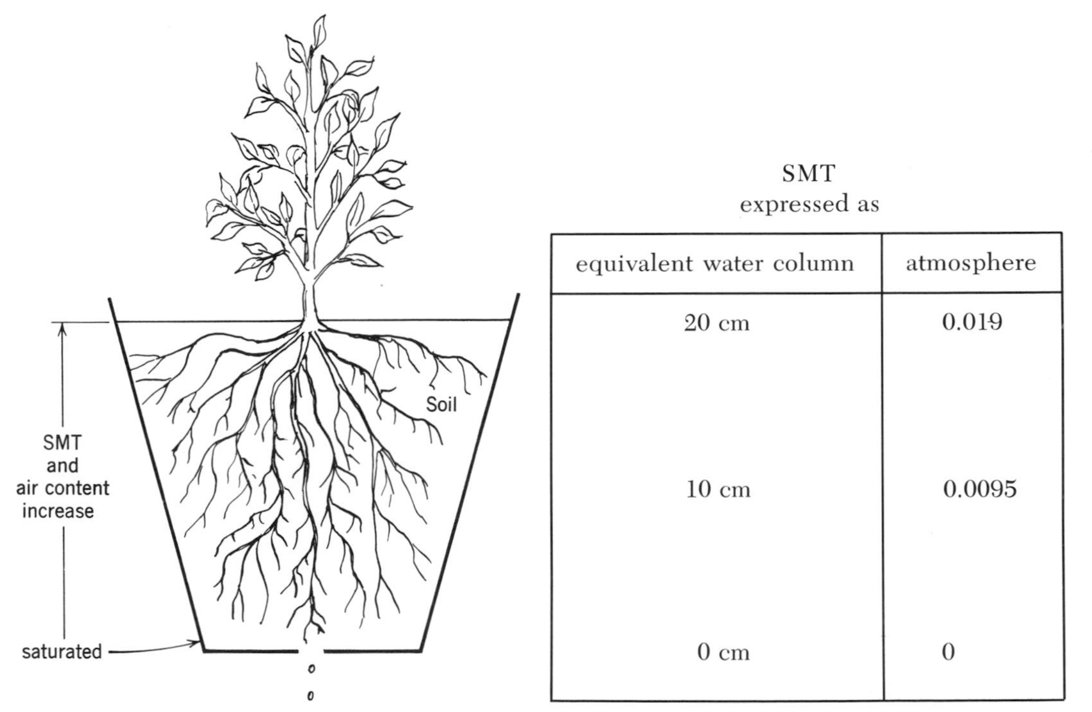


SMT
expressed as

| equivalent water column | atmosphere |
|---|---|
| 20 cm | 0.019 |
| 10 cm | 0.0095 |
| 0 cm | 0 |

**Fig. 4–6** Soil moisture tension and air relationships in a flower pot after watering and the excess water has stopped draining from the large hole (non-capillary size) in the bottom of the pot.

means that capillarity can not pull water out of the bottom of the pot. The soil at the very bottom of the pot will have an SMT of zero and the SMT will increase with distance upward into the soil. At the soil surface the SMT will be equivalent to a water column of 20 centimeters or 0.019 atmosphere. A very important consequence is that at the *very* bottom of the pot the soil is saturated and the air content is zero. The air content (and oxygen content) of the soil increases from the bottom to the top of the pot (see Fig. 4–6). Plants growing in pots only 10 centimeters tall would, in effect, have a less well aerated soil in which to grow. The same analogy applies for soil immediately above the water table and we shall consider this situation in more detail in the discussion of soil drainage.

## Measurement of Soil Moisture Tension

A vacuum gauge-type tensiometer, shown in Fig. 4–7, consists of a porous, fired clay cup, which is attached to a vacuum gauge by a water-filled pipe. If the porous cup is buried in soil where the water has a tension greater than zero, water will move from the cup into the soil. At equilibrium the tension created within the tensiometer and registered by the vacuum gauge is the SMT. This means that a force or pull (suction) *greater* than this tension must be exerted to remove some water from the soil. Wetting the soil releases the tension and water then moves from the soil into the porous cup of the tensiometer. Tensiometers work up to the range of about 0.7 to 0.8 atmosphere. This range is important in terms of plant growth making the tensiometer a useful instrument to determine when to apply irrigation water.

Measurement of SMT at higher tensions is made with the pressure chamber in the range 1 to 50 atmospheres. A membrane with very fine pores that are water filled rests on a screen in the bottom of the pressure chamber. The soil is placed on the membrane, the chamber is sealed, and air pressure is applied. Water is forced out of the soil and flows through the pores of the membrane. At equilibrium the SMT equals the air pressure that was applied. The pressure chamber is an invaluable research tool and is used to obtain the kind of data plotted in Fig. 4–8.

At an SMT of 346 centimeters of water, for all three soils shown in Fig. 4–8, the plants would need to exert the same amount of force to remove water from each soil, but the water content of the clay at this tension is about 7 times greater than that of the sand. This point brings out a most important fact, the ability of plants to remove water from soils is primarily related to SMT and not water content. It is the SMT and not the water content that indicates when to irrigate and this

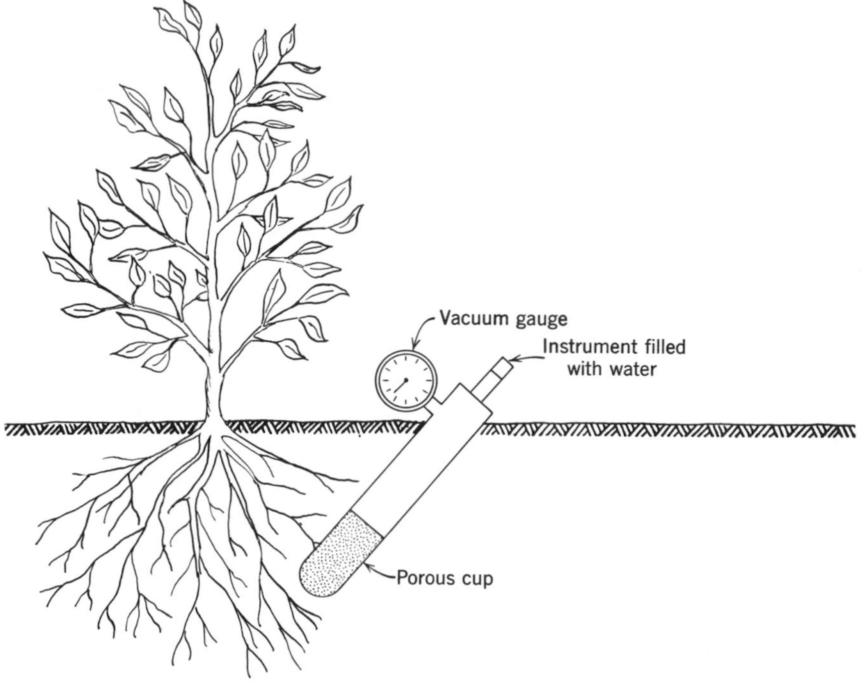


**Fig. 4–7** Vacuum-gauge tensiometer used to measure soil moisture tension. Tensiometer gauges can be wired to automatically turn on and off irrigation systems in response to changes in SMT. (Drawn after Richards and used with permission from *Soil-Plant Relationships,* C. A. Black, 2nd edition, p. 77, John Wiley & Sons Inc., New York, 1968.)

shows the importance of tensiometers (and other SMT measuring devices) in irrigation agriculture.

### Summary Statement

Adhesion water in soils:

1. Is held by strong surface (electrical) forces existing between soil particles and water molecules.
2. Is mostly "crystalline," exhibits little or no movement, and possesses a low energy content
3. Exists as a film on the surface of soil particles, several molecular layers thick.
4. Is unavailable to plants.
5. Exists on surfaces of dry soil particles that occur as dust in the air, but can be removed by drying the soil in an oven.

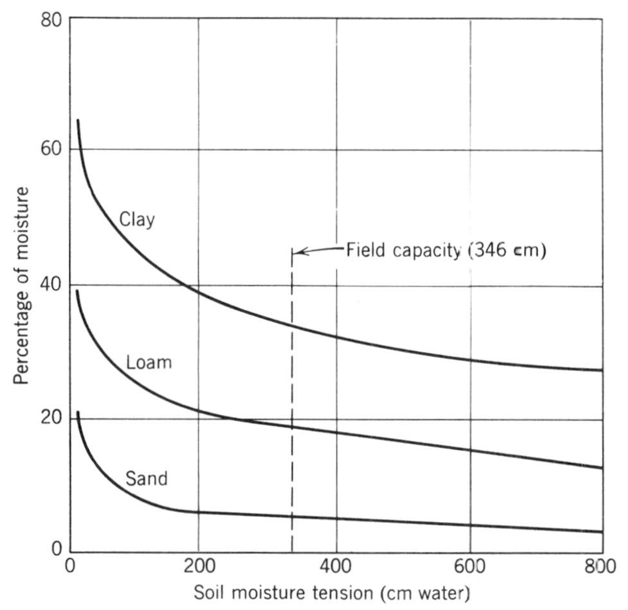


**Fig. 4–8** The moisture contents of soils of different textures at various moisture tensions. (Data from "The Usefulness of Capillary Potential to Soil Moisture and Plant Investigators," L. A. Richards, *J. Agr. Research*, Vol. 37, No. 12, 1928, p. 732.)

Cohesion water in soils:

1. Is held by the attraction of water molecules for each other through H bonding.
2. Exists in the liquid state in the water films around soil particles and in micropores.
3. Is the major source of water for plant growth.
4. Has a higher energy content than adhesion water.
5. Moves very slowly by film adjustment from areas where the films are thickest and SMT is lower to areas where the water films are thinner and the SMT is higher.

Gravitational water in soils:

1. Exists in the macropores.
2. Is either free water or under very low tension.
3. Moves freely through macropore spaces in response to very low water pressure differences or gravitation.

In saturated soils water movement is rapid in macropore spaces in response to gravity and the soil lacks oxygen for biological activities.

In unsaturated soils the capillary forces govern water movement, which occurs very slowly over very small distances and the macropore spaces contain air. As the water is removed by plants in unsaturated soil, the SMT increases and this results in greater resistance to movement of water and uptake by plant roots. The role of SMT in utilization of soil water by plants will be examined in the following section dealing with plant–soil water relations.

## PLANT–SOIL WATER RELATIONS

In the course of a summer day, it is not unusual for a plant to transpire an amount of water equal to many times its weight. Since the water that plants absorb from soils is not free flowing but diffuses slowly into plant roots by osmosis, an enormous area of contact between roots and soil particles is required. The total length of the root system of a corn plant could easily equal the distance from New York to San Francisco and back. Our concern in this section centers around the ability of plants to satisfy their water requirements from the soil.

### Gravitational Water and Field Capacity

When rain falls on the soil, water is pulled or sucked down into the soil by capillary action. If free water builds up on the surfaces of the soil, water may flow freely down through large macrospores. The macropores, however, must be open at the top of the soil. Where a typical loamy soil is unprotected by vegetation, rain drop impact will break apart soil aggregates and fine soil particles will float over the top of the soil creating a "dense" layer that will be composed mainly of microspores (capillary sized openings). Then, the water moves into the soil by capillarity. We shall be more correct if we visualize water being pulled into the soil surface during a rain by capillarity instead of thinking that water is running into the soil through "large" holes. The situation at the soil surface is the reverse of a capillary tube standing in a pan of water. A capillary tube fills from the bottom to the top, while by contrast, capillary-sized pores, in the surface of the soil, pull water downward into the soil during a rain.

Water in excess of the ability of the soil to retain adhesion and cohesion water exists in the large or noncapillary pore space and moves downward in response to *gravity* and the *suction* or *pull* of the underlying soil pores. This "excess" or gravitation water moves downward and moistens drier soil below. We see, then, that water that was considered gravitational at one level becomes capillary (nongravitational) water at a lower level. Under these conditions, the water moves downward as a front as shown in Fig. 4–9. A sharp line

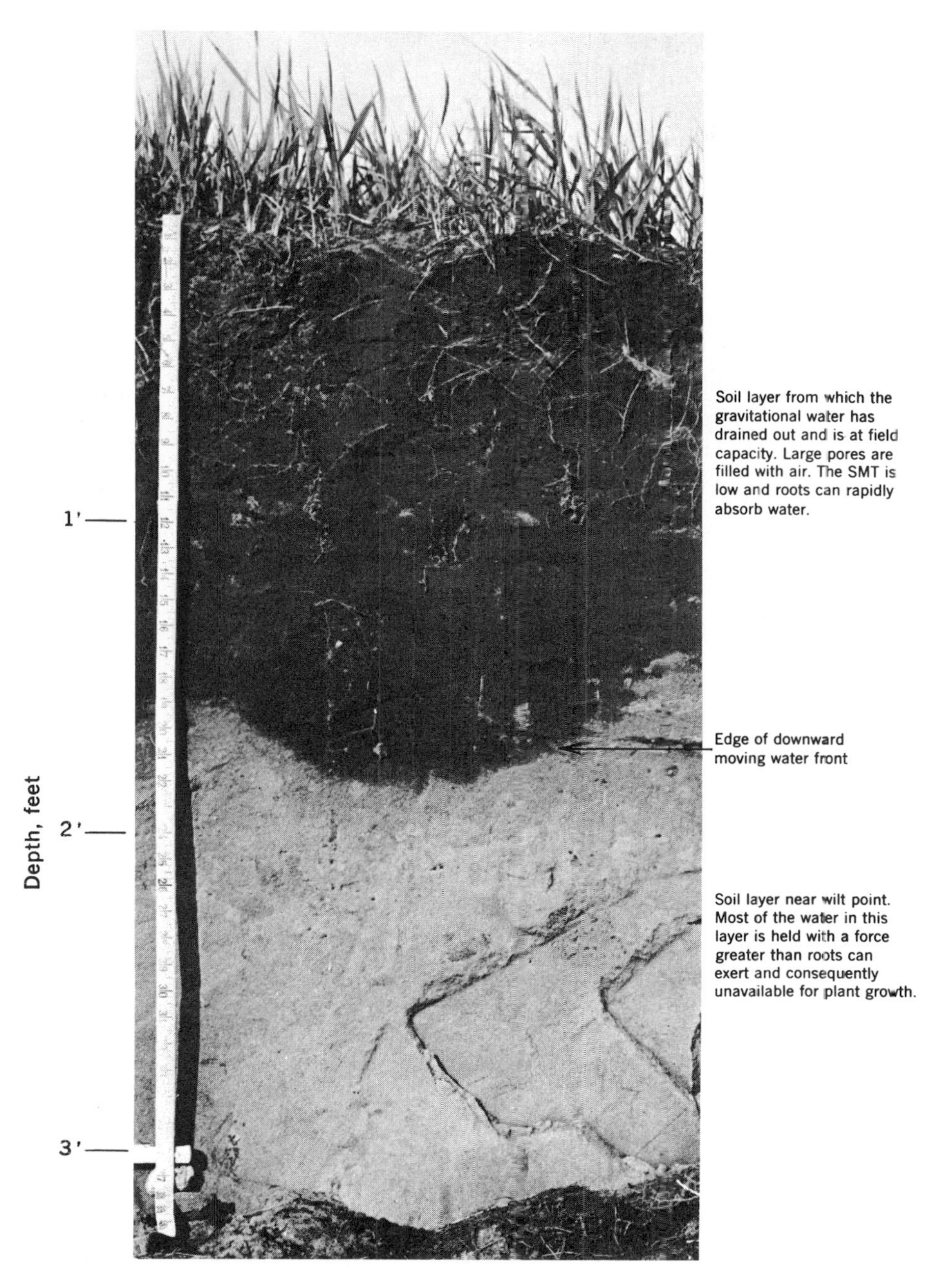


**Fig. 4–9** Moisture relationships one day after a rain when the soil was near the wilt point to a depth of 3 feet or more. The sharp line of demarcation between the moist upper and drier lower layer should be observed.

of demarcation is formed between the moist upper layer and the drier lower layer, and this line may persist for days. The upper moist soil layer is at *field capacity,* which is the water content of soil in the field after the gravitational water has drained out.

Two questions present themselves. First, "Why doesn't the water continue to move with "reasonable" speed from the moist upper layer into the drier lower layer?" Second, "What is the soil moisture tension in the moist soil (at field capacity)" as shown in Fig. 4–9?

The answers to these two questions are related. Consider, again, the capillary tube. What determines how high water rises in a capillary? Obviously, it is the size or diameter of the capillary. The same situation is true of the soil in regard to the ability of soil pores to pull water downward. The question then becomes, "What is a reasonable or typical capacity of soil pore spaces to pull water into the soil surface or of the soil pores to pull water out of a moist soil layer" as shown in Fig. 4–9? On a theoretical basis it has been calculated that water may be lifted from a free water surface to a height of 1½ feet by coarse sand and to 150 feet by fine silt. Soils vary greatly in their kinds and sizes of soil pores and, thus, in the soil moisture tension at field capacity. It has become customary, however, to consider a value of ⅓ atmosphere (equivalent to a column of water about 11 feet high) to be equal to the typical pulling power of capillaries in well-moistened soil. Thus, the SMT at field capacity is generally considered to be ⅓ atmosphere or 346 centimeters of water (see Fig. 4–10). This means a suction or tension *greater* than ⅓ atmosphere must be applied to remove water or to cause water to move in moist soil. The water in the moist soil layer in Fig. 4–9 exhibits very little tendency to move into the underlying drier soil because the retention forces in the moist soil (equal to about ⅓ atmosphere) are "balanced" by the capacity of the pores in the drier soil to pull the water down. Soil pores are irregularly shaped at the wetting front and the small pores which hold water open into larger open pores in the dry layer. Large capillaries (pores) cannot pull water out of the smaller capillaries (pores). This accounts for the resistance to movement of water from the moist to the dry soil.

The water between saturation and field capacity (gravitational water) has a low tension and would be easily absorbed by plant roots. However, the gravitational water is of little value in most soils because it drains downward rather quickly. In addition, the presence of gravitational water excludes air that is needed for root respiration and many other biological activities. As seen in Fig. 4–10, the pore space is about half air filled and half water filled at field capacity.

| Soil moisture classification | Tension | | Approximate % pore space occupied by water |
|---|---|---|---|
| | Atmospheres or bars | Centimeters of water | |
| Oven dry | 10,000 | 10,000,000 | 0 |
| Hygroscopic water (unavailable to plants) | | | |
| Hygroscopic coefficient | 31 | 31,600 | 15 |
| Capillary water (unavailable to plants) | | | |
| Wilt point | 15 | 15,800 | 25 |
| Capillary water (available to plants) | | | |
| Field capacity | 1/3 | 346 | 50 |
| Gravitational water (subject to drainage) | | | |
| Saturation | 0 | 0 | 100 |

**Fig. 4–10** Diagram showing soil moisture classification soil moisture tension equivalents, and the approximate percentage of the soil pore space occupied by water at various tensions. The soil is assumed to be a well-aggregated, medium-textured plow layer.

In humid regions gravitational water migrates through the soil profile in most years and enters the water table. Data in Table 4–1 from a location in Ohio show that about 1/6 of the precipitation percolated through the soil. Significant amounts of plant nutrients were removed from the soil and their removal results eventually in the development of acid soils. Leaching losses of calcium and magnesium in Ohio were more than the amount required to produce an average crop of wheat but less than the amount contained in a three-ton crop of alfalfa (see Table 11–1).

### Water Absorption and the Wilt Point

At field capacity the SMT is low and plant roots can easily absorb water. As roots absorb water by osmosis, water near the roots will move slowly in the direction of the root by film adjustment (see Fig. 4–5). We have already observed that as the soil becomes drier, the SMT increases and the movement of water becomes slower. Furthermore, as the SMT increases, the cells' ability to absorb soil moisture decreases. Eventually, if no additional water is added to the soil, the plant will absorb water slower than water is lost by transpiration. A water deficit is developed inside the plant and eventually wilting occurs (unless the plant has some special adaptation such as found in many desert plants that can stop transpiration loss). To determine the wilt point, plants such as sunflowers or wheat are grown on the soil until the plants wilt and are unable to regain turgor when placed in a saturated atmosphere. A soil moisture tension of 15 atmospheres has been found to correspond generally with the wilt point. The wilt point, like the field capacity, is not a precise value and varies with soil and environmental conditions. High temperatures and strong winds could cause some plants to wilt with SMT as low as 2 atmospheres.

When the SMT throughout the root zone is low or near field capacity, roots will absorb water most rapidly from the upper part of the soil where oxygen is the most abundant and near the base of the plant because less resistance will be encountered in translocating the water through the large roots to the stem. As the soil dries and the SMT in the surface soil layers increase, water uptake will shift to deeper soil layers where the oxygen supply is less but the soil is moist and the SMT is low. In this way the root zone is progressively depleted of available soil moisture (in the absence of rain or irrigation water). In the case of young annual plants root extension into moist deeper layers must occur more rapidly than upper soil layers are brought to the wilt point if the plant is to obtain water for continued growth.

**Table 4–1** Plant Nutrient Losses in Lysimeter Percolates on Kenne Silt Loam During a Period of High Precipitation (1950) and Low Precipitation (1953), as Compared with the 16-Year Average (1940–1955), by Practice

| Practice and Period | Total Precipitation | Percolation | Nutrients Percolated Per Acre | | | | | |
|---|---|---|---|---|---|---|---|---|
| | | | Ca | Mg | K | N | Mn | S |
| | Inches | Inches | Pounds | Pounds | Pounds | Pounds | Pounds | Pounds |
| Conservation | | | | | | | | |
| 1950 (high precipitation) | 47.28 | 12.61 | 51.79 | 30.18 | 11.5 | 5.62 | 0.64 | 82.21 |
| 1953 (low precipitation) | 28.20 | 3.54 | 5.88 | 0.74 | 2.77 | 0.64 | 0.13 | 8.28 |
| 16-year average (1940–55) | 37.16 | 6.29 | 29.35 | 17.61 | 9.74 | 3.47 | 0.30 | 31.42 |
| Poor | | | | | | | | |
| 1950 (high precipitation) | 47.28 | 13.40 | 30.81 | 18.69 | 20.56 | 2.87 | 0.94 | 51.18 |
| 1953 (low precipitation) | 28.20 | 5.04 | 4.59 | 0.68 | 3.61 | 1.18 | 0.19 | 12.51 |
| 16-year average (1940–55) | 37.16 | 7.20 | 21.04 | 12.29 | 13.41 | 4.13 | 0.37 | 20.89 |

From *Technical Bulletin* 1179, U.S. Department of Agriculture, p. 142, 1958.

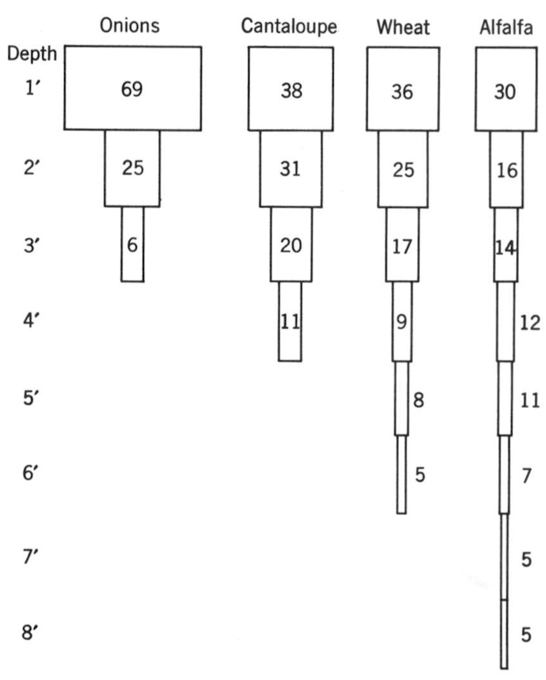


**Fig. 4–11** Percentage of water used from each foot of soil when produced with irrigation in Arizona. Onions are a shallow rooted annual crop that were grown in the winter and used a total of 17.5 inches of water. Alfalfa, by contrast, is a deep-rooted perennial crop, which grew the entire year and used a total of 74.3 inches of water. (Data from *Arizona Agr. Exp. Sta. Tech. Bul.* 169, 1965.)

When the upper soil layers are rewetted by rain or irrigation water, water absorption may shift back toward the surface soil layers near the base of the plant. This pattern of water utilization results in: (1) more deeply penetrating roots in dry years than in wet years and in arid regions as compared to humid regions and (2) a greater use of water from the upper soil layers than from the lower soil layers as shown in Fig. 4–11.

## SMT and Plant Growth

We have already noted that at very low SMT the lack of air may limit plant growth. The rate of plant growth is at or near a maximum at field capacity because there is adequate oxygen accompanied with low SMT for rapid water absorption. As soil moisture is absorbed, the moisture films become thinner, the SMT increases, and the rate of water absorption decreases. Generally, increasing SMT between field capacity and the wilt point is associated with a reduced rate of photosynthesis and growth. Thus, two important facets of plant growth are associated with SMT, namely, lack of oxygen at low SMT and slow rate of water absorption at high SMT. Forest tree growth is commonly limited by low water supply. It is interesting to note that maximum growth of sugar maple in the north central states was found to occur when SMT was in the range of 1/3 to 3 atmospheres.[1]

### Role of Water in Nutrient Absorption

Plant roots do not engulf and absorb the soil solution containing nutrients as animals drink water containing soluble material. Rather, the water enters the roots as pure water without regard to the intake of any of the materials dissolved in it. The entrance of dissolved substances is entirely a separate process. Nutrients dissolved in the soil solution move with it, and so when moisture flows by capillarity toward the roots to replace that which has been taken up by plants, a supply of nutrients is moved near the roots. Although this action takes place through short distances only, the net result in the course of a growing season may add materially to the nutrient supply of the crop.

### Available Water Holding Capacity of Soils

Water held in the soil between field capacity and the wilt point is water that in general comprises the soil solution and is the major source of water for plant growth. The capacity of the soil to hold water is related to both surface area and pore-space volume. Water-holding capacity is therefore related to structure as well as to texture. It can be seen in Fig. 4–12 that fine-textured soils have the maximum *total* water-holding capacity but that maximum *available* water is held in medium-textured soils. Research has shown that available water in many soils is closely correlated with the content of silt and very fine sand.

### Other Soil Moisture Coefficients

The soil still contains water at the wilt point which is considered to be unavailable to plants (see Fig. 4–10). To remove the remaining water (excluding water of hydration) the soil is dried in an oven for 24 hours at 110 degrees C. The soil is then brought to the *oven-dry* state. If oven-dry soil is placed in a water-saturated atmosphere, water will be adsorbed by the soil. At equilibrium the soil will contain

[1] "A Field Study of the Soil-Nutrient Status for Sugar Maple," R. L. Heninger, MS Thesis, Mich. Tech. Univ., 1969.

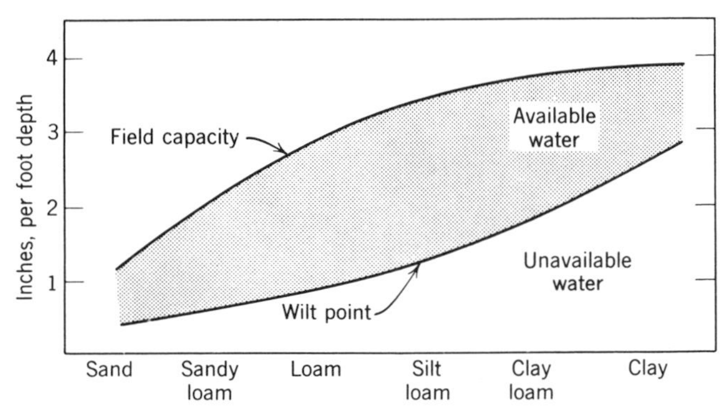


**Fig. 4–12** Typical water-holding capacities of different textured soils. Note that maximum available water-holding capacity occurs in the silt loam soil. (Adapted from *USDA Yearbook* 1955, p. 120.)

an amount of water described as the *hygroscopic* coefficient. The hygroscopic coefficient has little relevancy for plant growth but is a qualitative measure of the surface area in the soil.

## WATER REQUIREMENT OF CROPS

The total quantity of water required to produce a crop, excluding that retained in the plant, has two components: water lost by evaporation from the soil surface and water lost by evaporation from plant surfaces (transpiration). The word, *evapotranspiration* refers collectively to these two processes. The inches of water lost by evapotranspiration in the production of a crop is the *consumptive use.*

### Loss of Water by Transpiration

The amount of water transpired to produce one pound of dry matter was studied intensely by Briggs and Shantz shortly after the turn of this century. They grew plants in large galvanized pots which had tight fitting covers with openings only for the stems of the plants. They measured the amount of water utilized by the plants by weighing the pots and harvested the plants to determine the quantity of dry matter produced. Some of their findings, presented in Table 4–2, show that plants commonly transpire 500 pounds or more of water for each pound of dry matter produced. Furthermore, differences existed in the amount transpired between crops.

As transpiration is simply the evaporation of moisture from plant surfaces, it is influenced by the same factors that affect the evaporation of water from any moist surface; exposure to direct sunlight, air temperature, humidity, wind movement, and atmospheric pressure are among the most important. Since these are variable from year to year, so should the amount of water transpired vary (Table 4–2).

### Consumptive Use—Amount of Water Used to Produce a Crop

Consumptive water use is that amount of water lost by evaporation from the soil and plants during the time a crop is grown. The quantity varies widely from less than 10 inches for a quickly maturing crop

**Table 4–2** Water Transpired by Plants

| Pounds of Water Transpired Per Pound of Dry Plant Tissue Produced | | | | | |
|---|---|---|---|---|---|
| Crop | 1911 | 1912 | 1913 | Greatest Variation | Average |
| Wheat | 468 | 394 | 496 | 102 | 452.7 |
| Oats | 615 | 423 | 617 | 194 | 551.7 |
| Corn | 368 | 280 | 399 | 119 | 349.0 |
| Sorghum | 298 | 237 | 296 | 61 | 277.0 |
| Alfalfa | 1,068 | 657 | 834 | 411 | 853.0 |

From "Relative Water Requirements of Plants," L. J. Briggs and H. L. Shantz, *J. Agr. Research,* Vol. 3, No. 1, p. 56, 1914.

in a humid region to more than 70 inches for a long-season crop in an arid environment. Soybeans commonly require about 13 to 23 inches per season which averages about 0.14 to 0.18 inch per day. During periods of maximum use, the rate commonly rises to a fourth or third of an inch per day.

Evaporation of water requires energy, making the climatic environment the major factor in determining the amount of water used. This is shown in Table 4–2 where the amount of water transpired to produce one pound of dry matter varied greatly between years. In fact, studies made with an experimental setup like that shown in Fig. 4–13 reveal that a green actively growing crop that completely covers the ground may lose water more rapidly than a free water surface (open pan), if the soil moisture tension is low.

On any given day, many different kinds of crops require essentially the same amount of water when the cover is complete, they are green, and the soil moisture tension is the same. Great differences, however, exist in the total amount of water crops utilize because they have different maturation periods or they grow at different seasons of the year as shown in Fig. 4–14 (see also Fig. 4–11).

**Fig. 4–13** Experimental set-up for studying the effect of various factors on consumptive use of water. When rain starts to fall, the sheds automatically cover the plots and return to their former position when rain ceases. This permits the control of the amount of water added to the plots.

### Water Use Efficiency

Plants growing in a medium containing relatively small quantities of nutrients appear to grow slower and transpire more water per pound of plant tissue produced than those growing in a medium containing an abundance of plant-nutrient materials. Since it has been shown that water loss is mainly dependent on the environment, any management practice that increases the rate of plant growth will tend to result in more dry matter produced per pound of water used. The data in Table 4–3 show that the yield of oats was increased from 2.4 to 4.0 bushels for each inch of water utilized by the addition of fertilizer. In the humid regions it has been commonly observed that fertilized crops are more drought resistant. This may be explained on the basis that increased top growth results in increased root growth and penetration so that the total water consumed may be greater and is also used more efficiently.

Fertilizer used in a subhumid region may occasionally decrease yields. If fertilizer causes a crop to grow faster early in the season, the greater leaf area at an earlier date results in greater loss of water

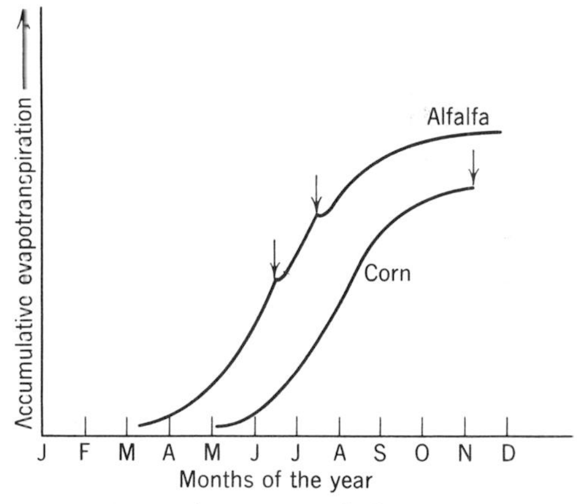


**Fig. 4–14** The seasonal distribution of water use by alfalfa and corn. The arrows indicate harvest dates. (From "Agricultural Water Use," D. E. Argus, *Advan. in Agron*, Vol. 11, p. 20, 1959. Used by permission.)

by transpiration. If no rains occur or no irrigation water is applied, plants could run out of water near harvest and a serious reduction in grain yield could occur. Work in Kansas has shown this to be the case for grain sorghum. Forage, when fertilized, does not depend on a supply of water to the end of the growing period, and can be produced, the same as in the humid region, when fertilized, with less water per pound. The forage will stop growing when the water is exhausted after having used the water "efficiently" as long as it lasted.

This concept may have far-reaching consequences in the future for crop production. Allocation of water to those situations resulting in the most efficient use of the water, in terms of crop yield per unit of water consumed, may be required if supplies become sufficiently limited.

**Table 4–3** Effect of Nitrogen Fertilizer on Water Utilization

| | Bushels of Oats Per Inch of Water Utilized | |
|---|---|---|
| Year | Low Nitrogen | High Nitrogen |
| 1949 | 2.1 | 4.4 |
| 1950 | 2.7 | 3.7 |
| Average | 2.4 | 4.0 |

From "Water Consumption by Plants as Influenced by Soil Fertility," R. J. Hanks and C. B. Tanner, *Agron. J.*, **44**:99, No. 2, 1952.

### Cultivation and Loss of Water by Evaporation

So far we have emphasized that water in soils between field capacity and the wilt point is immobile for practical purposes. Plant roots must extend into moist soil layers to obtain water. The idea of water moving up in the soil like kerosene moving up a wick is invalid in unsaturated soils. In Fig. 4–9 it was shown that a dry soil layer cannot pull water downward out of a moist layer. If a dry soil layer overlies a moist layer, neither will the dry surface layer pull water upward out of the moist underlying layer. If the surface soil layer is air dry, water cannot move to the soil surface by capillarity and be lost by evaporation. Once the surface soil dries, water can move from the underlying moist soil to the dry soil surface only in the vapor phase. This vapor movement is so slow as to be largely discounted. This produces a "capping" feature, which has great significance for water conservation. Once the surface of the soil has become dry after a rain (with or without cultivation), the water in the underlying moist layers is largely protected from loss by evaporation when the time considered is a few weeks or months. If another rain occurs the soil may be moistened to a greater depth, but, when the surface of the soil again becomes dry, the water is again "trapped" in the soil.

In regions of limited rainfall the land is commonly kept bare or free of vegetation (fallow) for a year so that the amount of water stored in the soil can be increased. During the second year the stored water plus the natural rainfall is generally sufficient to produce a "good" crop where there would be a crop failure in most years with continuous cropping. Weeds must be eliminated during the fallow year as shown in Fig. 4–15. Much of the subhumid plains in western United States and Canada are used extensively for wheat production by fallowing half of the land and growing wheat on the other half. The fallowing system is not 100 percent efficient because the self-mulching effect does not work perfectly and some runoff and evaporation does occur. A good estimate is that about 25 percent of the rainfall during the fallow period will become stored in the soil for use in crop production. This extra quantity of water, however, has a great effect on yields.

## SOIL DRAINAGE

About 1200 A.D. farmers in the Netherlands became engaged in an interesting water control problem. Small patches of fertile soil affected by tide and flood waters were enclosed by dikes and drained. Windmills provided the energy to lift gravitational water from drainage

**Fig. 4–15** A scene in the central Great Plains where the annual precipitation is about 15 inches and land is fallowed for wheat production. Cultivation kills the plants and prevents them from using soil moisture. Rainfall during the fallow year contributes to an increase in soil moisture, which supplements the next years' rainfall for the production of wheat.

ditches into higher canals for return to the sea. Now, 800 years later, about one-third of the Netherlands is protected by dikes and kept dry by pumps and canals. The use of drainage systems to remove water from soil is highly important throughout the world where the water table is close enough to the surface so that saturated soil occurs within the root zone of plants, foundations of engineering structures, or septic tank drain fields. We shall discuss the principles underlying these problems in this section.

### Effects of Poor Drainage on Plant Growth

Plant roots tend to suffocate in poorly drained soils. This suffocation is related to the oxygen diffusion rate, which is about 10,000 times more rapid in air than in water. In unsaturated soil there are thin films of water around the root surfaces producing a short diffusion path that enables roots to obtain sufficient oxygen. The enormously longer diffusion path in water-saturated soil, coupled with the very low diffusion rate of oxygen in water, produces oxygen deficiencies as shown in Figs. 2–2 and 3–12. Flooding of a field during the growing season produced the effect seen in Fig. 4–16.

Root tips are regions of rapid cell division and elongation and have a high oxygen requirement. Typically, roots of most crop plants do not

**Fig. 4–16** Crop damage resulting from a temporary ponding of water. Note the larger plants on the slightly higher ground in the rear on both sides of the field. (U.S.D.A. photo.)

penetrate water-saturated soil because of oxygen deficiency. The response of roots to the depth of the water table is shown in Fig. 4–17. Paddy rice, as well as many other plants, have stems through which oxygen can diffuse from the atmosphere to the roots growing in saturated soil and are not dependent on soil oxygen for root respiration.

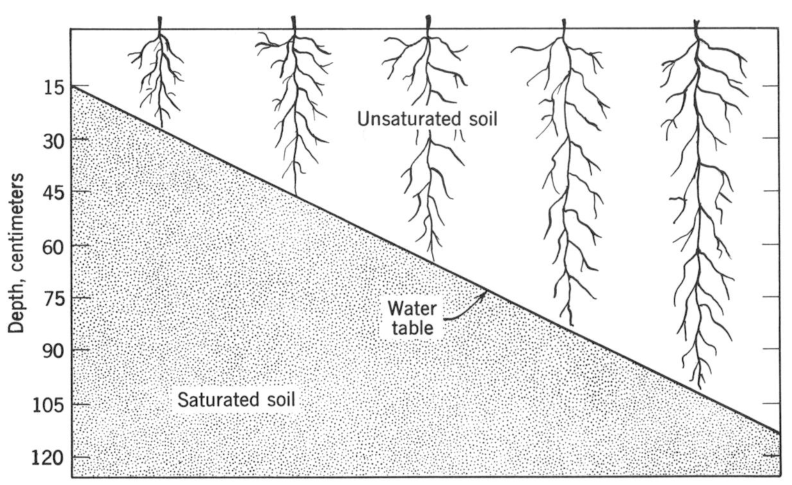


**Fig. 4–17** The effect of depth to water table on the length of millet roots. Roots did not enter the saturated soil. (Based on data from "Effect of Water Table Depth and Flooding on Yield of Millet," R. E. Williamson *et al.*, *Agron. Jour.*, Vol. 61, p. 312, 1969.)

**Fig. 4–18** Clay tile lain at the base of the foundation of the walls of a basement. Cracks between adjacent tile are covered with a durable black material to keep soil from entering the tile.

## Drainage Systems

Ditches can be quickly and inexpensively made to remove gravitational water. Drainage ditches, however, require periodic cleaning and are inconvenient for the use of machinery. There are also many situations where ditches are not satisfactory, as in the case of water removal around the walls of the basement of a building as shown in Fig. 4–18. The drainage tiles shown in Fig. 4–18 are made of fired clay and are laid side by side with a small crack between adjacent tiles. When the soil surrounding the tile is saturated with water, water seeps into the tile laid on a grade and the water eventually reaches an outlet where it is disposed.

Drain tile are installed in fields with trenching machines (Fig. 4–19). Long plastic tubes with holes are sometimes used for draining organic soils (muck) where the low supporting capacity of the soil results in unequal settling and misalignment of the shorter tile sections.

**Fig. 4–19** Installation of 7-inch drainage tile in a field. (U.S.D.A. photo.)

### Depth and Spacing of Tile Lines

Unless some unusual condition prevails, tile should be laid at least 2½ feet deep. The depth of the outlet has some bearing on the depth at which the drains may be placed and still provide a satisfactory fall. A fall of at least $1/10$ foot per 100 feet is considered necessary, and more is desirable. The water level is lowered by the lines as shown in Fig. 4–20. In order to have a sufficiently low water table between tile lines to permit crops to develop an adequate root system, it is essential that the tile be 2½ or more feet deep.

The freedom with which water moves through the soil determines the distance apart that drains may be placed and still afford adequate

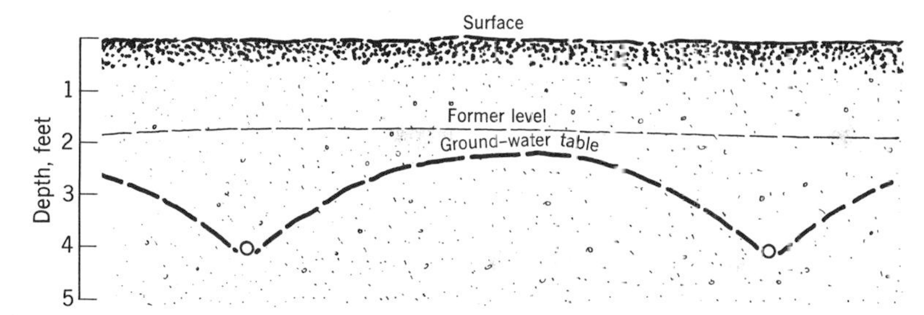


**Fig. 4–20** The effect of tile lines in lowering the water table or ground-water level. In very heavy soil it takes several years for the water table to reach the proper adjustment. The benefit of the drainage is first evident immediately over the tile lines and gradually spreads to the soil area between them.

drainage. If the soil through which the water must percolate to reach the tile contains a high percentage of clay and the structural condition is such as to make water movement slow, the tile lines should not be more than 70 feet apart. In more permeable soil a spacing of 80 feet is permissible, and in soil through which water drains rapidly the drains may be placed at 100 feet intervals.

### Drainage or Water Movement in Stratified Soil

In the Columbian Basin of northwestern United States some of the best agricultural soils consist of two feet of fine sandy loam soil with sand and gravel underneath. These soils are noted for their great water-holding capacity. One's first reaction is that soils underlain by sand and gravel would be very droughty. But this is readily understood if one realizes that when a downward moving water front (from rainfall or irrigation) encounters a coarse textured layer, water movement will stop, at least temporarily Water movement by capillarity through the upper 2 feet of fine sandy loam soil will stop when the sand and gravel layer is encountered as shown in Fig. 4–21. The sand layer has "large" pores and very low suction capacity. Water will not be pulled down into the sand layer until the water content of the overlying finer-textured layer is increased so that an SMT less than the suction capacity of the pores in the sand has been attained. This is clearly shown in Fig. 4–21. The soil just above the sand interface will become nearly saturated before water continues to move down into the sand. This explains the high water-holding capacity of the Columbian Basin soils.

When a moving water front, encounters a finer-textured layer or horizon, as a Bt horizon, the water front continues to move. The finer pores pull the water downward. The low transport capacity of the

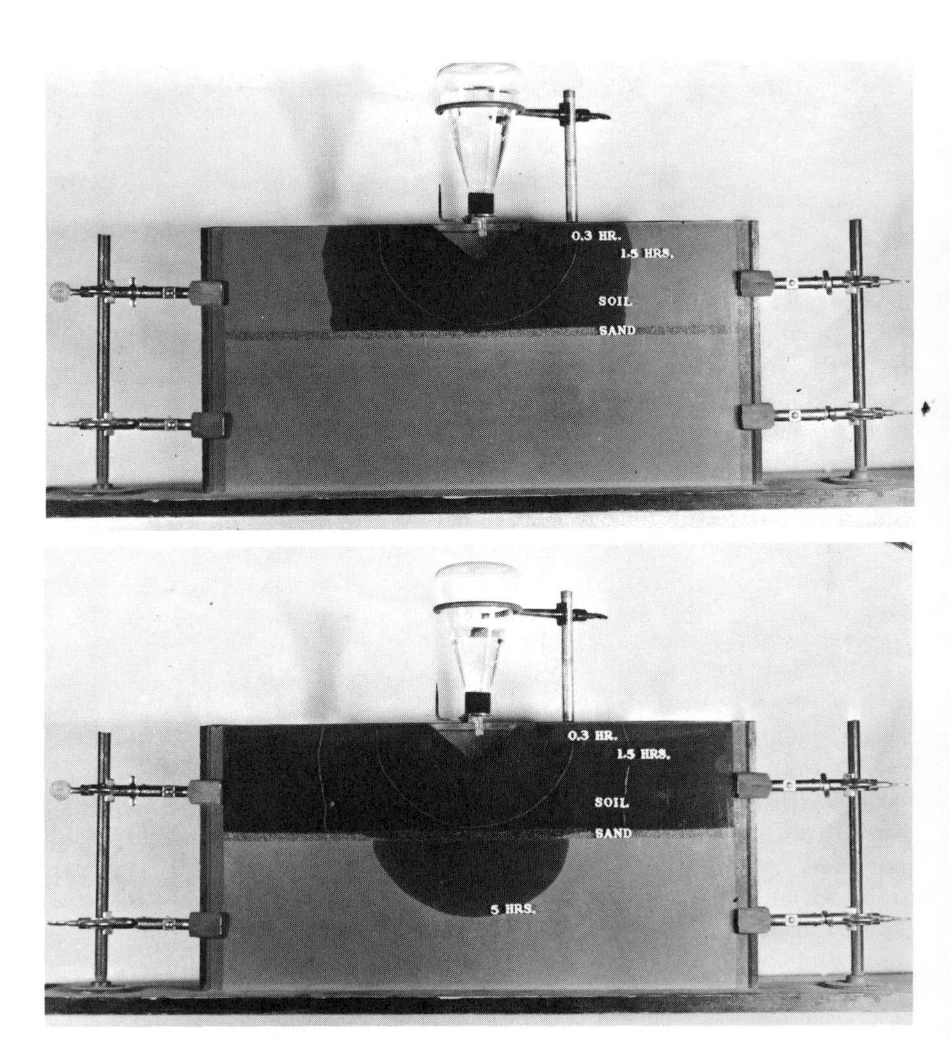


**Fig. 4–21** Photographs illustrating water movement in stratified soil where the water under conditions of unsaturated flow is moving from a finer-textured soil into a coarser-textured layer. In the upper photo it can be seen that the downward movement of water is inhibited by the sand layer. The lower photo shows that sometime after an elapsed time of 1.5 and 5.0 hours the SMT just above the sand layer was lowered sufficiently so that the pores in the sand layer could pull the water downward by capillarity. (Photos courtesy W. H. Gardner of Washington State University, Pullman, Washington.)

finer pores, however, may cause a build up of water above the finer-textured layer. During wet seasons this commonly produces water-saturated soil overlying well-developed Bt horizons.

## Drainage of Septic Tank Effluent Through Soils

Farmers who reside beyond the limits of municipal sewer lines have used septic tank sewage disposal systems for many years. The recent rapid expansion of rural residential areas and the development of summer homes near lakes and rivers has greatly increased the role of the soil in waste disposal. The sewage enters a septic tank where solid material is digested and the liquid effluent flows out of the top of the septic tank and into the tile lines of a filter field as shown in Fig. 4–22. The seepage lines are laid in a bed of gravel from which the effluent seeps into and drains through the soil.

A major factor influencing the suitability of the soil for filter field use is permeability. Soil permeability of less than one inch per hour (or a percolation rate of more than 60 minutes per inch) is too slow. Sand may have percolation rates in excess of 10 inches per hour and may be too permeable. When sewage effluent moves too rapidly through soil, there is danger that shallow water supplies may become contaminated.

## Soil Drainage Classes

Well-drained soils are located on the landscape in positions where the soil is never saturated with water. Other soils are poorly drained

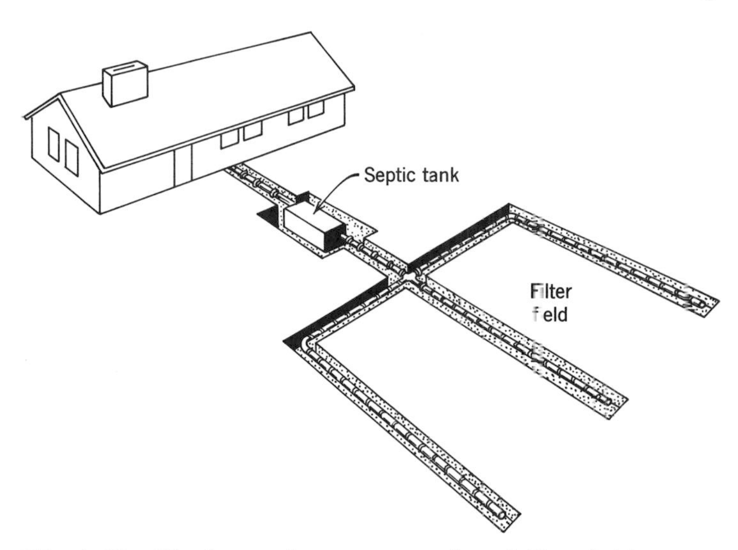


**Fig. 4–22** The layout for a septic tank and filter field.

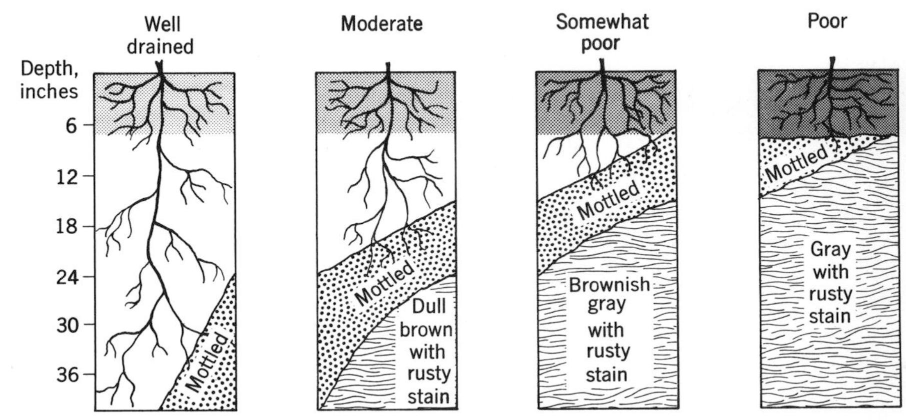


**Fig. 4–23** Soil properties and root extension in relation to soil drainage classes. (From "Soil Survey of Genessee County," New York, p. 11, USDA and Cornell Agriculture Experiment Station, 1969.)

because they are located in low wet areas where the soil is saturated all or most of the year. Obviously, the natural drainage greatly influences soil use. Well-drained soils are ideal for fruit orchards. Poorly drained soils need artificial drainage for production of most agricultural crops. Septic tank filter fields should never be installed in poorly drained soils because the soil will become saturated with water and there is a danger that disease organisms in the effluent will contaminate work or play areas. The basements of homes constructed on poorly drained soils are likely to become flooded.

Except in the very young soils, the natural drainage of the soil is reflected in the soil's properties. The drainage classes and the most important associated features are presented in Fig. 4–23. In summary, the following happens as the soil becomes more poorly drained.

1. Rooting depth decreases.
2. Mottled soil colors occur nearer the top of the soil.
3. Subsoil colors change from bright browns (and yellows and reds) to grays with iron stains.
4. The organic matter content increases and the color becomes darker in the surface soil layer.

The soil properties are a reliable guide for interpreting soil drainage classes at any time of the year. This is true whether the soil is wet or dry at the time of inspection, which accounts for the fact that the soil properties associated with drainage classes are very important in determining wise land use.

# 5

# Soil Organisms

The soil is the home of innumerable forms of plant and animal life. Some of the fascination and mystery of this underworld has been described by Peter Farb.

"We live on the rooftops of a hidden world. Beneath the soil surface lies a land of fascination, and also of mysteries, for much of man's wonder about life itself has been connected with the soil. It is populated by strange creatures who have found ways to survive in a world without sunlight, an empire whose boundaries are fixed by earthen walls."[1]

Life in the soil is amazingly diverse, ranging from microscopic single-celled organisms to large burrowing animals. As is the case with the organisms above the ground, there are well-defined food chains and intense competition for survival.

Historically, the soil has been a dumping ground for all kinds of refuse. "Dust thou art and to dust thou shall return."[2] We are coming to realize, however, that the soil organisms have limited capacity to degrade and detoxify materials. Recent concern about environmental quality and recognition of the role of the soil has created a new awareness and appreciation for soil organisms.

[1] "Living Earth," Peter Farb, Harper and Brothers Publishers, 1959.

## KINDS AND NATURE OF THE SOIL FLORA

The many forms of plant life in the soil range in size from those so small they can only be studied with a microscope to mushrooms and the roots of large trees. The original concept of the soil population as predominantly bacterial has been expanded through years of investigation to include other plant groups as fungi, actinomycetes, and algae as well as many micro- and macroscopic animals. A summary of the more important groups of soil microflora are presented in Table 5–1. Bacteria and fungi are the most important flora in soils concerned with decay and nutrient recycling and we shall discuss the bacteria first.

### Bacteria—The Most Abundant Soil Organisms

Bacteria are single-celled plants and exceed all other soil organisms in numbers and kinds. A gram of fertile topsoil may contain over one billion bacteria. The most common soil bacteria are rod-shaped, a

**Table 5–1** A Brief Classification of Important Soil Microflora

- I. Bacteria
  - A. Heterotrophic
    1. Nitrogen fixers
       - *(a)* Symbiotic
       - *(b)* Nonsymbiotic
    2. Those requiring fixed nitrogen
  - B. Autotrophic
    1. Nitrite formers
    2. Nitrate formers
    3. Sulfur oxidizers
    4. Iron oxidizers
    5. Those that act on hydrogen and various hydrogen compounds
- II. Fungi
  - A. Yeasts and yeast-like fungi
  - B. Molds
  - C. Mushrooms
- III. Actinomyces
- IV. Algae
  - A. Blue-green
  - B. Grass-green
  - C. Diatoms

micron (1/25,000 of an inch) or less in diameter and up to a few microns long (see Fig. 5–1). Researchers have estimated that the live weight of bacteria per acre may exceed 1000 pounds.

Soil bacteria may be divided broadly into two large groups, based on their energy requirements: (1) the *heterotrophic* bacteria, which obtain their energy and carbon from complex organic substances; (2) the *autotrophic* bacteria, which can obtain their energy from the

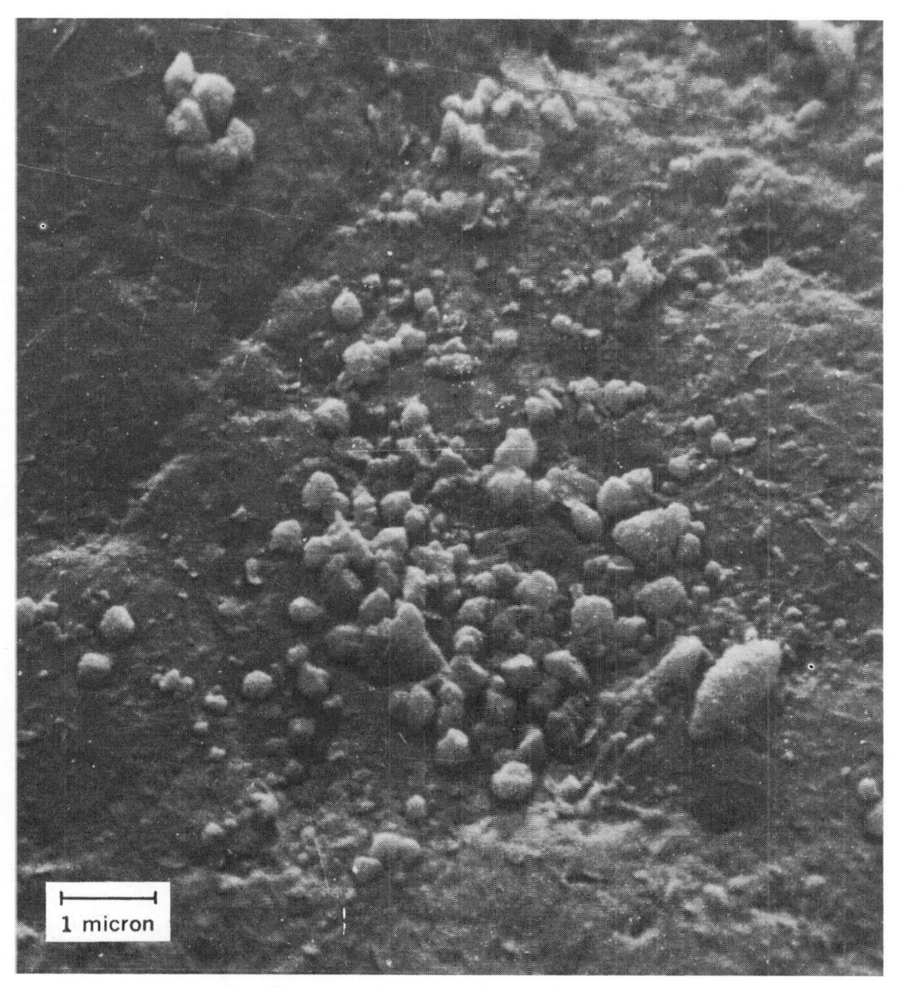


**Fig. 5–1** Bacterial cells in a colony on the surface of a sand grain as photographed through an electron microscope. (From "Stereoscan Electron Microscopy of Soil Microorganisms," T. R. G. Gray, *Science* **155:** p. 1668, 1967. Copyright 1967 by Am. Assoc. Adv. Sci.)

oxidation of inorganic elements or compounds, their carbon from carbon dioxide, and their nitrogen and other minerals from inorganic compounds (see Table 5–1). In the autotrophic group are found such organisms as the nitrite formers, the nitrate formers, the sulfur-oxidizing bacteria, the iron oxidizers, and those that act on hydrogen and its compounds.

Most of the soil bacteria require oxygen from the soil air and are classified as *aerobes*. Some aerobic bacteria can adapt to living where the soil air is devoid of oxygen; they are *facultative aerobes*. Other bacteria cannot live in the presence of oxygen and are *anaerobes*. The soil bacteria also differ considerably in their nutrition and in their response to environmental conditions. Consequently, the kinds and abundance of bacteria depend both on the available nutrients present and on the soil environmental conditions.

## Fungi

Fungi are heterotrophic plants that vary greatly in size and structure from single-celled yeasts to molds and mushrooms. Fungi typically grow from spores by a threadlike structure that may or may not have crosswalls. Individual threads are *hypha*, and a mass of extensive threads is the *mycelium*. The mycelium is the working structure that absorbs nutrients, continues to grow, and eventually produces special hyphae that produce reproductive spores. The average diameter of hyphae is about 5 microns or about 5 to 10 times the diameter of a typical bacterium.

It is difficult to accurately determine the number of fungi per gram of soil since mycelium are easily fragmented. It has been observed that a gram of soil commonly contains 10 to 100 meters of mold filament per gram. On the basis of the amount of filament, researchers have concluded that live weight of fungal tissue exceeds or equals that of the bacterial tissue in most soils.

All of us have seen mold mycelia growing on bread, clothing, or leather goods. Some mold colonies growing on plant leaves produce a white-cottonish appearance called downy mildew disease. Many fungi have morphological features that resemble higher plants. *Rhizopus*, shown in Fig. 5–2, has rootlike absorptive structures called *rhizoids*, which penetrate the substrate on which the mold is growing. Hyphae elongating over the substrate are called *stolons*. Stalk or stemlike hyphae originating from the stolons bear spore cases. Unlike higher or vascular plants, however, fungi have no specialized xylem or phloem-conducting tissue.

Fungi are important in all soils and their tolerance of acidity makes

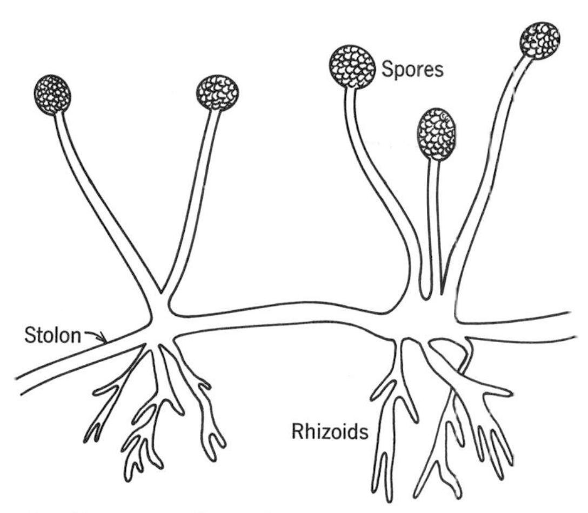


**Fig. 5–2** Hyphae of *Rhizopus* fungi, a common soil mold.

them particularly important in acid forest soils. The woody residues of the forest floor provide an abundance of food for fungi which are effective decomposers of lignin (Fig. 5–3).

Yeast fungi are found in soils only to a limited extent and are believed to be of no great importance in soil development or the growth of higher plants.

**Fig. 5–3** Mushroom fungi *(Agaricales)* growing on soil, wood, and bark. The mushroom fruiting bodies form after the hyphae growing in soil, wood, or other substratum have accumulated sufficient food and the moisture and temperature are favorable.

## Actinomycetes—the "Fungilike" Bacteria

The actinomycetes occupy a position between the bacteria and fungi from the morphological viewpoint. They are frequently spoken of as *ray fungi* or *thread bacteria.* The actinomyces resemble bacteria in that they are unicellular and of about the same size in cross section. They resemble the filamentous fungi in that they produce a branched filamentous network. Many of these organisms reproduce by means of spores, and these spores appear very much like bacterial cells.

These organisms are present in great abundance in soil, making up as much as 50 percent of the colonies that develop on plates containing artificial media inoculated with a soil extract. The numbers of actinomyces may vary between 0.1 million and 36 millions per gram of soil. In actual weight of live substance per acre, they may exceed bacteria but as a rule will not equal fungus tissue. Under optimum soil conditions the total weight of actinomyces on the average may exceed 700 pounds per acre of soil.

## Algae—the Simplest Chlorophyllous Plants

Algae exhibit great diversity in form and size, ranging from single-celled organisms, with a diameter about 5 to 10 times greater than that of a bacterium, to kelps of the ocean that are over 100 feet in length. Although algae are the most important plants living in water, algae are of only minor importance in most soils. The most common soil algae are single-celled or are small filaments. Algae are universally distributed in the surface layer of soils wherever moisture and light are favorable. A few algae are found below the soil surface in the absence of light and appear to function heterotrophically.

The common forms of algae inhabiting soils are (1) blue-green, (2) green, and (3) diatoms. The blue-green are the most abundant in soils and to the extent that they fix carbon, they contribute to the organic matter content of soils. Their photosynthetic ability accounts for their growth on numerous exposed surfaces, including rocks and soils. Some algae grow in close association with fungi in a form known as *lichens.* In the initial weathering of rock exposures and the formation of soils from freshly exposed parent material, lichens play an important role in the early accumulation of organic matter. Further, the ability of some blue-green algae to fix atmospheric nitrogen also helps plant communities become established on freshly exposed rocks and parent materials. The nitrogen fixed by algae living in the water of rice paddies is of great importance in rice production.

## INFLUENCE OF SOIL CONDITIONS ON THE MICROFLORA

As has been indicated, the environmental conditions determine the nature of the microbial population present at any given time in the soil. In general, the fertile, fine-textured soils high in organic matter contain many more microbes than the coarse-textured soils low in organic matter. If we examine the changes which occur in the microbial population of soils, we find certain causes for these variations. Let us consider briefly some of these causes.

### Nutrient and Energy Supplies

Soil organisms have, in general, the same nutrient-element requirements as higher forms of life. For their growth and development they all require supplies of energy in addition to the several essential elements, including carbon, hydrogen, oxygen, nitrogen, phosphorus, potassium, sulfur. With the exception of algae all the important soil microbes are devoid of chlorophyll, and they must obtain their energy either from the oxidation of simple inorganic substances, as do the autotrophic bacteria; or from complex organic substances, as do most bacteria (heterotrophic organisms), all the fungi, and all actinomycetes. Thus, for the great bulk of organisms, the soil organic matter is the source of energy and nutrients.

The invasion of a soil region by an elongating root is followed by a large increase in microbial activity near the root surface. Excretory products and sloughed-off root tissue serve as readily available sources of energy and nutrients. Bacterial activity appears to be stimulated more than that of actinomycetes or fungi. The zone near the root where the abundance and composition of microbes is greatly affected is the *rhizosphere*.

### Temperature

Temperature regulates the reacting velocities of chemical and biological changes occurring in the soil. Within a rather narrow range the rate of biological reactions increases two to three times for each increase in temperature of 10°C. Roughly, the limits of microbiological functions are reached with a temperature of 80°C. For the majority of the soil organisms the optimum temperature is about 35°C, although they can grow at rather wide temperature ranges and may adapt themselves readily to gradual changes in temperature.

## Moisture

Another major factor affecting the numbers and activities of soil microorganisms is soil moisture. The influence of moisture depends to a large extent on the nature of the soil and the nature of the organisms concerned. The optimum amount of water for most soil organisms is between 50 and 70 percent of the water-holding capacity of the soil, about the same as for most higher plants. Most of the microbes are aerobic and perhaps only a few bacteria can tolerate water-saturated soils. Actinomycetes are particularly able to remain active as the soil dries out and can tolerate the greatest range of moisture conditions. Organic matter decomposition occurs about half as fast at the wilt point than when the soil is at field capacity.

## Aeration

Soil aeration is governed primarily by fluctuations in soil moisture. Aeration increases with a decrease in soil moisture, whereas an excess of water tends to encourage anaerobic conditions. The development and activities of soil organisms are greatly affected by the concentration and rate of supply of certain gases (particularly oxygen, carbon dioxide, and nitrogen) in the air. Oxygen is used for oxidation processes, carbon dioxide as a source of carbon for autotrophic organisms, and nitrogen for the nitrogen-fixing organisms. Abundant oxygen favors the activities of the nitrite and nitrate formers, the nitrogen fixers, fungi, actinomyces, and other organisms which oxidize organic matter. Sandy soils frequently are well aerated, and this condition restricts the accumulation of organic matter, whereas fine-textured soils frequently are insufficiently aerated. Poor aeration favors reduction processes and the accumulation of soil organic matter owing to restricted organic matter decomposition.

## Acidity and Alkalinity

The degree of acidity or alkalinity of the soil is of particular importance in influencing the activities and relative abundance of the different groups of soil organisms. It is frequently noted that the proportion of fungi to bacteria and actinomyces is greater in acid than in neutral soils; thus it appears that an acid soil favors the development of fungi but is unfavorable to the development of other forms. As a rule, the actinomyces prefer a reaction of 7.0 to 7.5, the bacteria from 6.0 to 8.0, and the fungi from 4.0 to 8. In strongly acid soils legume bacteria fail to develop and function normally, and as a result poor inoculation is obtained and the organisms do not persist in the soil for any great length of time. The nitrifying organisms are also sensitive to a highly acid condition. Thus it becomes evident that the optimum in acidity for the majority of the soil population (especially the more desirable groups) is essentially the same as for most higher plants. The tolerance of soil organisms for acidity, as of higher plants, is influenced considerably by other conditions like nutrient supply and favorable moisture content and temperature.

### Vertical Distribution of Microflora in the Soil Profile

The surface of the soil is the interface between the lithosphere and the atmosphere. At or near this interface the quantity of living matter is greater than at any region above or below. As a consequence, the A horizon contains more organic debris that serves as food for the microflora than the B or C horizons. Although other factors influence the activity and numbers of microflora besides nutrient and energy supplies, the greatest number of microbes as a rule occurs in the A horizon or surface layers, as shown in Fig. 5–4. In forest and meadow soils the greatest number of microbes are likely to be in or very close to the surface of the soil, although the number at the very surface of cultivated soils may be relatively low because of a lack of moisture and the germicidal action of sunlight.

## IMPORTANT ROLES OF SOIL MICROFLORA

The great diversity of soil microflora accounts for the numerous roles they play in the ecosystem. Perhaps their most important role is the decomposition of dead plants and animals and the recycling of nutrients. The intimate relationship of microflora and roots in the

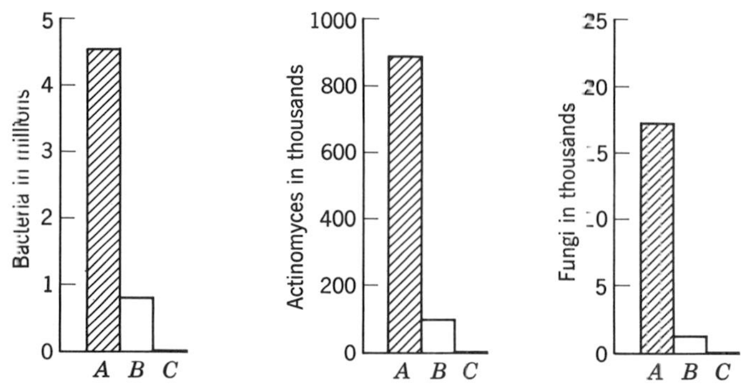


**Fig. 5–4** Distribution of microflora in the A, B, and C horizons of a cultivated Prairie (Mollisol) soil. All values refer to the number of organisms per gram of air-dry soil. (Data from Iowa Research Bull. 132.)

rhizosphere accounts for many interactions between microflora and higher plants. Some microbes establish symbiotic relationships with higher plants that result in mutual benefit. Other microflora produce disease. These and other important ecological roles of the soil microflora will be considered.

### Organic Matter Decomposition and Nutrient Recycling

The vast majority of bacteria, actinomycetes, and fungi are saprophytic and function as organic matter decomposers. These organisms bring about the hydrolysis and oxidation of organic compounds through enzymes. Simpler and simpler chemical compounds are produced until at last the carbon, hydrogen, and oxygen appear as carbon dioxide and water. Other nutrients contained in the organic matter also appear in inorganic form. The conversion of nutrients in organic matter into the mineral inorganic form is termed *mineralization.*

Figure 5–5 shows that the organic matter that is added to the soil consists of a variety of compounds. These include fats, carbohydrates, proteins, lignins, and others. Incorporation of these organic compounds into the soil stimulates to the greatest extent those organisms benefited the most. As decomposition proceeds, the most easily digested materials disappear first. All groups can effectively break down and utilize carbohydrates and proteins, but the fungi are the most effective in decomposing the lignin.

While digesting the plant residues, the microbes utilize some of the carbon, energy, and other nutrients for their own growth. In time the synthesized microbial tissue dies and becomes the substrate for further decomposition. Figure 5–5 indicates this by the subcycle where constitutents in the living organisms are temporarily unavailable or *immobilized.* The immobilization of nutrients refers to the use and incorporation of nutrients into living matter by both microflora and higher plants. The immobilized nutrients are again mineralized when the organisms die. In time even the most resistant materials succumb to the enzymatic attack of the microflora. The net effect is the release of energy as heat, formation of carbon dioxide and water, and the appearance of nitrogen as ammonium ($NH_4^+$), sulfur as sulfate ($SO_4^=$), phosphorus as phosphate ($PO_4^{-3}$), and many other nutrients as simple metallic ions ($Ca^{++}$, $Mg^{++}$, $K^+$). Most of these forms are available to living organisms for another cycle of growth.

The annual recycling of nutrients resulting from decomposition of most plant residues within a year means that a considerable amount of the nutrients absorbed by higher plants can be reused each year for growth. For example the data in Table 5–2 show that 67 to 86 percent

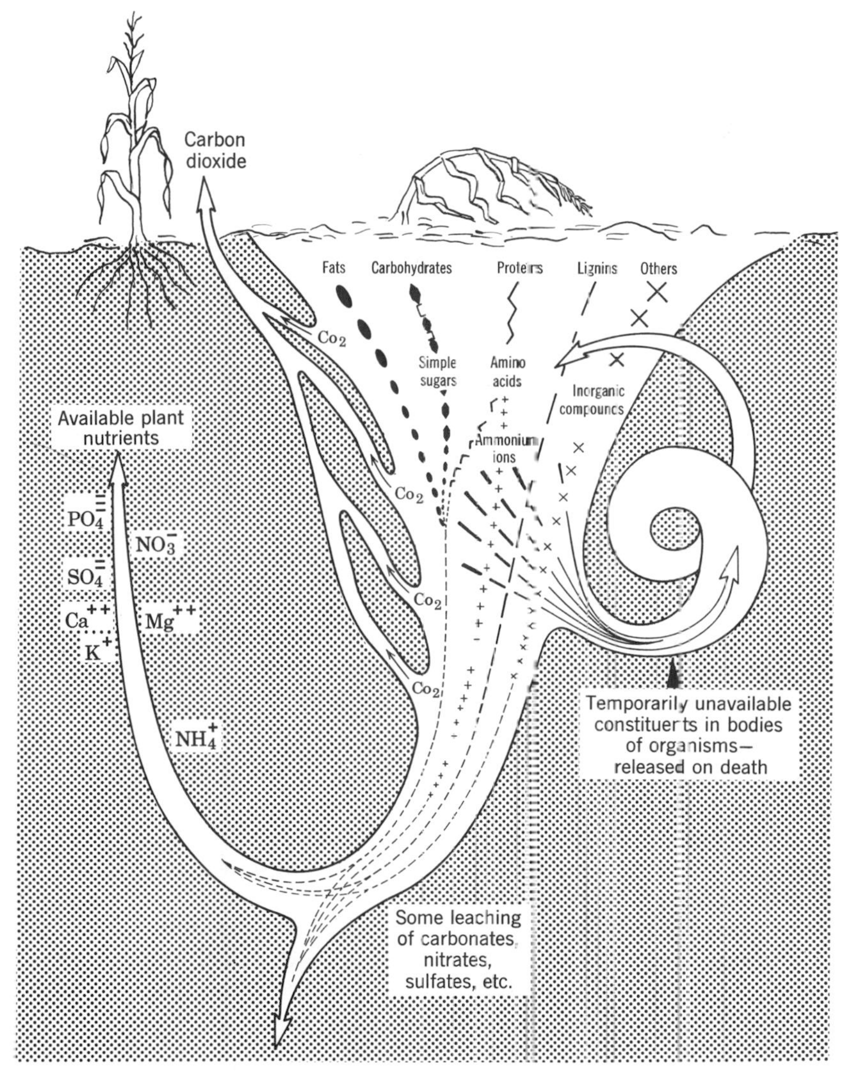


**Fig. 5–5** Schematic diagram of organic matter decomposition and nutrient recycling. (Courtesy of Dr. Burns R. Sabey.)

of 4 major nutrients taken up by pine and beech trees was returned to the soil each year in the litter (leaves and wood). Nutrient recycling accounts for the existence of enormous forests on some very infertile soils. In these situations most of the available nutrients are in organic matter.

As soon as some of ions or elements released in organic matter

**Table 5–2** Annual Uptake, Retention and Return of Nutrients to the Soil in Pine and Beech Forests

| | Nutrients, pounds per acre | | | | | | | |
|---|---|---|---|---|---|---|---|---|
| | Scotch Pine | | | | Beech | | | |
| | N | P | K | Ca | N | P | K | Ca |
| Uptake from soil | 40 | 4 | 6 | 27 | 45 | 11 | 13 | 86 |
| Stored in wood or lost from soil | 9 | 1 | 2 | 9 | 9 | 2 | 4 | 12 |
| Returned to soil in litter | 31 | 3 | 4 | 18 | 36 | 9 | 9 | 74 |
| Percent recycled | 77 | 75 | 67 | 67 | 80 | 82 | 70 | 86 |

(Data of Dengler and cited in "Fertilizers in Forestry," H. F. Arneman, *Agronomy*, **12**, p. 174, 1960.)

decomposition appear, other specialized organisms oxidize some of them. Some of the reactions and the bacteria involved are as follows:

$$NH_4^+ + 1\tfrac{1}{2}O_2 \rightarrow NO_2^- + 2H^+ + H_2O + \text{energy (Nitrosomonas)} \quad (1)$$
$$NO_2^- + \tfrac{1}{2}O_2 \rightarrow NO3 + \text{energy (Nitrobacter)} \quad (2)$$
$$S + 1\tfrac{1}{2}O_2 + H_2O \rightarrow H_2SO_4 + \text{energy (Thiobacillus)} \quad (3)$$

These transformations are beneficial to the extent that the oxidized forms are more readily used by higher plants. This is the case for the oxidation of sulfur to sulfate. The oxidation of iron and manganese makes them less soluble and, therefore, less available to higher plants. In this way microflora contribute to nutrient deficiencies in higher plants. Specialized bacteria in the rhizosphere oxidize manganese which produces a manganese deficiency on some oat plants growing on soils where nonsusceptible oat varieties do not develop manganese deficiency. Manganese deficiency on oats produces the disease called "gray speck."

When soils become anaerobic, still other microbes become active. The oxidized forms of plant nutrients are reduced. Sulfate is reduced to hydrogen sulfide or elemental sulfur. Perhaps one of the most important reactions in anaerobic soil is the reduction of nitrate and the formation of nitrogen gas, which escapes from the soil (denitrification). Loss of nitrogen by dentrification normally involves about 10 percent of the nitrogen turnover per year.

Under the anaerobic conditions of flooded rice paddy soils, the ammonium produced during organic matter decomposition is not oxidized to nitrate. Interestingly, the paddy rice plant shows a preference for ammonium nitrogen over the nitrate form. Not only is the ammonium not oxidized in flooded rice paddy soil, many other compounds are not oxidized as well. Sulfur accumulates as hydrogen sulfide and methane ($CH_4$) is formed instead of carbon dioxide and water. Hydrogen sulfide, methane, and other toxic materials build up in rice paddies requiring an occasional draining to permit aeration of the soil and the oxidation and elimination of reduced forms of materials that are toxic to higher plants.

The biological transformations of nitrogen in the soil are sufficiently important to summarize them in the form of a diagram, as shown in Fig. 5–6. The nitrogen in the soil essentially all comes from the atmosphere and most of the nitrogen enters the soil as a result of biological nitrogen fixation. We shall consider this important topic next.

## Fixation of Atmospheric Nitrogen

Nitrogen in its elemental form is a colorless, odorless gas and very inert. There is an inexhaustible supply of it in the air, but it is in the free state and does not easily combine with other elements. There are about 34,500 tons of nitrogen in the atmosphere for every acre of land area, the atmosphere being approximately 79 percent nitrogen. Regardless of the fact that nitrogen is so inert in the free state, certain groups of soil organisms have the ability to take nitrogen out of the air and utilize it in the building of their cells. The nitrogen of the air is thereby changed to a "fixed" form in which it can be of subsequent use to higher plants. This changing of atmospheric nitrogen into nitrogen compounds in the soil by microorganisms is known as *nitrogen fixation*. The process is accomplished mainly by two groups of bacteria, symbiotic and nonsymbiotic There are several species of each.

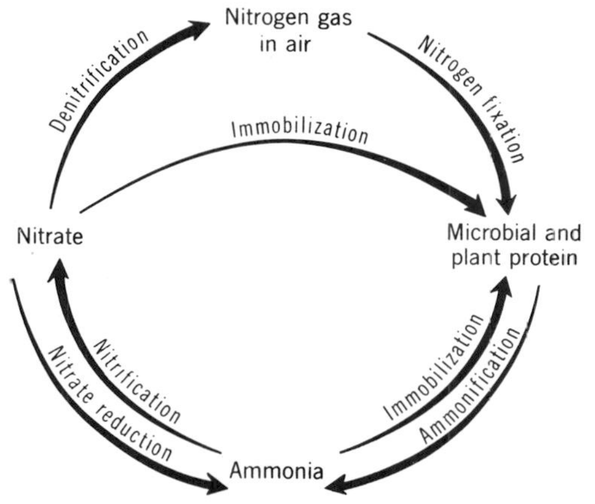


**Fig. 5–6** The soil nitrogen cycle. (USDA Yearbook of Agr. 1957, p. 154.)

**Nonsymbiotic Nitrogen Fixation.** There are certain groups of bacteria living in the soil independently of higher plants that have the ability to use atmospheric nitrogen in the synthesis of their body tissues. Since these bacteria do not grow in association (mutual relationship) with higher plants, they are termed nonsymbiotic. A dozen or more different bacteria have been found that fix $N_2$ nonsymbiotically. However, the two organisms that have been studied the most belong to the genus *Azotobacter* and the genus *Clostridium.*

*Azotobacter* are widely distributed in nature. They have been found in soils (of pH 6.0 or above) in practically every locality where examinations have been made. The greatest limiting factor affecting their distribution in soils appears to be the soil reaction. These organisms may exist in soils below pH 6.0, but as a rule they are not active, as far as nitrogen fixation is concerned, under such conditions. *Azotobacter* are favored by good aeration, abundant organic matter (particularly of a carbonaceous nature), the presence of ample available calcium and sufficient quantities of available nutrient elements, especially phosphorus, and proper moisture and temperature relations.

The anaerobic bacteria, *Clostridia*, are much more acid tolerant than most members of the aerobic group and perhaps for that reason are more widespread. It is believed that these organisms can be found in every soil and that under suitable conditions they fix some nitrogen. It is not necessary that soils be waterlogged in order for anaerobic bacteria to function. A soil in good tilth may contain considerable areas within the granules favorable for the activities of these anaerobic nitrogen-fixing bacteria.

A question that naturally comes to mind is, how much nitrogen is fixed per acre per year by these nonsymbiotic microorganisms under favorable field conditions? This question cannot be answered definitely because of the many difficulties encountered in making such a measurement under field conditions. From laboratory studies it is known that the nitrogen fixers utilize nitrate and ammonium nitrogen, which are normally present in the soil. To the extent that soil nitrogen is used, symbiotic fixation of atmospheric nitrogen is inhibited. Further, large amounts of carbohydrates are used in relation to the amount of nitrogen fixed. It is estimated that only 5 to 20 pounds of nitrogen is fixed per 1000 pounds of organic matter decomposed. Considering the amount of decomposing organic matter available in soils, the large number of competing organisms, and the inhibitory effect of the available soil nitrogen, it appears that nonsymbiotic nitrogen fixation is a minor or unimportant factor in crop production. For the natural ecosystem, the small quantity fixed each year over thousands of years is undoubtedly important.

**Symbiotic Nitrogen Fixation.** The most important bacteria, from the agricultural point of view, capable of utilizing the free nitrogen of the air are those that cause the formation of nodules on the roots of legumes. These organisms, when growing in the nodules of legume plants, derive their food and minerals from the legume, and in turn they supply the legume with some of its nitrogen. This growing together for a mutual benefit is called *symbiosis*, and hence the organisms are designated symbiotic nitrogen-fixing bacteria. It has been estimated that nearly 2,000,000 tons of nitrogen are fixed annually by legume bacteria in the United States.

**Nodulation and Nitrogen Fixation.** Legume plants form a symbiotic relationship with heterotrophic bacteria of the genus *Rhizobium*. The root of the host plant appears to secrete a substance that activates *Rhizobium* bacteria. When the bacteria make contact with a root hair, the root hair curls. An infection thread is formed in the root through which the bacteria migrate to the center of the root as shown in Fig. 5–7. Once inside the root the bacteria rapidly multiply and are transformed into swollen irregular shaped bodies called *bacteroids*.

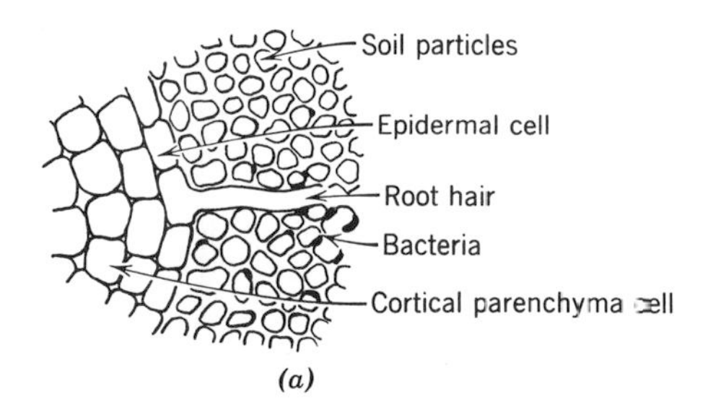


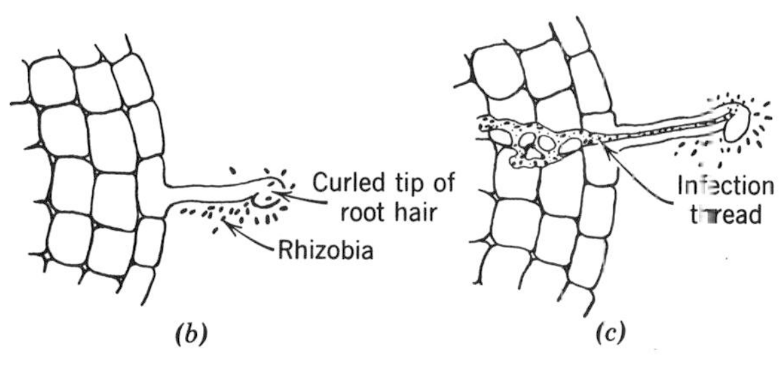


**Fig. 5–7** Early stages in the formation of a nodule. (a) Response of the bacteria to a product of the host plant; organism moves toward root hair. (b) Curling of the root hair. (c) Early penetration of the infection thread. (Courtesy of P. W. Wilson and the University of Wisconsin Press, *The Biochemistry of Symbiotic Nitrogen Fixation*, 1940.)

An enlargement of the root occurs and eventually a gall or nodule is formed. The bacteroids receive food, nutrients, and probably certain growth compounds from the host plant. The legume host plant is benefited by the $N_2$ fixed in the nodule. Some of the fixed nitrogen is transported from the nodules to various parts of the host plant. Nonnitrogen-fixing associated plants may benefit from the fixed nitrogen when nodules disintegrate and decompose, since most nodules do not live longer than one year. Under some conditions nitrogen is excreted from the nodules and is available to all organisms.

**Quantity of Nitrogen Fixed Symbiotically.** The quantity of nitrogen added to the soil through growth of legumes varies greatly according to conditions, such as the kind of legume, the nature of the soil, the effectiveness of the bacteria present, and seasonal conditions. It appears that the intimate relations existing between nodule bacteria and their host plants are determined mainly by the carbohydrate supply in the host plants. Any environmental condition affecting the production of carbohydrates in the plant would automatically affect the quantity of nitrogen fixed by good strains of legume bacteria. It has also been found that symbiotic nitrogen fixation is inhibited by an abundance of available soil nitrogen.

One method used to study the quantity of nitrogen fixed is to compare the amount of nitrogen in nodulated and nonnodulated plants. Weber used this method and found that as much as 142 pounds of nitrogen was fixed when soybeans were grown. This is shown in Fig. 5–8, which also shows that symbiotically fixed nitrogen represented 75 percent of the total nitrogen in the tops of the soybean plants. Further, as the amount of available nitrogen in the soil was increased by the addition of fertilizer, the total amount and percentage of nitrogen fixed by bacteria was markedly decreased. From these and other data it is reasonable to conclude that 50 to 200 pounds of nitrogen are commonly fixed per acre per year when properly inoculated legume crops are grown.

In actual farm practice the amount of nitrogen added to a soil by legume bacteria is determined by the methods of disposing of the legume crop. If the crop is turned under as a green manure, the total quantity of nitrogen taken from the air is added. If the crop is cut for hay and sold off the farm, little or no gain is realized; with some legumes there may even be a net loss of nitrogen. And, if the crop is cut for hay and fed on the farm, about one-half the nitrogen that was taken from the air by the legume bacteria can be returned to the soil if special care is exercised in handling the manure to prevent loss. It is generally assumed (although not necessarily true for all legumes) that the amount of nitrogen in the roots and stubble equals the amount of nitrogen taken from the soil; this would mean that the quantity

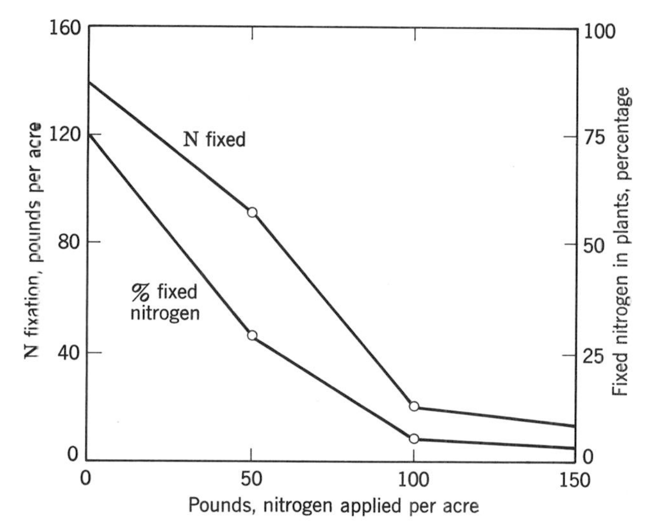


**Fig. 5–8** Amount of total nitrogen fixed and percent of nitrogen in soybean plants from symbiotic fixation in relation to the application of nitrogen fertilizer. (Data from "Nodulating and Nonnodulating Isolines," C. R. Weber, *Agronomy Journal*, **58**, p. 47, 1966.)

of nitrogen removed in the harvested crop is equal to the nitrogen obtained from the air.

## Mycorrhizae or Fungus Roots

Some fungi form an association with higher plants that results in the formation of structures known as *mycorrhizae* or fungus roots. The mycorrhizal fungi grow in the close proximity of plant roots and invade the roots. Two types are recognized. The *ectotrophic* fungi penetrate plant roots with mycelium existing largely between cell walls. The mycelium of the *endotrophic* fungi invade the cells of the plant roots and appear to be parasitic in some cases. The fungi are heterotrophic and are believed to obtain food and, perhaps, other growth factors from the host plant, while the extensive mycelium or fungus roots act as an extension of the root system. The effective root system of the host plant is increased many fold. This correlates with greater development of mycorrhizae in soils low in nitrogen and phosphorus. Many plants, particularly forest trees, are greatly benefited, as shown in Fig. 5–9.

## Production of Disease

The soil frequently contains a rather large number of organisms that cause diseases either in plants or animals. Some of these organisms live in the soil only temporarily, and others use it as a permanent

**Fig. 5–9** The uneven growth of these Red Pine seedlings is attributed to nonuniform inoculation with mycorrhiza fungi. This area was a new tree propagation site and required mycorrhiza inoculation for good seedling growth.

habitat. The soil may harbor organisms that cause such bacterial diseases as wilt of tomatoes and potatoes, soft rots of a number of vegetables, leaf spots, and galls. Some of the most destructive parasites are the disease-causing fungi such as those that cause damping-off of seedlings, cabbage yellows, mildews, blights, certain rusts, wilt diseases, scab, dry rot of potatoes and many others. Certain species of *Actinomyces* may cause diseases like scab in potatoes and sugar beets and pox in sweet potatoes. The catastrophic potato famine in Ireland in 1845–1846 was caused by a fungus that produced potato blight.

## SOIL MICROFLORA AND ENVIRONMENTAL QUALITY

During the billions of years of evolution of living organisms there evolved organisms that could decompose all compounds formed directly or indirectly from photosynthesis. This led to the concept of the "infallibility of soil organisms." This concept is being challenged today because man has become an important contributor of synthetic compounds to the environment. New questions have been raised, including: "Can the soil organisms destroy any compound that man can synthesize?" The topics of pesticide degradation, nitrate contamination of ground water, and disposal of sewage effluent will be considered as they relate to environmental quality.

### Nitrate Contamination of Ground Water

About 90 to 95 percent of plant tissue is composed of carbon, hydrogen, and oxygen that the plant obtains from air and water. The remainder of the elements in plants come from the soil, and of these nitrogen is generally the most abundant. Thus, nitrogen is an element required in large amounts and frequently is deficient for crop production. The low cost of nitrogen fertilizer has encouraged its use in quantities that approach or exceed the immobilization capacity of the soil. In the humid regions nitrate nitrogen is carried downward by water during those periods when precipitation exceeds evapotranspiration. When nitrate has migrated beyond the depth of biological activity, nitrates are not immobilized or dentrified. As a result the nitrate moves with percolation water to the water table and persists as nitrate. The continued use of nitrogen at rates of 150 pounds or more per acre per year caused an accumulation of nitrate nitrogen in the Marshall silt loam in Missouri when the land was used continuously for corn production (see Fig. 5–10). The zero and 100 pound rate produced similar amounts of nitrate nitrogen in the soil profile. It is natural for some nitrate nitrogen to move to the ground water in the humid regions, but excessive use of nitrogen fertilizer can pollute the ground water with harmful quantities. Drinking water standards of the Public Health Service give 10 parts per million as the amount of nitrate nitrogen that should not be exceeded for infants. Excess nitrate contributes to the development of methemoglobemia; an oxygen shortage in the blood.

Although excessive use of nitrogen fertilizer can cause nitrate pollution of ground water and many persons credit fertilizer use for extensive pollution, the evidence at the present time suggests that use of fertilizers contribute little to ground water pollution. In fact, studies in Missouri and Colorado show that nitrate pollution of ground water is more related to numbers of livestock. Nitrate nitrogen is produced during the decomposition of manure in the same ways that organic matter decomposition in soils produces nitrate. Where animals congregate in large numbers, as in feedlots, growing plants are absent and the nitrate produced from manure decomposition is not immobilized. Under these conditions rain water can leach the nitrate out of the feedlot and transport the nitrates to the ground water table. The same situation applies to corrals.

There has always been a natural amount of nitrate in ground water resulting from the natural reactions of the nitrogen cycle. Since an-

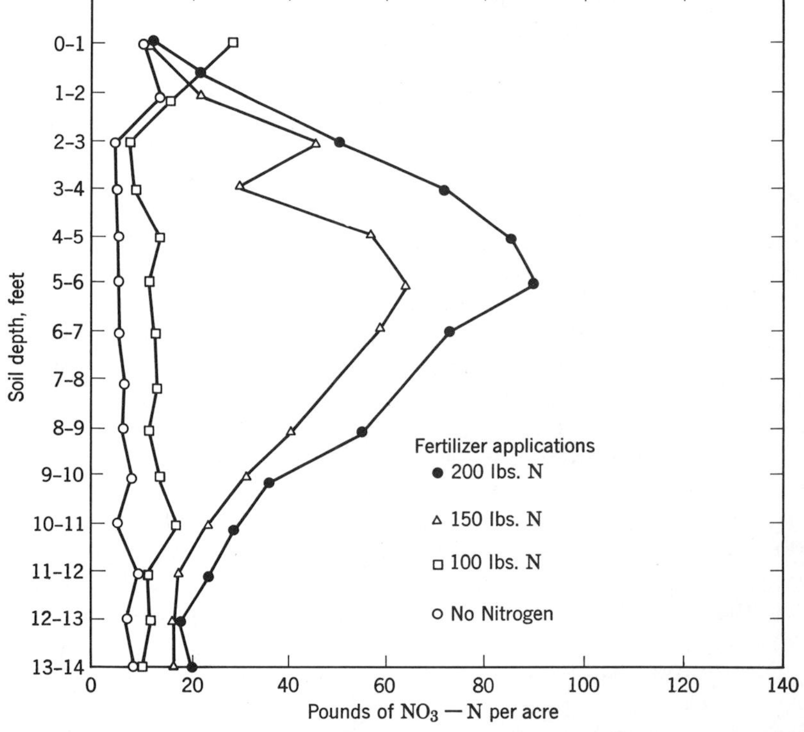


**Fig. 5–10** Nitrate nitrogen in the Marshall silt loam after seven annual applications of nitrogen fertilizers in a continuous corn program. (From "Residual Nitrate in Missouri Soils," K. W. Linville, MS Thesis, University Missouri, 1968.)

cient times the manner in which human and other animal wastes were disposed created varying degrees of nitrate pollution of water supplies. Today, however, the problem is different because of the magnitude of nitrogen used as fertilizer and the great concentrations of cattle that occur on large dairy and beef farms.

## Pesticide Degradation in Soils

Pesticides include those substances used to control or tc eradicate insects, disease, organisms, and weeds. One of the first and most successful pesticides was DDT, used to kill mosquitoes for malaria control. Today, about 30 years later, there is evidence to indicate that some DDT exists in the cells of "all" living animals. This has dramatized the resistance of DDT to biodegradation and the "fallibility" of the soil microflora. Evidence supports the view that the structure of

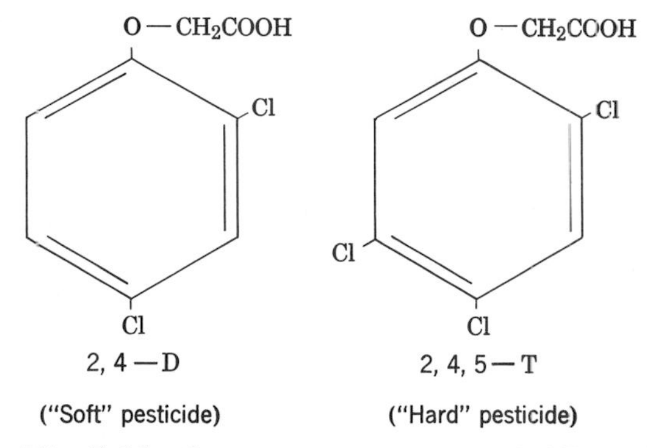


**Fig. 5–11** Structures representing 2,4-D on the left and 2,4,5-T on the right. The structures are very similar except for the additional Cl at the meta position of 2,4,5-T which is metabolized with great difficulty, if at all, by soil microorganisms.

DDT is different from any naturally occuring compound and, as a consequence, no soil organism has developed an enzyme system that degrades DDT. In general, it is believed that those pesticides with structures similar to those found in naturally occuring compounds are degradable and those pesticides with "new" structures not naturally found are persistent. This is illustrated by comparing 2,4-D(2,4-dichlorophenoxyacetic acid) with 2,4,5-T (2,4,5-trichlorophenoxyacetic acid). The two compounds have very similar structures, as shown in Fig. 5–11, except that 2,4,5-T has an extra chlorine at the meta position of the ring. While 2,4-D is readily decomposed, in fact some organisms can use 2,4-D as their only source of carbon, 2,4,5-T is very resistant. A chlorine on the meta position is metabolized with difficulty, if at all, and consequently the 2,4,5-T is a very persistent or is a "hard" pesticide.

We have become very dependent on pesticides and the likelihood of eliminating their use is remote. The challenge for scientists is the development of pesticides that can perform their useful function and disappear from the ecosystem without any undesireable side effects.

## Soil as a Living Filter for Sewage Effluent Disposal

A relatively small community of 10,000 people may produce about 1 million gallons of waste water per day from its sewage treatment plant. This is approximately 40 acre inches of sewage effluent a day. The effluent looks much like ordinary tap water and when chlorinated is safe for drinking. Discharging the effluent into a nearby stream does not create a health hazard. The effluent, however, is enriched with nutrients and may contain "hard" detergents. When the effluent is discharged into streams or lakes a menacing foam may appear from the detergents and plant growth may be increased by the addition of nutrients. Increased plant growth in the water increases the consumption of oxygen because more organic matter decomposition will eventually result. As a consequence, oxygen levels in the water may become too low for fish. Weed growth may be stimulated by nutrients and weeds interfere with boating and swimming. The problem is more acute today because many cities have grown rapidly while the amount of water available for diluting and carrying away the effluent has remained about the same. At the same time many cities are depleting their underground water supplies, which is a major source of water for use in sewage disposal.

To combat these problems researchers at Pennsylvania State University set up experiments to use the soil as a living filter for sewage effluent disposal. The researchers expected microorganisms to degrade the detergents (similar to the problem of pesticide degradation) and microorganisms and higher plants to immobilize nutrients as the effluent water slowly percolated through the soil. By applying water in excess of the potential evapotranspiration they expected the "filtered" water to migrate downward and eventually recharge the aquifer for reuse. The water was applied with a sprinkling system (see Fig. 5–12).

The experiment has been a success. For a community of 10,000 people producing 1 million gallons of effluent a day, the researchers found that an application of 2 inches per week on 129 acres of land was satisfactory.[2] Under these conditions over 80 percent of the effluent water migrated deep enough to be considered aquifer recharge water. After three years the water recovered at a depth of four feet showed that over 90 percent of the hard detergents had been removed and there was almost complete removal of the nitrogen and phosphorus (two major nutrients that have been associated with eutrophication). Further, the trees and crops growing on the land were greatly stimulated.

## SOIL ANIMALS

Perhaps higher plants could grow and provide us with food and the microflora could recycle all the nutrients without the aid of animals. However, animals play an important role in organic matter decomposition. Furthermore, soil animals forage on higher plants, move

[2] "Waste Water Renovation by the Land-A Living Filter," L. T. Kardos, *Agriculture and the Quality of Our Environment,* Am. Assoc. Adv. Sci. Pub. 85, 1967.

**Fig. 5–12** Water application of sewage effluent in a forest with a sprinkler system in winter at Pennsylvania State University. (Photo courtesy USDA.)

considerable quantities of soil, and produce disease. Before we discuss these activities we shall look briefly at the kinds and extent of soil animals.

## Kinds and Abundance of Soil Animals

Some of the most important soil animals are listed in Table 5–3. Protozoa are single-celled animals and are the smallest and most numerous of the soil animals. Soil protozoa live in the films of water surrounding soil particles and in a sense are aquatic animals. When the soil dries out, food supplies become short, or conditions are harmful, protozoa encyst and become active again when conditions become favorable. Soil protozoa are largely predators, feeding on soil bacteria, although some protozoa also feed on fungi, algae, or dead organic matter.

Numerous worms inhabit the soil ranging from the microscopic nematodes that invade plant roots to the larger earthworms. Arthropods comprise a group of animals that have segmented bodies and jointed

**Table 5–3** Some Important Soil Animals

| | | |
|---|---|---|
| Protozoa | Amoebae | |
| | Ciliates | |
| | Flagellates | |
| Worms | Nematodes | |
| | Earthworms | |
| Mollusks | Snails | |
| | Slugs | |
| Arthropods | Wood lice | Ants |
| | Spiders | Termites |
| | Mites | |
| | Millipedes | |
| | Centipedes | |
| | Springtails | |
| | Insect larvae | |
| Vertebrates | Reptiles | |
| | Moles | |
| | Gophers | |

legs. Arthropods are very mobile and large numbers inhabit the litter layer of forest soils. The vertebrates are largely burrowing animals and may spend only part of their life in the soil. As with the soil flora, the smaller the organisms, the more numerous they are.

### Role in Organic Matter Decomposition and Nutrient Recycling

The importance of microflora in organic matter decomposition and nutrient recycling has been stressed. The role of the smaller soil animals is of about equal importance. In fact, the microflora and fauna work together as a team. Consider for a moment that a leaf falls on the forest floor. Both microflora and animals attack the leaf. Holes made in the leaf by springtails and mites facilitate the entrance of microflora inside the leaf. Soil animals ingest bacteria when feeding and the bacteria continue to function in the digestive tract of the small animals. The excrement of animals is attacked by both the flora and fauna. The entire decomposing mass of stuff along with mineral soil particles may be ingested by earthworms, thereby, producing an intimate mixing of organic and mineral matter. The net result is the humification of organic matter with both microflora and fauna playing vital roles. In the process the major role of the animals is the fragmentation and mixing that greatly increase the surface area and prepare the organic matter for the microflora. Major credit for the mineralization and recycling of mineral elements goes to the microflora.

## Activities of Nematodes

Nematodes are worms that are mainly microscopic in size and are the second most abundant animals in soils. On the basis of their food requirements three groups are distinguished: (1) those that feed on decaying organic matter; (2) those that feed on earthworms, other nematodes, plant parasites, bacteria, protozoa, and the like; and (3) those that infest the roots of higher plants, passing a part of their life cycle embedded therein. Nematodes are sometimes called eel worms; they are round or spindle-shaped and usually have a pointed posterior.

Members of the first group are in greater abundance in most soils than those of the other two groups, although the last group is the most important from the agricultural point of view. The roots of a great number of plants, such as the English pea, cowpea, tomato, and carrot, are entered by certain species of the third group, and a great deal of damage is often done. Investigations are indicating that nematode damage is much more extensive than originally thought. Nematodes are a very serious problem for pineapple production in Hawaii where the soil for new plantings is routinely fumigated to control nematodes. Nematodes may also become serious pests in greenhouse soils unless special care is taken to avoid infestation. Not only do the nematodes injure the plant roots themselves, but also by puncturing the plant they prepare an entrance for other parasites. Not all the activities of nematodes are harmful to the growth of higher plants, for they aid in bringing about an intimate mixture of the mineral and organic matter in soil and in decomposing organic materials.

## Earthworms and Their Activities

Perhaps the best known group of larger animals inhabiting the soil is the common earthworm, of which there are several species. These organisms prefer a moist environment with an abundance of organic matter and a plentiful supply of available calcium. Consequently, earthworms are found most abundantly, as a rule, in fine-textured soils which are high in organic matter and are not strongly acid, and occur only sparingly in acid sandy soils low in organic matter.

Obviously, then, the number and activity of the earthworms vary greatly from one location to another and, as with other soil organisms, figures indicating numbers are merely suggestive. The number of earthworms in the plowed layer of an acre may range from a few hundred or even less to more than a million. It has been estimated that between 200 and 1000 pounds of earthworms are present in an acre of soil.

The common earthworm, *Lumbricus terrestris*, was imported into

**Fig. 5–13** Clay loam soil before (left) and after (right) being thoroughly worked by earthworms. Note the almost complete granulation resulting from passage of the soil through the digestive tract of the worms.

the United States from Europe. *Lumbricus terrestris* makes a shallow burrow and forages on plant material at night. Some of the plant material is dragged into the burrow. Other kinds of earthworms exist by ingesting organic matter that exists in the soil. Excrement or castings are deposited both on and in the soil. The intimate mixing of soil materials, creation of channels, and production of castings leaves the soil more open and porous (see Fig. 5–13).

Darwin made extensive studies of earthworms and showed that they may annually deposit 10 or more tons of castings on the surface or build up a one-inch layer in 10 years. This activity produces thicker than normal dark-colored surface layers in some forest soils and buries stones and artifacts that are laying on the top of the soil. The burying of artifacts is of importance to archeologists.

Earthworms don't add anything to the soil that wasn't already there. As a consequence, their effect on plant growth is minimal or of no importance. Attempts to increase plant growth by increasing earthworm activities in soils have been disappointing. Earthworms are objectionable when they produce an abundance of casts on lawns or golf courses.

## Ants and Their Activities

While most persons appear to be more conscious of earthworms and their activities, the activities of ants are perhaps of greater importance. Harvester ants are a pest in many places including the southwestern part of United States. Harvester ants denude the area surrounding

**Fig. 5–14** Denuded areas surrounding Red Harvester ant mounds in an alfalfa field. (Photo courtesy USDA.)

their nests to distances as great as 10 feet or more. Thorp[3] estimated 20 ant hills per acre with denuded areas ranging from 6 to 20 feet in diameter. Assuming an average denuded area of 13 feet diameter, about 6 percent of the land surface would be denuded. Such harvesting of vegetation by ants can be of economic importance as seen in Fig. 5–14. On range lands the forage for wildlife and cattle is reduced and the bare denuded areas are more subject to erosion. Harvester ants also gather seeds for food which retards the reseeding of natural grasslands.

Ants transport large quantities of material from within the soil and deposit the material on the surface. Some of the largest ant mounds are a few feet high and more than 10 feet in diameter. The effect of this transport is comparable to that of earthworms in creating thicker dark-colored A horizons and burying objects laying on the surface. A study of ant activity on a prairie in southwestern Wisconsin showed that ants brought material to the surface from depths greater than 5 feet and built mounds about 6 inches high and over 1 foot in diameter (see Fig. 5–15). Furthermore, it was estimated that 1.7 percent of the land

[3] "Effects of Certain Animals That Live in Soils," J. Thorp, *Sci. Monthly*, **68**:180–191, 1949.

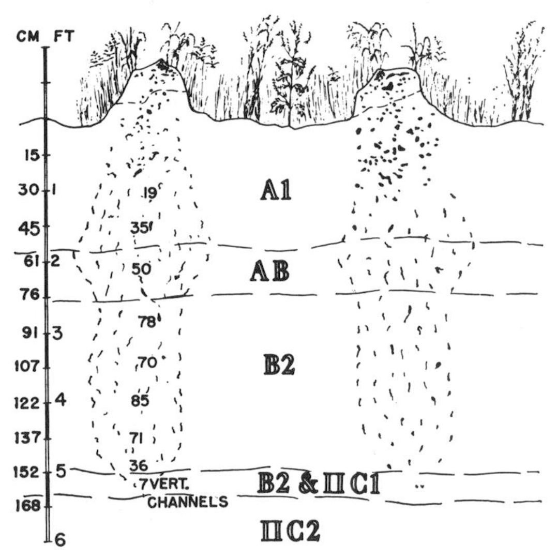


**Fig. 5–15** Ant *(Formica cinera)* in a Prairie soil in southwestern Wisconsin. *Upper* photo shows ant mounds over six inches high and over a foot in diameter. The *lower* sketch shows soil horizons and location of ant channels with numbers referring to the number of channels observed at the depths indicated. (Courtesy F. D. Hole, Soil Survey Division, Wisconsin Geological and Natural History Survey, University of Wisconsin.)

was covered with mounds. Assuming the average life of a mound as 12 years, the entire land surface would be reoccupied every 600 years. The researchers believe this evidence supports the view that the incorporation of subsoil material (Bt horizon) into the A horizon has helped produce a thicker dark-colored A horizon with a greater than "normal" clay content. The increase in clay content is supported by the fact that the clay contents of A horizons in a nearby forest are only 10 percent compared to 22 percent clay in the A horizons on the prairie.

Termites bring about similar changes in soils as ants. In summary, ants and termites create channels in soils and transport soil materials that tend to alter or obliterate soil horizons. A concentration of nutrients builds up where the mounds are located because plant materials are stored and fecal material accumulates there. Some farmers in southeast Asia recognize this and make use of the higher fertility of areas occupied by mounds.

# 6

# Soil Organic Matter

Almost all of the life in the soil is dependent on organic matter for energy and nutrients. For thousands of years man has recognized the importance of organic matter in food production. The story of how the Indian, named Squanto, helped the Pilgrims raise corn by burying a dead fish near each hill is well known. Perhaps, the most poetic expression of the effects of organic materials on plant growth was expressed by Omar Khayyam.

> "I sometimes think that never blows so red
> The Rose as where some buried Caesar bled."

Although organic matter in soils is very beneficial, Liebig pointed out over 100 years ago that soils composed entirely of organic matter are naturally very infertile. The major objective of this chapter is to clarify the role and importance of the soil organic matter.

## HUMUS FORMATION AND CHARACTERISTICS

In the previous chapter we discussed the decomposition of plant residues and the synthesis of many compounds by soil organisms. As a result of these activities the soil contains an enormous number or organic compounds in various states of decomposition. *Humus* is the word used to refer to the organic matter that has undergone extensive decomposition and is quite resistant to further alteration.

## Humus Formation

The organic residues added to soils are not decomposed as a whole, but the chemical constituents are decomposed independently of one another. In the formation of humus from plant residues there is a rapid reduction of the water-soluble constituents, of the celluloses, and of the hemicelluloses; a relative increase in the percentage of lignin and lignin complexes; and an increase in the protein content. The new protein is believed to be formed for the most part through the synthesizing activities of the microorganisms. The lignin in humus originates mostly from plant residues with perhaps certain chemical modifications. The major differences between the composition of typical mature plant tissue and soil organic matter are shown in Table 6–1.

**Table 6–1** Partial Composition of Mature Plant Tissue and Soil Organic Matter

| | Percent | |
|---|---|---|
| Component | Plant Tissue | Soil Organic Matter |
| Cellulose | 20–50 | 2–10 |
| Hemicellulose | 10–30 | 0–2 |
| Lignin | 10–30 | 35–50 |
| Protein | 1–15 | 28–35 |
| Fats, waxes, etc. | 1–8 | 1–8 |

The high content of lignin and protein in humus or soil organic matter is of particular importance. Normally, proteins are readily decomposed in soils and the increase in proteins during humus formation can be explained by the fact that the nitrogenous complexes are rendered resistant to further rapid decomposition. The exact mechanisms involved are not known, but two have been purposed. First, there is reason to believe that the protein molecules can be adsorbed on the surface of clay minerals and rendered resistant to decomposition. Second, enzymes that decompose proteins may also be adsorbed by clay minerals so that the proteins are less susceptible to decomposition. That the clay plays an important role is supported by the fact that soils high in clay tend to have a high organic matter content. This slow rate of decomposition of humus is obviously of considerable practical importance. It offers a means whereby nitrogen can be stored in the soil and released gradually.

A general schematic summary of the processes leading to humus

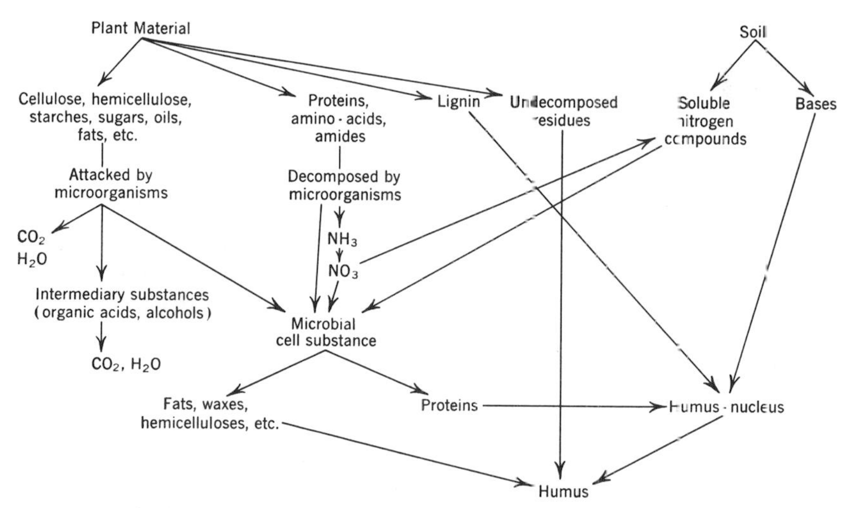


**Fig. 6–1** Schematic representation of the mechanism of the formation of humus in the decomposition of plant residues in soil. (After Waksman, from *Humus*, by permission of Williams and Wilkins Co.)

formation is presented in Fig. 6–1. The material commonly referred to as humus includes the mass of plant residues undergoing decomposition, together with the synthesized cell substance and certain intermediary and end products. It is constantly changing in composition. It is better, therefore, to speak of humus not as a single group of substances but rather as a state of matter, which is different under varying conditions of formation.

## Characteristics and Properties of Humus

During humus formation some plant constituents become completely decomposed, others are modified to a greater or lesser extent, and others are only slightly attacked. These processes are accompanied by the synthesis of new complexes (fats, carbohydrates, and proteins) by the microorganisms. It is in order next to point out that humus as a whole is characterized by certain specific physical, chemical, and biological properties that distinguish it from other forms of organic matter in nature.

Humus is practically insoluble in water, although some of it may go into colloidal suspension in pure water. To a large extent it is soluble in dilute alkali, and certain of the humus constituents may dissolve in acid solutions.

One of the most important and characteristic properties of humus is its nitrogen content, which usually varies from 3 to 6 percent, although the nitrogen concentration may be frequently lower or higher than these figures. The carbon content is usually 55 to 58 percent. According to Waksman,[1] the average theoretical carbon content of soil humus is 56.24 percent, and the average nitrogen content is 5.6 percent. This gives a theoretical ratio of carbon to nitrogen of 56.24:5.6, or 10.04, which is very close to the ratio commonly found in the humus of soils. This ratio varies with the nature of the humus, the stage of its decomposition, the nature and depth of the soil, and climatic and other environmental conditions under which it is formed.

Another important property of humus is its high cation-exchange capacity. Cation exchange is associated with several chemically active groups in both living and dead organic matter. One of the most important groups is *carboxyl* (—COOH). During humification of organic matter, lignin is altered in such a way that there is a decrease in noncation-exchanging groups as methoxyl ($-OCH_3$) and an increase in the cation-exchanging carboxyl groups. As a result the cation-exchange capacity of humus is many times greater than that of the organic residues originally added to the soil. Cation exchange sites adsorb cations such as Ca, Mg, K, and, in so doing, humus acts similarly to clay in retaining available nutrients against leaching and maintaining the nutrients in a form available to higher plants and microorganisms. The cation-exchange phenomenon of humus (and other kinds of soil organic matter) is illustrated by the following equation:

$$-R-C\begin{matrix} \diagup\!\!\!\!\diagup O \\ \diagdown OH \end{matrix} + KCl \longrightarrow -R-C\begin{matrix} \diagup\!\!\!\!\diagup O \\ \diagdown OK \end{matrix} + HCl \qquad (1)$$

The equation shows how water-soluble potassium chloride reacts with carboxyl groups of humus. The potassium (K) is exchanged for the H of the carboxyl group. The K is adsorbed with enough energy to retard its loss from the soil by leaching, but the K is still readily available for plant use.

Humus absorbs large quantities of water and exhibits the properties of swelling and shrinking. It does not exhibit so pronounced properties of adhesion and cohesion as do the mineral colloids and is less stable because it is subject to microbial decomposition. It has already been shown that soil humus is an important factor in aggregation (structure formation). Humus possesses other physical and physiochemical properties which make it a highly valuable soil constituent.

[1] *Humus*, S. A. Waksman, Williams and Wilkins Co., 2nd edition, p. 182, 1938.

## AMOUNT AND DISTRIBUTION OF ORGANIC MATTER IN SOILS

As the rocks and minerals of the earth's crust decomposed, mineral elements were made available to plants; and, as supplies of nitrogen in usable chemical combinations were produced from the store of nitrogen in the air, plants grew, died, and contributed their remains to the soil. Thus organic matter began to accumulate. As the supply of available plant nutrients in the soil increased, the accumulation of soil organic matter increased accordingly. This condition continued until an equilibrium was reached at which the rate of organic-matter accumulation was equal to the rate of decompositon. This section will discuss factors that influence the amount of organic matter in soils including climate, vegetation, drainage conditions, cultivation and soil texture.

### Influence of Climate and Vegetation on Organic Matter Content of Soils

Generally, as the quantity of organic residues added annually to soils is increased, there is an increase in the total organic matter content. One would expect the soils in deserts to contain very little organic matter because the annual additions of organic matter from plant growth are very small. With increasing precipitation and an accompanying increase in the annual production of organic matter there is an increase in the organic matter content of soils. On the plains in the United States from eastern Colorado to Indiana the annual precipitation increases from about 15 to 35 inches. This is accompanied by a shift from widely spaced bunch and short grasses to tall grass and an increase in the organic matter content of soils from about 80 to 160 tons per acre to a depth of 40 inches (see Fig. 6–2). Similar changes occur in Argentina from the Andes Mountains to Buenos Aires and in the southern part of the Soviet Union from a south to north direction.

Eastern United States is a forested region and the soils have considerably less organic matter than nearby soils developed under tall grass. Further, with increasing average annual temperature in the forested area from north to south, the organic matter content of the soils decreases (Fig. 6–2). A major cause is the increased rate of microbial activity and decomposition of organic matter with increasing temperature (also see Fig. 9–5).

### Organic Matter in Forest Versus Grassland Soils

The settlers that colonized America adapted to farming on soils developed under forest vegetation. By the early 1800s the settlers had

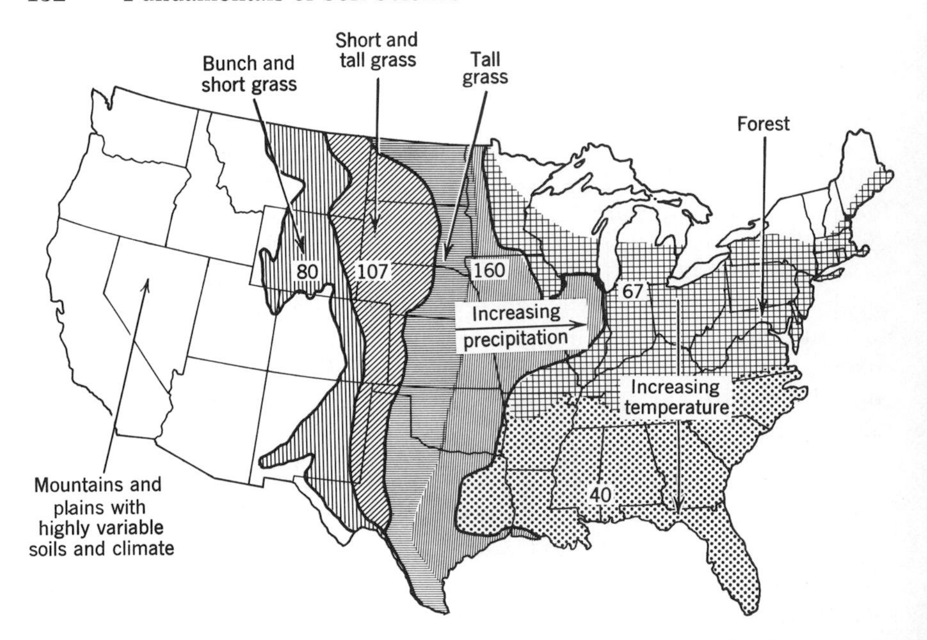


**Fig. 6–2** Generalized map showing organic matter content of soils (tons per acre to 40 inches) as related to climate and vegetation. (Adapted from "Soil Nitrogen," O. Schreiner and B. E. Brown, *USDA Yearbook*, 1938, pp. 365 and 366.)

spread westward to the tall grass prairie lands of central United States. The tough sod made it difficult to plow the land and new tools and techniques had to be developed. At first, the prairie lands were avoided but once they were broken, the superiority of grassland soils over forest soils was readily apparent. Now, the great productivity of the American prairies and the Argentine pampa are well known. One reason for the high productivity of these grassland soils is related to the amount and distribution of organic matter. Studies show that grassland soils, as compared to nearby forest soils, have (1) about twice as much organic matter in the soil profile and (2) a more gradual decrease of organic matter with increasing soil depth (Fig. 6–3).

**Fig. 6–3** The distribution of organic matter in forest (white oak, black oak) and prairie (big bluestem, Indian grass) ecosystems in south central Wisconsin. (Adapted from "A Study of the Natural Processes of Incorporation of Organic Matter into Soil in the University of Wisconsin Arboretum," G. A. Nielsen and F. D. Hole, Wisconsin Academy of Sciences, *Arts and Letters:* **52**, pp. 213–227, 1963.)

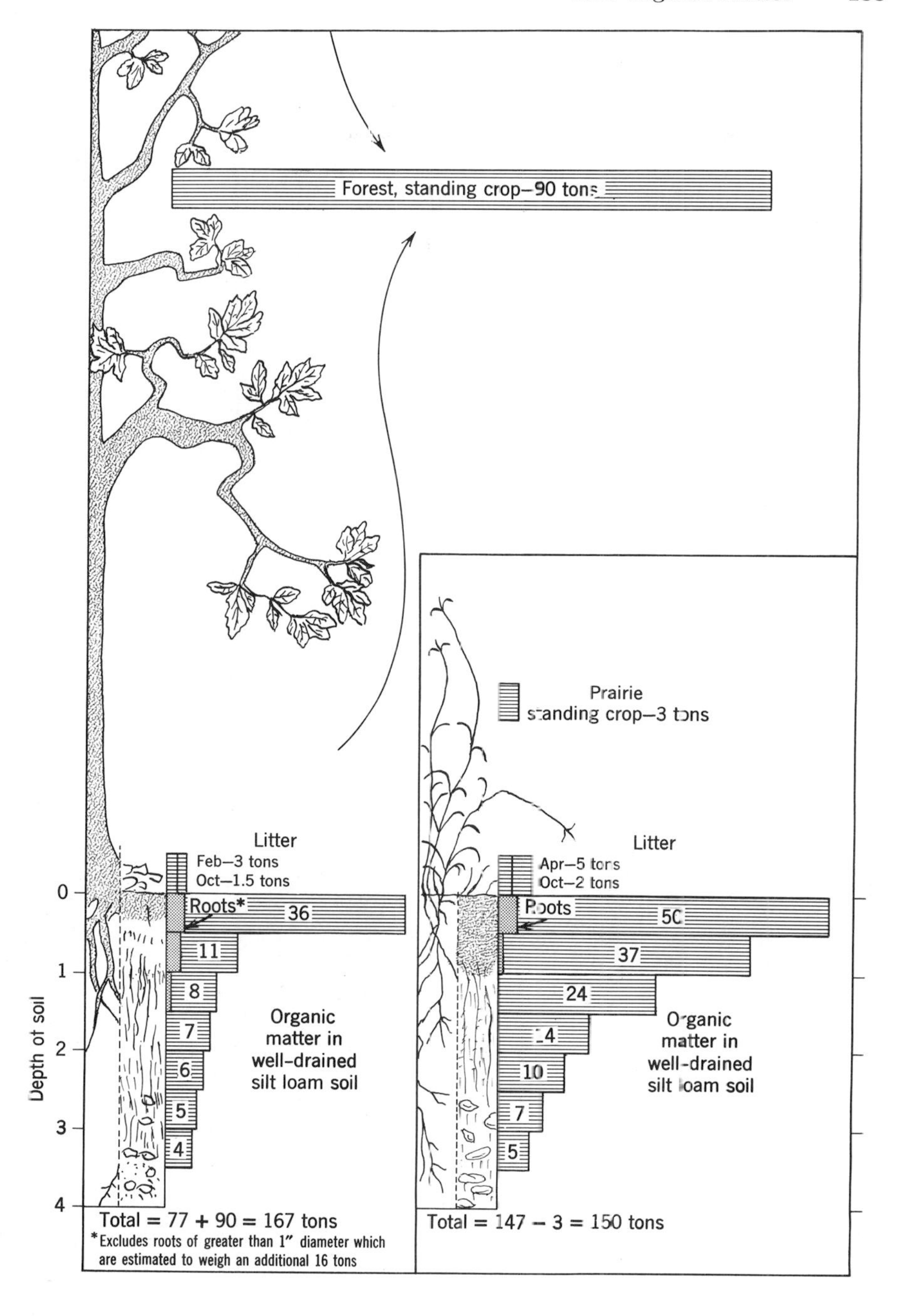
Forest, standing crop–90 tons
Prairie
standing crop–3 tons
Litter
Feb–3 tons
Oct–1.5 tons
Roots*
36
11
8
7
6
5
4
Organic
matter in
well-drained
silt loam soil
Litter
Apr–5 tons
Oct–2 tons
Roots
50
37
24
14
10
7
5
Organic
matter in
well-drained
silt loam soil
Depth of soil
0
1
2
3
4
Total = 77 + 90 = 167 tons
*Excludes roots of greater than 1" diameter which
are estimated to weigh an additional 16 tons
Total = 147 + 3 = 150 tons

The explanation for the differences in amount and distribution of organic matter in forest and grassland soils is related to differences in the growth of the plants and how the plant residues become incorporated into the soil. The roots of grasses are short-lived and each year the decomposition of dead roots contributes to the quantity of humified organic matter. Further, the quantity of roots decreases gradually with increasing soil depth (see Fig. 3–7 for the root system of oats). In the forest, by contrast, the roots are long-lived and the annual addition of plant residues is largely as leaves and dead wood that fall onto the surface. Some of the residues decompose on the surface but small animals transport and mix some of the surface litter with a relatively thin layer of top soil. In the hardwood forest in southern Wisconsin (where earthworms are active) it was found that 36 tons of organic matter per acre existed in the upper 6 inches of soil (A1 horizon) and only 11 tons per acre in the next deeper 6-inch layer of soil (A2 horizon).

Another interesting fact shown in Fig. 6–3 is that there is a similar amount of total organic matter in each ecosystem, but in the forest most of the organic matter exists in the *standing trees* while in the prairie ecosystem over 90 percent of the organic matter exists within the *soil*. When settlers cleared forests they burned or harvested the trees and, in so doing, removed about half of the organic matter. Breaking of the prairie land, by comparison, left virtually all of the organic matter in the soil, even if the grass was burned off before plowing. The differences in amount and distribution of organic matter is one of several explanations for the larger crop yields on the grassland soils. Even today, with good soil management, the average yields of corn on well-drained grassland soils is about 10 to 20 bushels per acre greater than for well-drained forest soils in central United States. Let us consider next the changes in soil organic matter that were produced when the established equilibrium level of organic matter was disturbed by putting the land into cultivation.

## Organic Matter Changes by Cultivation

Even on nonerosive land that is brought under cultivation, rapid losses of organic matter usually occur. It has been observed that the losses are most rapid immediately after farming is started, and thereafter the rate of disappearance is decreased; ultimately the organic content of the soil will reach a new equilibrium level.

It has been found at the Missouri Agricultural Experiment Station that, as a result of cultivation over a period of 60 years, soils in a noneroded condition lost over one-third of their organic matter, the losses being much greater during the earlier than the later periods. The organic matter losses amounted to about 25 percent the first 20 years, about 10 percent the second 20 years, and only about 7 percent the third 20 years. In other words, a new equilibrium level was almost attained after about 30 years (Fig. 6–4).

Soils in arid regions naturally have very low organic matter contents. Irrigating arid-region land and producing crops result in large increases in the amount of organic matter returned to the soil each year. As a consequence, irrigating arid lands and growing crops results in the establishment of a new equilibrium organic matter level much higher than the original level.

## Maintenance and Restoration of Soil Organic Matter

Although there is a rapid depletion of the organic matter in the soil immediately after virgin lands of humid regions are brought under cultivation, there is some consolation in the fact that this high rate does not continue indefinitely. It has been emphasized that after a period of heavy loss, a fairly constant level is attained during a long period of continued cultivation; the level is determined by the environment associated with a particular soil. Once the organic content has reached a low level, restoring the organic matter to its original level would require that the original vegetation be reestablished. In time a new equilibrium level of organic matter content would be achieved that would be the same or similar to the level that existed before the land was cultivated. While the land is being farmed it is virtually impossible and much too expensive to maintain the organic matter content similar to that which existed in the virgin soil. It is,

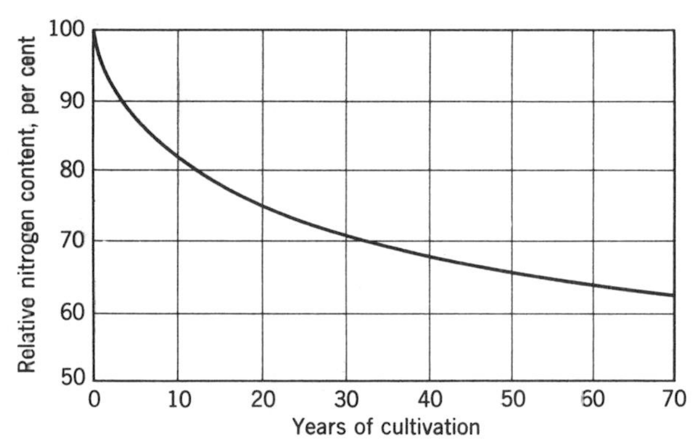


**Fig. 6–4** Decline of soil nitrogen (or organic matter) with length of cultivation period under average farming practices in the Middle West. (After Jenny, *Missouri Agr. Exp. Sta. Bull.* **324**, 1933.)

therefore, usually unwise and uneconomical to maintain the organic matter above a level consistent with good crop yields. Attention, then, should be directed toward the frequent additions of small quantities of fresh organic materials rather than to practices of maintaining the organic-matter content at any particularly high level.

In a consideration of the maintenance of soil organic matter, the amount of crop residues that must be returned to maintain a given organic-matter content depends on soil and climatic conditions. It is interesting to note that 3600 pounds of crop residues were needed annually to maintain the organic-matter content of Blackland soils near Dallas, Texas (Fig. 6–5). Cropping systems that returned more than this amount showed an increase in the organic-matter content after twelve years, whereas those returning less than this amount caused a decline in the organic-matter content. Only the tops were measured and root residues in addition to top residues were returned to the soil. It would be reasonable to allow 1000 pounds for roots, making the total weight of residues added per year to maintain the organic-matter content of these soils equal to about 4600 pounds.

## The Feasibility of Attempting to Maintain a High Soil Organic Content

The maintenance of an adequate supply of organic-matter in soil is rendered more difficult because of the large quantity that is dissipated each year. Since the rate of organic-matter loss from soils increases rapidly as the organic content is raised, the maintenance of

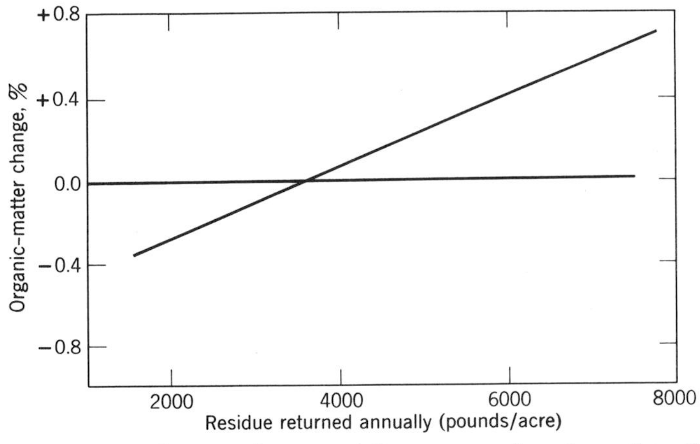


**Fig. 6–5** Relation of crop residue returned to the soil to the change in soil organic-matter content. (From "Farming Systems for Soil Improvement in the Blacklands," W. Derby Laws, *Tex. Res. Found. Bull.* **10**, 1961.)

a high organic level is not only difficult but also expensive. It is wise, therefore, to maintain the organic matter at a level that will result in good crop yields.

The "pool" of organic matter in a soil can be compared to a lake. Changes in the level of water in a lake depend on the difference between the amount of water entering and leaving the lake. This idea applied to soil organic matter is illustrated at the top of Fig. 6–6. Soil organic matter, being largely lignified material, decomposes in mineral soils at a rate equal to about 1 to 4 percent per year. Assuming a 2 percent rate and 40,000 pounds of organic matter per acre furrow slice, 800 pounds of soil organic matter would be lost or decomposed

**General Equation for the Loss or Gain in Soil Organic Matter**

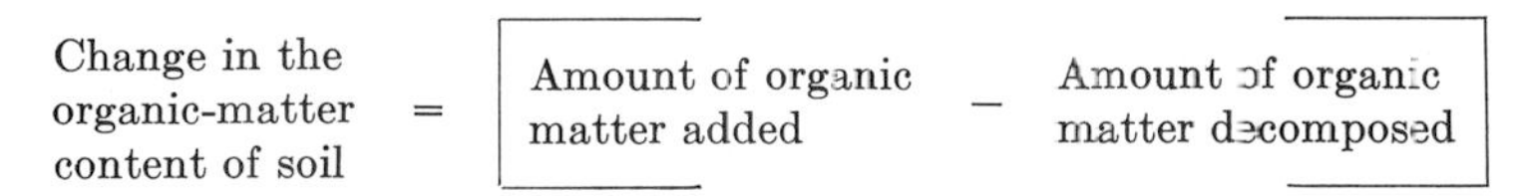


**A Case Where the Loss and Gain of Soil Organic Matter Is in Equilibrium**

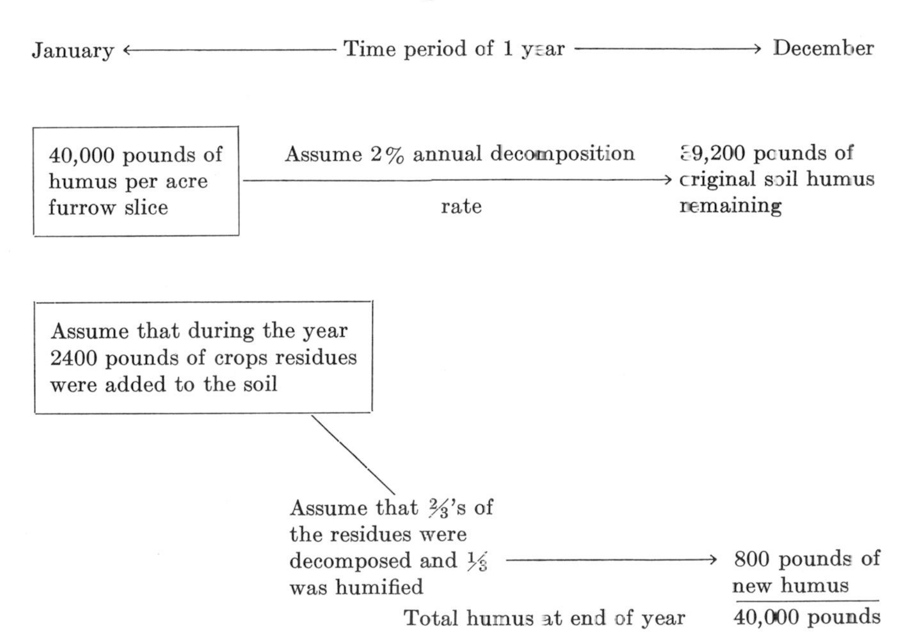


**Fig. 6–6** Schematic illustration of the equilibrium concept of soil organic matter as applied to a representative plow layer (2,000,000 pounds) containing 2 percent organic matter.

each year. On the other hand, if 800 pounds of humus was formed from the residues added to the soil, the organic matter content of the soil would remain the same from one year to the next and the soil would be at the equilibrium level as shown in Fig. 6–6.

## Organic Matter Content Versus Soil Texture

Locally, there tends to be a correlation between the clay content of the soil and the content of organic matter. The greater combined supply of water and nutrients favors the production and accumulation of more organic matter in the finer textured soils. Clay also adsorbs decomposing enzymes that become inactivated. Organic molecules adsorbed on clays are partially protected from decomposition by microorganisms. As the content of organic matter in the soil increases, the content of nitrogen and phosphorus increase since they are important constituents of organic matter. The content of nitrogen and phosphorus in some New York soils is used to illustrate the general relationship between soil texture and organic matter content in Fig. 6–7.

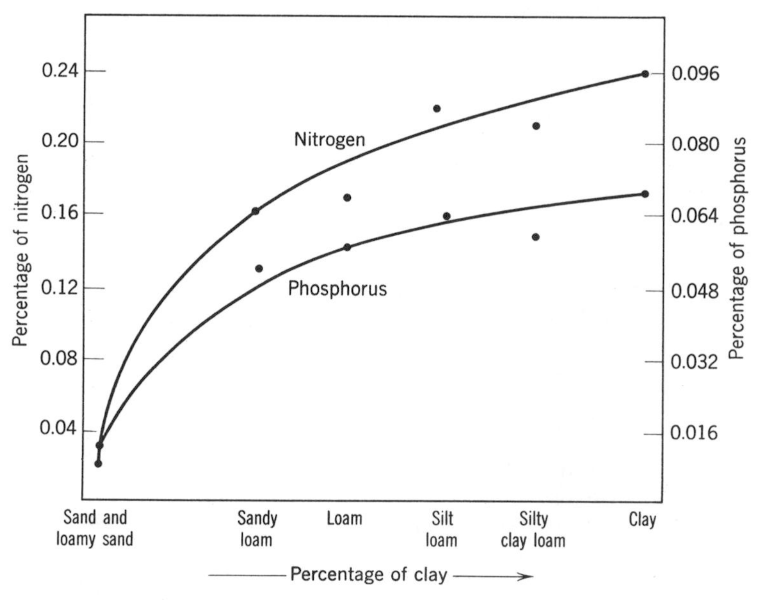


**Fig. 6–7** Nitrogen and phosphorus contents of several New York soil classes illustrate the geographically local trend of increasing organic matter content with increasing clay content of the soil.

## Organic Soils

In the shallow water of lakes and ponds, plant residues accumulate rather than decompose under anaerobic conditions in the water. As a consequence, soils consisting almost entirely of organic matter develop. Plant tissue and pollen grains in peat can be easily identified. The remarkable preservation capacity of some swamp and bog waters can be illustrated by the fact that human bodies up to 3000 years old have been found in peat bogs. One of the best preserved is the 2000-year-old Tollund man found in 1950 in Denmark.[2] The facial expression at death and the bristles of the beard of the Tollund man were well preserved. Excellent finger prints were made and an autopsy revealed that his last meal consisted mainly of seeds, many of them weed seeds. When found, the Tollund man was buried under 7 feet of peat that had formed in the 2000 years after his burial.

In Chapter 3 organic soils were distinguished from mineral soils by having 20 to 30 percent organic matter (depending on the texture of the mineral soil material). A characteristic feature of organic soils is a stratification or layering that represents changes in the kind of plants that produced the organic matter as a result of changes in climate or water level. Such a series of horizons that developed in Sweden is shown in Fig. 6–8. The woody peat layer is indicative of a dry period and was preceded and followed by wet periods when the vegetation was sphagnum moss. The 15 feet of peat accumulated in 9000 years or at the rate of one foot each 600 years. About 1 foot every 200 to 800 years is the usual range.

Organic soils must be drained before they can be used for crop production. The upper soil layers then become aerobic and the peat begins to decompose. This converts undecomposed peat into well-decomposed muck. In time the entire organic soil above the underlying mineral soil may disappear as a result of decomposition. This is a serious problem in warm climates, as in southern Florida, where the high annual temperature stimulates rapid decomposition.

# SOME ORGANIC MATTER MANAGEMENT CONSIDERATIONS

Organic matter plays many important roles in soils. Since soil organic matter orginates from plant remains, soil organic matter orginally contained all nutrients needed for plant growth. Organic matter per se influences soil structure and tends to promote a desirable

[2] "Lifelike Man Preserved 2,000 Years in Peat," P. V. Glob, *National Geographic Magazine*, 105, pp. 419–430, 1954.

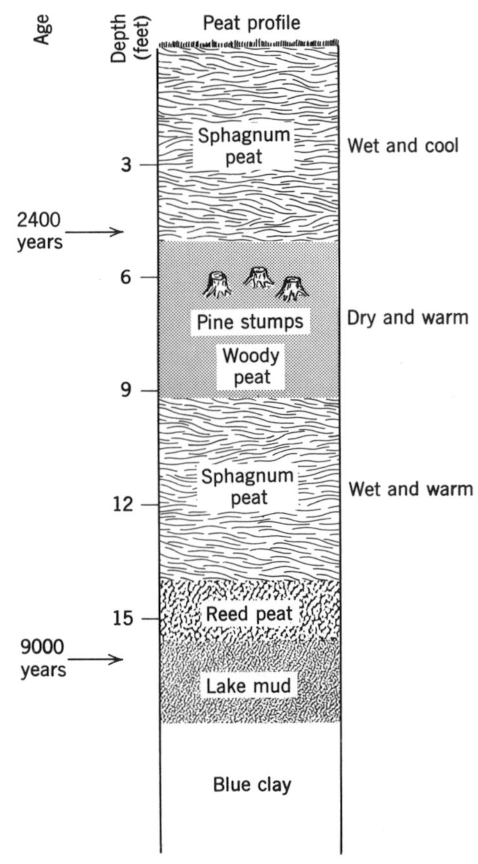


**Fig. 6–8** Stratification found in a peat soil in Sweden where changes in climate produced changes in the kind of vegetation that formed the peat. (From "Organic Soils," J. F. Davis and R. E. Lucas, *Mich. Agr. Exp. Sta. Spec. Bull.*, **425**, 1959.)

physical condition. Soil animals depend on organic matter for food and contribute to a desirable physical condition by mixing soil and creating channels. Naturally, there is much interest in managing organic matter to make soils more productive. A discussion of these kinds of considerations will help place in better perspective the value of organic matter in soils.

## Carbon–Nitrogen Ratio and the Decomposition of Organic Residues

The soil microflora are the primary agents for organic matter decay, and being plants themselves, have certain dietary requirements. Of major concern from a practical standpoint is the amount of carbon relative to nitrogen in the decomposing organic matter. A problem arises when the nitrogen content of decomposing organic matter is small because the microflora may become deprived of nitrogen and compete with the higher plants for whatever available nitrogen exists in the soil. Since the carbon content of organic materials is relatively constant between about 40 to 50 percent, while the nitrogen content varies many fold, the carbon–nitrogen ratio is a convenient way to express the relative content of nitrogen. Thus, the carbon–nitrogen ratio of organic materials is a indication of the likelihood of a nitrogen shortage and competition between microflora and higher plants for whatever nitrogen is available in the soil.

The carbon–nitrogen ratios of some organic residues that are frequently added to soils are given in Table 6–2. They range from 10 to 12 for humus and immature sweet clover tissue to 400 for sawdust. Materials with small or narrow ratios are relatively rich in nitrogen, while those with higher or wider ratios are relatively low in nitrogen.

**Table 6–2** The Carbon–Nitrogen Ratio of Some Organic Materials

| Material | C : N Ratio |
|---|---|
| Soil humus | 10 |
| Sweet clover (young) | 12 |
| Barnyard manure (rotted) | 20 |
| Clover residues | 23 |
| Green rye | 36 |
| Cane trash | 50 |
| Corn stover | 60 |
| Straw | 80 |
| Timothy | 80 |
| Sawdust | 400 |

Data is taken from several sources. The values are approximate only, and the ratio in any particular material may vary considerably from the values given.

Organic residues with a narrow ratio, less than 15 to 20, usually have enough nitrogen in them to satisfy the requirements of the decomposing microflora. In fact, as the residues decompose, there will likely be nitrogen in excess of that needed by the microflora which will be released as ammonia. This nitrogen can be utilized by higher plants as decomposition proceeds. By contrast, the nitrogen content of sawdust and straw is so low that when they are incorporated into the soil nitrogen must be supplied from some other source to enable the nitrogen requirements of the decomposing microflora to be satisfied. When these materials are incorporated shortly before planting a crop, competition for nitrogen between the crop and microflora may occur (see Fig. 6–9).

Several things may be done to prevent competition for nitrogen between the microflora and higher plants. Straw and similar residues may be burned rather than incorporating them into the soil. Such a practice deprives the soil of a source of organic matter, and the organic matter content of the soil will be less than where straw is periodically returned to the soil. Secondly, nitrogen fertilizer can be added if crops are to be grown immediately after the turning under of wide carbon–nitrogen ratio materials. This nitrogen can be used early in the year for the decomposition of residues of a previous crop,

**Fig. 6–9** Paper mill sludge reduced the growth of young corn plants grown in the greenhouse when mixed with soil in quantities of 1/4 or 1/2 of the total soil volume. The sludge contains about 40 percent paper fiber (cellulose) and little if any nitrogen, so it has a wide or high carbon-nitrogen ratio. Addition of the sludge reduced the supply of available nitrogen for the corn.

and a month or so later, when the decomposition of residues is largely completed, the nitrogen can be used by the crop.

The *nitrogen factor* is a convenient term to express the extent to which a material is deficient in nitrogen for decomposition. It is defined as the number of units of inorganic nitrogen that must be supplied to 100 units of organic material in order to prevent a net immobilization of nitrogen from the environment. A factor of approximately 0.9 represents straw. The nitrogen factor can be calculated for straw as follows. Assume that one hundred pounds of straw contains about 40 pounds of carbon and 1/2 pound of nitrogen. Assuming that 35 per cent of the carbon will be assimilated by the microflora and that one-tenth as much nitrogen will be assimilated as carbon, the nitrogen factor will be 0.9:

$$40 \times 0.35 = 14 \text{ pounds of carbon assimilated}$$

$$14/10 = 1.4 \text{ pounds of nitrogen assimilated}$$

$$1.4 - 0.5 = 0.9 \text{ pound of nitrogen deficient}$$

The addition of 0.9 pound of inorganic nitrogen to the soil at the time 100 pounds of straw are incorporated should prevent the immobilization of nitrogen from the soil environment and prevent competition for nitrogen between a crop and the microflora. The above calculations assumes that 65 percent of the carbon in the straw is converted to carbon dioxide during respiration in the decomposition process.

Many urban people have organic matter residues with wide carbon–nitrogen ratios such as tree leaves, grass clippings, or other plant wastes from a garden. The carbon–nitrogen ratio of these materials can be lowered by *composting*. Composting consists of storing the organic materials in a pile while maintaining favorable moisture, aeration, and temperature relationships. As the organic matter is decomposed, much of the carbon, hydrogen, and oxygen are released as carbon dioxide and water. Nutrients, like nitrogen, are continually reused by the microflora and are conserved. Thus, while there is a loss of carbon, the amount of nitrogen remains about constant resulting in a narrowing of the carbon–nitrogen ratio. There is also general enrichment of all plant nutrients. The rotted material is easily incorporated into the soil or can be used as an organic matter mulch.

The low nitrogen content of the composting materials may greatly retard the rate of decomposition and for this reason most composters add some nitrogen fertilizer. The quality of compost can further be improved by adding other materials. Some recommendations of the United States Department of Agriculture are given in Table 6–3.

**Table 6–3** Materials Recommended for Making Compost

| | Cups per tightly packed bushel |
|---|---|
| **For general purposes, including acid-loving plants** | |
| Ammonium sulfate | 1 |
| Superphosphate (20 percent) | 1/2 |
| Epsom salt | 1/16 |
| or: | |
| 10-6-4 fertilizer | 1 1/2 |
| **For plants not needing acid soil** | |
| Ammonium sulfate | 1 |
| Superphosphate (20 percent) | 1/2 |
| Dolomitic limestone or wood ashes | 2/3 |
| or: | |
| 10-6-4 fertilizer | 1 1/2 |
| Dolomitic limestone or wood ashes | 2/3 |

From Soil, USDA Yearbook, 1957, page 674.

## Effects of Green-Manuring

One of the oldest agricultural practices is the growing of legumes for soil improvement. The yields of nonleguminous crops are usually greater when grown after legumes, like alfalfa or clover, because of an increased nitrogen supply. In these cases the legume crop is harvested and the benefit to crops grown later is a by-product. In green-manuring a crop is planted just to be plowed under to add some organic matter to the soil. This is particularly beneficial for sandy soils very low in organic matter content. In these soils little nitrogen is mineralized from soil organic matter and added nitrogen fertilizer may be leached out of the soil before the crop has utilized it. In such cases the green-manure crop is planted after harvest in the late summer or fall and plowed under just before planting the next crop in the spring. The gradual decomposition of the plowed-under crop provides plant nutrients, particularly nitrogen, for some weeks after planting. The major effect of the green-manure crop in this case is to increase the supply of nitrogen (and other nutrients) instead of providing a significant increase in the organic matter content of the soil. Other benefits commonly cited are protection of soil from erosion and reduced loss of nutrients by leaching. However, it has been difficult to attribute economic benefits of the practice to any other effect than that of increased nitrogen supply. Where green-manure crops deplete soil moisture and contribute to a droughty situation, green-manure crops may cause a reduction in crop yields.

## Use of Peats

Peats used for soil amendments are generally classified as moss peat, reed-sedge peat, and peat humus. Moss peat forms from moss vegetation, reed-sedge peat from reeds, sedges, cattails, and other associated plants, while peat humus is any peat that has undergone considerable decomposition. The peats are used largely for mulches and greenhouse-soil mixes.

**Table 6–4** Characteristics of Common Horticultural Peats

| Type | Range in Nitrogen[a] (percent) | Range of Water-Absorbing Capacity[a] (percent) | Range in Ash Content[a] (percent) | Range in Volume Weights[a] (lb./cu. ft.) | Range in pH |
|---|---|---|---|---|---|
| Sphagnum moss peat | 0.6–1.4 | 1500–3000 | 1.0– 5.0 | 4.5– 7.0 | 3.0–4.0 |
| Hypnum moss peat | 2.0–3.5 | 1200–1800 | 4.0–10.0 | 5.0–10.0 | 5.0–7.0 |
| Reed-sedge peat (low lime) | 1.5–3.0 | 500–1200 | 5.0–15.0 | 10.0–15.0 | 4.0–5.0 |
| Reed-sedge peat (high lime) | 2.0–3.5 | 400–1200 | 5.0–18.0 | 10.0–18.0 | 5.1–7.5 |
| Decomposed peat | 2.0–3.5 | 150–500 | 10.0–50.0 | 20.0–40.0 | 5.0–7.5 |

[a] Oven-dry basis. (From "Peats for Soil Improvement and Soil Mixes," *Ext. Bull.*, 516, Michigan State University.)

Some properties of common horticultural peats are given in Table 6–4. The nitrogen content (wide C/N ratio) of sphagnum peat indicates that its incorporation into soil may temporarily lower the available soil nitrogen supply. The low pH of sphagnum peat, however, makes it desirable as a mulch for acid-requiring plants such as azaleas and rhododendrons. Peat moss makes a neat looking surface that "sets off" plants and protects the soil from the disruptive force of rain drips, thereby keeping the soil porous so that water rapidly enters the soil when irrigated.

## Nutrient Accumulation Under Shifting Cultivation

When one flies over the jungles of the humid tropics, one can see small clearings where food crops are grown. These people have few animals that produce manure for the fields, chemical fertilizers are not available, and the highly weathered tropical soils are very infertile. "How can these people produce the subsistence food crops they need?" They use a system of cultivation known as *shifting cultivation.*

We have already noted the relatively efficient recycling of nutrients that can occur in forests (Chapter 5) and the potential for organic matter accumulation in trees. The shifting cultivator utilizes the nutrients in the forest (trees, vines, leaves, etc.) through controlled cutting and burning that kills most of the trees but does not destroy all of the organic matter. Crops are planted among the few remaining living trees, stumps, and fallen trees. Nutrients are made available to the crops as organic matter decomposes. After about 1 to 5 years, nutrients are depleted and weed and diseases invade the cultivated land resulting in extremely poor yields. The land is abandoned and the forest quickly reestablishes itself. Perhaps 10 to 20 years are needed before enough nutrients have accumulated in the trees to permit another short period of cultivation. Nutrient accumulation by a regenerating

**Table 6–5** Nutrient Accumulation (or Immobilization) in Forest Fallow in the Congo

| Age of Forest Fallow | Nutrients Immobilized in Vegetation, pounds per acre | | | | |
|---|---|---|---|---|---|
| | N | P | S | K | Ca+Mg |
| 2 years | 168 | 20 | 33 | 166 | 143 |
| 5 years | 505 | 29 | 92 | 406 | 375 |
| 8 years | 516 | 31 | 90 | 748 | 595 |
| 18–19 years | 625 | 96 | 175 | 535 | 732 |

Adapted from "Shifting Cultivation," C. E. Kellogg, *Soil Science*, **95**:221–230, 1963.)

forest-fallow in the Congo is given in Table 6–5. The 18- to 19-year-old forest vegetation contained about 5 times more nutrients than were contained in the 2-year-old forest fallow.

A common practice is to grow crops for 2 or 3 years and then use about 15 years for forest fallow. A farmer would need 17 or 18 parcels. Each year a new field would be brought into cultivation and a field would be abandoned to forest fallow (see Fig. 6–10). A shifting cultivation system may require as much as 50 acres per person.

Grasslands are not as effective nutrient accumulators as forests, since grasses are shallow rooted and do not have as much potential for storage of biomass as standing vegetation. It is an interesting situation that in the humid and subhumid tropics forest soils have greater productivity than grassland or savannah soils under shifting cultivation, while in the humid temperate regions the grassland soils are

**Fig. 6–10** Landscape in Assam showing land use under shifting cultivation. Various stages of forest fallow can be observed as well as a small burning in the right background. (Photo courtesy Charles E. Kellogg.)

usually considered the most productive. One cannot help but be impressed by the ingenuity that must have been required to perfect the shifting cultivation system. Today over 200 million people depend on the system for their livelihood.

## The Organic Gardening Myth

There have been extravagant claims made for the merits of soil organic matter. An extreme position has been taken by some who believe in the superiority of food produced "organically." Organically produced foods have been described as those that have been grown without the use of chemical fertilizer or poisonous sprays. Nutrients are added to the soil in the form of organic matter or naturally occurring mineral compounds. Some claim that the plants produced in this manner are so healthy that they require no insecticides to control insects which allows the food to be produced without use of poisonous spray. Some rather phenomenal claims for organically produced food have been made, and one important question is, "Do organically produced foods have superior nutritional quality?"

According to Dr. Boswell of the Crops Research Division of the USDA, "There is no evidence of any difference in nutritive value between crops grown with manure and compost and those grown with commercial fertilizers. Most soils need additions of organic matter, but the claim that organic matter alone is a cure-all and produced the pinnacle of quality is neither borne out by common observation nor demonstrated by adequate experiments. Muck and peat soils are high in organic matter, but they are so low in mineral elements that they do not produce the best quality of many crops. Manures, low in phosphorus, are unbalanced as sources of plant nutrients. It appears impossible economically to produce the amounts of the crops we need without commercial fertilizers."[3] This chapter has shown that organic matter is very important in soils, but the claims made by organic gardening enthusiasts appear to be unsupported by scientific facts.

[3] "The Great Organic-Gardening Myth," Harland Manchester, *Readers Digest*, p. 104, July 1962. Used with permission.

# 7

# Chemical and Mineralogical Properties of Soils

As recently as the eighteenth century, the soil was viewed as little more than a mixture of rock and organic matter particles. It was believed that plant roots injected soil particles and that these particles provided the sustenance of plants. Jethro Tull, an Englishman, believed that the swelling of plant roots caused a pressure that aided the entrance of soil particles into the roots. As a consequence, Tull invented the grain drill to plant grain in rows and a horse hoe (cultivator) to pulverize the soil when the grain was growing. Grain yields were increased and Tull thought this resulted from the effect of cultivation in loosening fine soil particles, which would make the particles more easily ingected. However, it was later learned that the grain grew better when cultivated because weed competition was reduced.

Shortly after 1800 de Saussure of Geneva used improved chemical techniques to discover the basic elements of photosynthesis and respiration. He showed that plants utilized carbon in the daytime and released carbon dioxide at night. The ash of the plant was composed of nutrients obtained from the soil. In spite of the excellent experimental work of de Saussure, many continued to believe the *humus theory*, which held that the plant's source of carbon was humus ingested by the roots. It remained for Liebig to bring a "quick" death to the humus theory through the publication of the book "Chemistry in Agriculture and Physiology" in 1840. Liebig, one of the foremost chemists of the nineteenth century, correctly viewed mineral weathering as a source of nutrients that were absorbed by roots as ions.

Roentgen's discovery of X rays in 1895 led to the development of the use of X-ray diffraction to study the arrangement of atoms in minerals. Using X-ray diffraction and more recently developed methods has resulted in great strides in understanding the role of mineral structure in the weathering and availability of plant nutrients.

Consideration of the characteristics of minerals found in soils and their transformation from one form to another is essential in understanding the nature of the soil's chemical properties and the origin of its fertility. Since the soil develops from material composed of rocks and minerals of the earth's crust, we will direct our attention first to the chemical and mineralogical composition of the earth's crust.

## CHEMICAL AND MINERALOGICAL COMPOSITION OF THE EARTH'S CRUST

About 92 chemical elements are known to exist in the earth's crust. When one considers the number of possible combinations of such a large number of elements, it is not surprising that about 2000 minerals have been recognized. Relatively few elements and minerals, however, are of real importance in soils.

### Chemical Composition of the Earth's Crust

About 98 percent of the crust of the earth is composed of 8 chemical elements (Fig. 7–1). In fact, two elements, oxygen and silicon, compose 75 percent of it. Many of the elements important in the growth of plants and animals occur in very small quantities. Needless to say, these elements and their compounds are not evenly distributed throughout the earth's surface. For example, in some places phosphorus compounds are so concentrated that they are mined, whereas in many other areas there is a deficiency of phosphorus for maximum plant growth.

### Mineralogical Composition of Rocks

Most of the elements of the earth's crust have combined with one or more other elements to form compounds called *minerals*. The minerals generally exist in mixtures to form the *rocks* of the earth. The mineralogical composition of igneous rocks, shale, and sandstone are given in Table 7–1. Limestone is also an important sedimentary

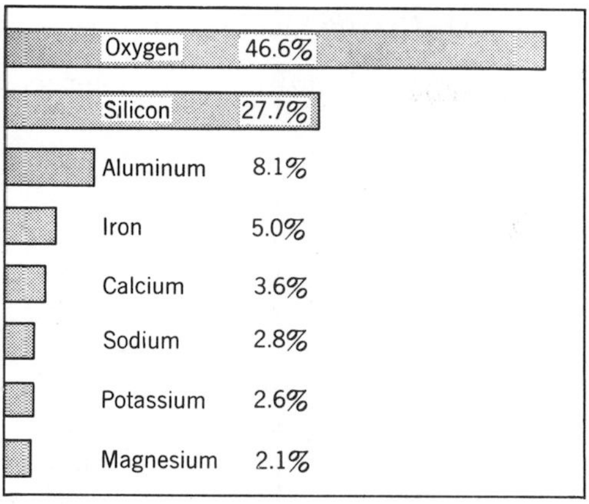

Oxygen 46.6%
Silicon 27.7%
Aluminum 8.1%
Iron 5.0%
Calcium 3.6%
Sodium 2.8%
Potassium 2.6%
Magnesium 2.1%

**Fig. 7–1** The eight elements in the earth's crust comprising over 1 percent by weight. The remainder of elements make up 1.5 percent.

rock and is composed largely of calcium and magnesium carbonates, with varying amounts of other minerals as impurities. The dominant minerals in these rocks are feldspar, amphibole, pyroxene, quartz, mica, clay minerals, limonite (iron oxide), and carbonate minerals.

**Table 7–1** Average Mineralogical Composition of Igneous and Sedimentary Rocks

| Mineral Constituent | Origin | Igneous Rock, percentage | Shale, percentage | Sandstone, percentage |
|---|---|---|---|---|
| Feldspars | Primary | 59.5 | 30.0 | 11.5 |
| Amphiboles and pyroxenes | Primary | 16.8 | . . . | [a] |
| Quartz | Primary | 12.0 | 22.3 | 66.8 |
| Micas | Primary | 3.8 | . . . | [a] |
| Titanium minerals | Primary | 1.5 | . . . | [a] |
| Apatite | Primary or secondary | 0.6 | . . . | [a] |
| Clay | Secondary | . . . | 25.0 | 6.6 |
| Limonite | Secondary | . . . | 5.6 | 1.8 |
| Carbonates | Secondary | . . . | 5.7 | 11.1 |
| Other minerals | . . . | 5.8 | 11.4 | 2.2 |

Data from Clarke cited by K. Lawton in "Chemistry of the Soil," *ASC Monograph* 126, p. 56, 1955. Edited by F. E. Bear.

[a] Present in small amounts.

## WEATHERING AND MINERALOGICAL COMPOSITION OF SOILS

Numerous examples of weathering abound and can be observed every day. Rusting of metal, cracking of sidewalks, and loss of mortar between bricks are a few examples. Weathering in soils results in the destruction of existing minerals and synthesis of new minerals. Nutrients are made available for plants and clay minerals are formed. In a real sense, all life on earth is "locked" in the minerals and, through weathering, nutrients essential to life are made available. It is interesting to contemplate the amount of food or lumber that could be produced from the nutrients contained in the rocks of the Rocky Mountains. Even life in the seas awaits nutrients released by weathering on the land and carried to the sea by rivers. This is manifested in the greater plant and fish populations that tend to occur at the mouths of rivers. For these reasons we need to consider weathering and the mineralogical composition of soils.

### Weathering—The Response of Rocks and Minerals to a New Environment

Rocks and minerals that are at or near equilibrium deep in the earth adjust to the greatly reduced pressure and temperature in the soil environment. The resulting adjustments or changes are called *weathering*. The changes are in the direction of a lower energy state

**Fig. 7–2** Unloading is the process whereby cracks and fissures are caused by expansion of rock that accompanies the removal of overlying layers by erosion or uplift. Limestone is the rock in this case.

and to a large extent are self-generating (exothermic). The response to reduced pressure is seen in the increased volume during *unloading*. Unloading is the removal of thick layers of sediment overlying deeply buried rocks by erosion or uplift. The release of pressure results in an accompanying bit of expansion that produces cracks and fissures (Fig. 7–2). Strains from temperature changes and the pressures of freezing water as well as the erosive action of water, wind, and ice also cause a slow and unceasing breaking up of hard rocks.

The response of minerals to reduced temperature is seen in the exothermic chemical reactions between minerals and the water, oxygen, and carbon dioxide in the soil. The abundance of water, oxygen, and carbon dioxide accounts for the fact that the major chemical weathering reactions are hydration, oxidation, and carbonation. An increase in volume accompanies these reactions and causes a peeling off of rock surfaces, producing exfoliation and spheroidal weathering. Hydration is considered to be very effective in producing spheroidal weathering (see Fig. 7–3).

## Crystal Structure—A Clue to Weathering Rate

Particle size, through its effect on specific surfaces, is an important factor influencing the weathering rate of minerals. Ionic bonding within the mineral crystal, however, is ultimately the major factor.

**Fig. 7–3** Hydration of minerals in the outer layer of a rock results in an increase in volume to cause a concentric spalling known as "spheroidal weathering." The grass in the center part of the photograph provides scale.

Sodium chloride exists as the natural mineral halite. Sodium and chlorine exist in the crystal as ions that attract each other. At the crystal faces the sodium and chloride ions respectively attract the negative and positive poles of water molecules. Adsorption of water molecules dislodges the sodium and chlorine from the crystal and greatly increases the solubility of the sodium and chlorine. The mineral dissolves easily in water and is readily leached from soils. This accounts for the general absence of halite in the soils of the humid regions.

We have noted that oxygen and silicon make up about 75 percent of the earth's crust on a weight basis, which explains the great abundance of silicate minerals in the crust. From Table 7–2, one can see that the large size of the oxygen atom, coupled with its abundance, causes oxygen to occupy over 90 percent of the volume of the earth's crust. It is productive to think of soil minerals as being composed of ping-pong balls, represented by oxygen atoms, and that marbles, representing the smaller metallic cations, occupy the interstices between the ping-pong balls. The cations are arranged on the basis of their size and valence to produce electrically neutral crystals. The larger the cation, the greater the number of oxygens surrounding it (coordination number) (see Table 7–2).

The silicon atom fits in an interstice formed by four oxygens. The covalent bonding between oxygen and silicon is directional and causes the oxygen and silicon to form a tetrahedron, as shown in Fig. 7–4. Silicon-oxygen tetrahedra are the basic units in the silicate minerals. Different silicate minerals exist depending on the way in which the tetrahedrons are linked together.

**Table 7–2** Size, Percent Volume in Earth's Crust, and Coordination Number of the Most Abundant Elements in Soil Minerals

| Element | Atomic Radius, A° | Volume, Percentage in Earth's Crust | Coordination Number with Oxygen |
|---|---|---|---|
| O | 1.32 | 93.8 | – |
| Si | 0.39 | 0.9 | 4 |
| Al | 0.57 | 0.5 | 6 (and 4) |
| Fe | 0.83 ($Fe^{+3}$ 0.67) | 0.4 | 6 |
| Mg | 0.78 | 0.3 | 6 |
| Na | 0.98 | 1.3 | 8 |
| Ca | 1.06 | 1.0 | 8 |
| K | 1.33 | 1.8 | 8 |

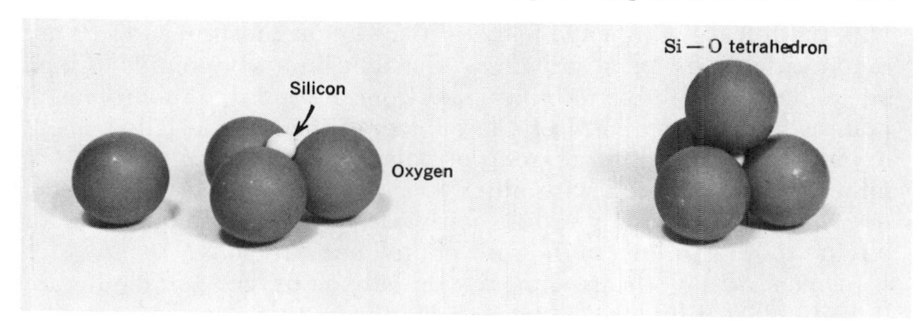


**Fig. 7–4** Models showing the tetrahedral (4-sided) arrangement of silicon and oxygen atoms in silicate minerals. The silicon-oxygen tetrahedron on the left has the apical oxygen set off to one side to show the position of the silicon atom.

Olivine, $MgFeSiO_4$, is composed of individual silicon-oxygen tetrahedra bound together by magnesium and iron. Each magnesium and iron atom is surrounded by six oxygens—three from each of two faces of adjacent tetrahedrons (see Fig. 7–5). The bonds between oxygen and the iron and magnesium are weak in comparison to the bonds in the tetrahedra between oxygen and silicon. During weathering, exposed magnesium along the crystal face react with hydroxyl ions to form magnesium hydroxide. The ferrous iron along the crystal face readily looses an electron and its size is reduced by about 20 percent (see Table 7–2). The smaller size of the ferric iron atom makes it easier for iron to leave the crystal and likely also to form a hydroxide. For these reasons the weathering of olivine can be visualized as shown in Fig. 7–5 as consisting of the separation of silicon-oxygen tetrahedra

**Fig. 7–5** Models representing the weathering of olivine. Olivine is composed of silicon-oxygen tetrahedra held together by iron and magnesium. Every other tetrahedron is "inverted" as shown by the light-colored tetrahedra in the olivine model on the left. During weathering the silicon-oxygen tetrahedra separate with the release of iron and magnesium.

with the release of iron and magnesium. The magnesium and iron are then available for plant growth. The magnesium compounds formed are sufficiently soluble to be leached from the soil, but the iron compounds formed may be insoluble enough to accumulate. The silicon-oxygen tetrahedra often regroup or polymerize and form some other mineral by combining with other constitutents in the soil solution. As with halite, olivine weathers fairly easily making it impossible for olivine to remain long in the soils of the humid regions.

In most of the silicate minerals the silicon-oxygen tetrahedra are linked together by the common sharing of some oxygen atoms. The number of cations other than silicon needed to neutralize the crystal is reduced to the extent that oxygen sharing occurs. Increased oxygen sharing by silicon generally causes greater resistance to weathering because a greater percentage of the crystal bonds are the strong covalent-like bonds that exist between silicon and oxygen. The ratio of oxygen to silicon in olivine is 4. It can be seen in Fig. 7–6 that in-

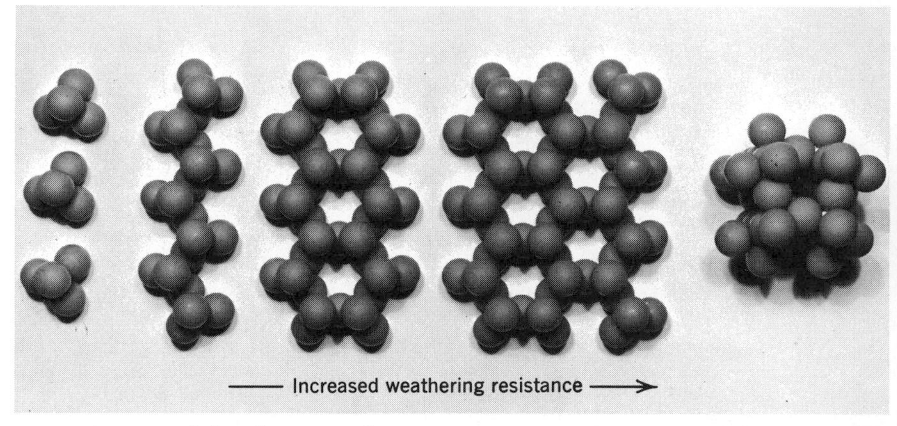


**Fig. 7–6** Models showing the common arrangements of silicon-oxygen tetrahedra in silicate minerals and their relation to weathering resistance.

**Arrangement of Si-O tetrahedra and representative minerals**

| Individual | Single chain | Double chain | Sheet | 3-dimensional |
|---|---|---|---|---|
| Olivine | Pyroxene augite | Amphibole hornblende | Biotite (mica) | quartz |
| | | **Oxygen-silicon ratio** | | |
| 4 | 3 | 2.7 | 2.5 | 2 |

creased oxygen sharing in the series olivine to pyroxene to amphibole to mica to quartz is associated with a decrease in the oxygen-silicon ratio from 4 to 2. The decreased ratio is associated with an increase in weathering resistance. In quartz ($SiO_2$) all of the oxygen atoms are neutralized by silicon, and quartz tends to persist in many soils as the more weatherable minerals disappear.

### Mineralogical Composition of Soils versus Weathering Stage

The differences in weathering resistance of minerals were used by Jackson and Sherman to establish 13 weathering stages that relate mineralogical composition of soils to weathering intensity. Representative minerals and typical soils associated with the weathering stages are given in Table 7–3.

Soils of weathering stage 1 may contain some gypsum and halite. Notice that soils containing significant amounts of olivine are representative of stage 3 and biotite mica of stage 4. This sequence is in agreement with our discussion of oxygen-silicon ratio and weathering resistance. Soils with minerals representative of stages 1 to 5 are considered to be in the early weathering stages. Such soils are frequently the youthful soils throughout the world, but are primarily the soils of arid regions where limited water restricts chemical weathering.

Soils of the intermediate weathering stages, 6 to 9, include most of the soils of the humid temperate regions. Quartz is often abundant in these soils and is representative of weathering stage 6. Here we notice minerals that we have not mentioned previously such as illite, vermiculite, and montmorillonite (see Table 7–3). These latter 3 minerals are typically synthesized in soils and increase in abundance as feldspars and micas weather and disappear. Some of the elements released in weathering recrystallize to form illite, vermiculite, and montmorillonite, which accumulate as fine-sized particles in the clay fraction.

Soils of the advanced weathering stages include the intensely weathered soils of the humid tropics. Soils dominated by minerals of weathering stages 10 to 13 may have lost almost all of the original minerals of the parent material and may consist mainly of stable minerals that have been synthesized during weathering. These minerals are kaolinite, gibbsite, hematite, and anatase ($TiO_2$). Thus, soils of the advanced weathering stages are usually characterized by extreme infertility, and, as was pointed out in Chapter 6, where these soils exist most of the nutrients in the ecosystem are circulating through the vegetation. In a recently developed soil classification system, which we shall consider in a later chapter, many of the intensely weathered soils of the humid tropics are classified as *Ultisols*, soils that have undergone the "ultimate in weathering and leaching."

**Table 7–3** Representative Minerals and Soils Associated with Weathering Stages

| Weathering Stage | Representative Minerals | Typical Soil Groups |
|---|---|---|
| | Early weathering stages | |
| 1 | Gypsum (also halite, sodium nitrate) | Soils dominated by these minerals in the fine silt and clay fractions are the youthful soils all over the world, but mainly soils of the desert regions where limited water keeps chemical weathering to a minimum. |
| 2 | Calcite (also dolomite, apatite) | |
| 3 | Olivine-hornblende (also pyroxenes) | |
| 4 | Biotite (also glauconite, nontronite) | |
| 5 | Albite (also anorthite, microcline, orthoclase) | |
| | Intermediate weathering stages | |
| 6 | Quartz | Soils dominated by these minerals in the fine silt and clay fractions are mainly those of temperate regions developed under grass or trees. Includes the major soils of the wheat and corn belts of the world. |
| 7 | Muscovite (also illite) | |
| 8 | 2:1 layer silicates (including vermiculite, expanded hydrous mica) | |
| 9 | Montmorillonite | |
| | Advanced weathering stages | |
| 10 | Kaolinite | Many intensely weathered soils of the warm and humid equatorial regions have clay fractions dominated by these minerals, they are frequently characterized by their infertility. |
| 11 | Gibbsite | |
| 12 | Hematite (also goethite, limonite) | |
| 13 | Anatase (also rutile, zircon) | |

From "Chemical Weathering of Minerals in Soils," M. L. Jackson and G. D. Sherman, *Advan. Agron.*, Vol. 5, 1953. Used by permission.

## ORIGIN, STRUCTURE, AND PROPERTIES OF CLAY MINERALS

The discussion so far has emphasized the degradational aspects of weathering. When something has been taken apart, the pieces remain. When minerals weather, these pieces are the atoms or groups of atoms set free from crystal degradation. We know that some of these atoms are used by plants; some are leached from soils, but many of the atoms regroup to form new minerals. Synthesis of minerals begins when a few atoms group together to form the nucleus of a crystal. The crystal grows and the mineral particle increases in size. This synthesis, or building up, accounts for the small particle size and great specific surface of the clay fraction as compared to silt and sand. The clay fraction of soils then tends to be composed of fine-sized secondary minerals formed in the soil called *clay minerals*. An understanding of their origin, structure, and properties is essential to an understanding of how soils retain soluble plant nutrients, the nature and importance of soil pH, and certain physical properties of soils.

## Origin of Clay Minerals

There are two major groups of clay minerals. The *silicate clays*, which include *illite*, *montmorillonite*, *vermiculite*, and *kaolinite*, and the *oxide clays*, which include primarily *iron and aluminum oxides*. Soils with clay fractions dominated by silicate clays are representative of weathering stages 7 to 10 and are widely distributed throughout the world. Soils with clay fractions dominated by oxide clays are representative of weathering stages 11 to 13 and are common in the humid tropics.

These secondary clay minerals are formed by alteration of existing minerals or by synthesis. Their origin is related to the kinds of minerals in the soil that can be altered and the amount and kinds of constituents that can recombine. The complexity of the processes makes a clear cut exposition of their origin impossible, but several general ideas will be presented.

Evidence indicates that micas, feldspars, and ferromagnesian minerals may weather directly to silicate clay minerals like *illite*, *montmorillonite*, *kaolinite*, or even hydroxide clays like *gibbsite* (hydrated aluminum oxide). The nature of the weathering environment plays an important role in determining when a given mineral will be formed. Illite formation is common in the temperate regions where weathering has not been intense. Alteration of mica minerals by the partial loss of structural potassium and hydration is a common mode of illite formation. Montmorillonite formation requires an abundant supply of magnesium and a neutral or only slightly acid environment. In the temperate regions illite can weather by alteration into montmorillonite.

Prolonged leaching of a soil in which montmorillonite has formed can lead to the development of high acidity and conditions favorable for the alteration or breakdown of montmorillonite and the formation of kaolinite. Kaolinite is frequently formed directly from primary minerals in soils of the humid tropics. Although kaolinite is very stable, it can weather to form gibbsite, $Al(OH)_3$ The diagram in Fig. 7–7

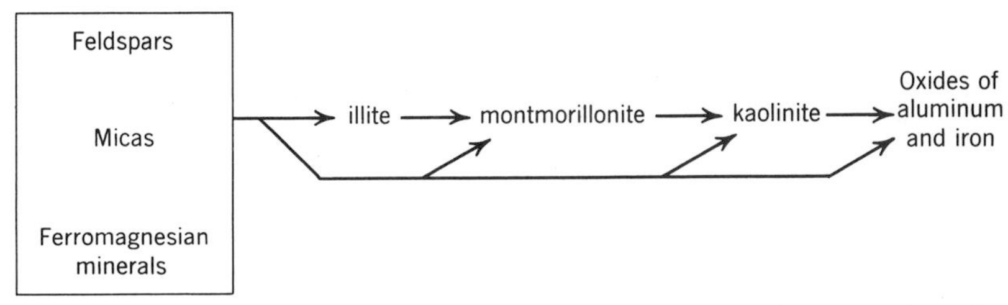


**Fig. 7–7** Some possible routes in the formation of silicate and oxide clays.

indicates some possible alteration routes in the formation of secondary minerals.

Kaolinite can also be formed by the resilication of aluminum oxide if the weathering zone is invaded by silica rich waters. The other silicate clays can also be formed by synthesis so that more than one method is available for their formation. This undoubtedly causes slight differences in their properties, depending on the mode of formation. It means that each kind of clay mineral category represents minerals with much similarity but within each group there is a range in properties.

Although much more information is needed before the origin of these minerals can be viewed in more certain terms, there is sufficient information about their characteristics to know they have pronounced effects on plant growth. A discussion of these characteristics follows.

## Structure of Silicate Clay Minerals of Soils

The silicate clay minerals consist of two basic components. One component is the silicon-oxygen tetrahedron (silicon in 4 coordination with oxygen) and the other component is the aluminum octahedron (8 sided with aluminum in 6 coordination with oxygen and/or hydroxyl). Silicon-oxygen tetrahedra in silicate clays are arranged in a sheet as shown in Fig. 7–6 (as in biotite). The upper or apical oxygens of the tetrahedral sheet have one bond satisfied by silicon and one bond of the oxygen unsatisfied. The formula for the tetrahedral sheet is $Si_2O_5^{-2}$.

We can visualize that the apical oxygens of the tetrahedral sheet can attract cations which in turn are surrounded by enough anions to neutralize all the cation charge. For simplicity, a pair of tetrahedra instead of a sheet are used to illustrate how the tetrahedral and octahedral sheets exist in kaolinite in Fig. 7–8. Items 1 and 2 of Fig. 7–8 show the apical oxygens, which have one free bond, and the position of the aluminum. Items 3, 4, 5, and 6 show the progressive

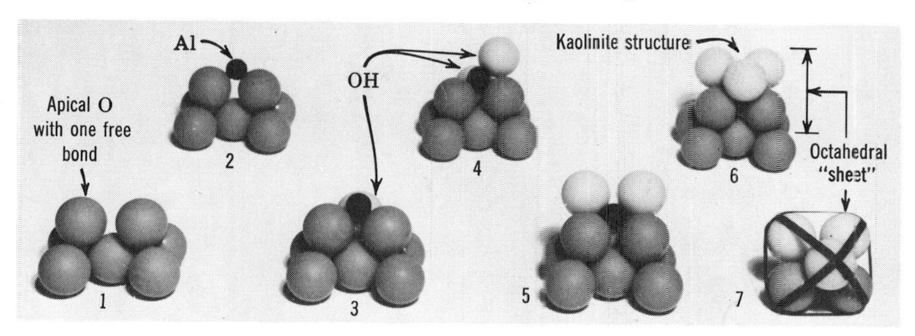


**Fig. 7–8** Models illustrating the structure of kaolinite. Item 1 shows that the apical oxygen of the tetrahedral sheet have a free bond that is neutralized by Al as seen in item 2. Items 3 to 6 show the progressive addition of 4 OH to complete the octahedron or put Al in 6 coordination. Item 6 represents the kaolinite structure, and item 7 shows that the Al is surrounded by 2 oxygen and 4 hydroxyl. Item 7 is a good example of an octahedron—4 faces are seen and 4 faces are hidden.

addition of OH to complete the aluminum octahedron. Model 6 is representative of the structure of kaolinite consisting of 1 tetrahedral and 1 octahedral sheet. For this reason kaolinite is called a 1:1 clay (ratio of tetrahedral to octahedral sheets). Item 7 of Fig. 7–8 shows that the apical oxygens are common to both the tetrahedral and octahedral sheets. In the octahedron, each Al neutralizes half a valence bond from each of the surrounding 6 anions, which results in a neutral structure.

The properties of kaolinite are derived from its structure. The tetrahedra and octahedra sheets form layers. Many layers are stacked one on top of the other as shown in Fig. 7–9. A kaolinite particle about 2-microns wide would have the length of about 20,000 oxygens side by side. If the particle was 1/10 as high as wide, being 0.2 micron high, it would contain about 50 layers stacked on top of each other. As an analogy, a clay particle is like a stack of sheets of plywood.

The hydroxyls of the aluminum octahedral layer are adjacent to the oxygens of the next layer above in kaolinite. This results in hydrogen bonding that firmly holds successive layers together (see Fig. 7–9). This bonding has two effects. First, it tends to favor the formation of large particles. Many of the kaolinite particles are 2 microns or larger in diameter and soils with a high content of this clay may have surprisingly high permeability. Secondly, water cannot permeate successive layers or units of the particles to cause expansion and contraction with wetting and drying. To carry the analogy of the structure further, it is similar to a stack of plywood sheets glued together. The

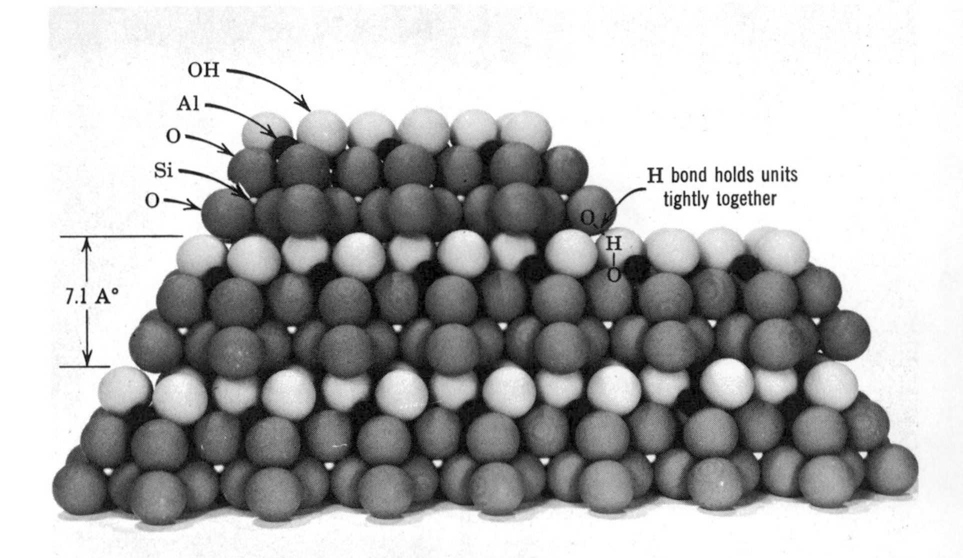


**Fig. 7–9** Model of the 1:1 clay mineral kaolinite showing three units stacked one on top of the other and held together tightly by H bonding (attraction of H of the hydroxyl for the O of the next adjacent unit). Each unit is 7.1 angstroms high.

electron micrograph in Fig. 7–10 shows the hexagonal shape and platy nature of kaolinite particles.

Montmorillonite has a 2:1 lattice. An aluminum octahedral sheet is sandwiched between two silica tetrahedral sheets to form the basic layer (see Fig. 7–11). The stacking of layers on top of each other, results in a side by side arrangement of oxygen atoms from adjacent tetrahedral sheets so that hydrogen bonding between layers does not exist. In fact, the adjacent planar oxygen surfaces have a natural tendency to repel each other. The layers readily expand and contract with wetting and drying to give an "accordian" effect. Particles also tend to be smaller than those of kaolinite, and soils with a high content of this clay will crack markedly upon drying and tend to be impermeable when wet. It gives the soil highly plastic characteristics. It also means that the interiors as well as the exteriors of the particles are available for the adsorption of water and nutrients (Fig. 7–12).

**Fig. 7–10** An electron micrograph of kaolinite particles magnified 35,821 times. Note the flaky and hexagonal-like shape of the particles. (Courtesy Mineral Industries Experiment Station, Pennsylvania State University.)

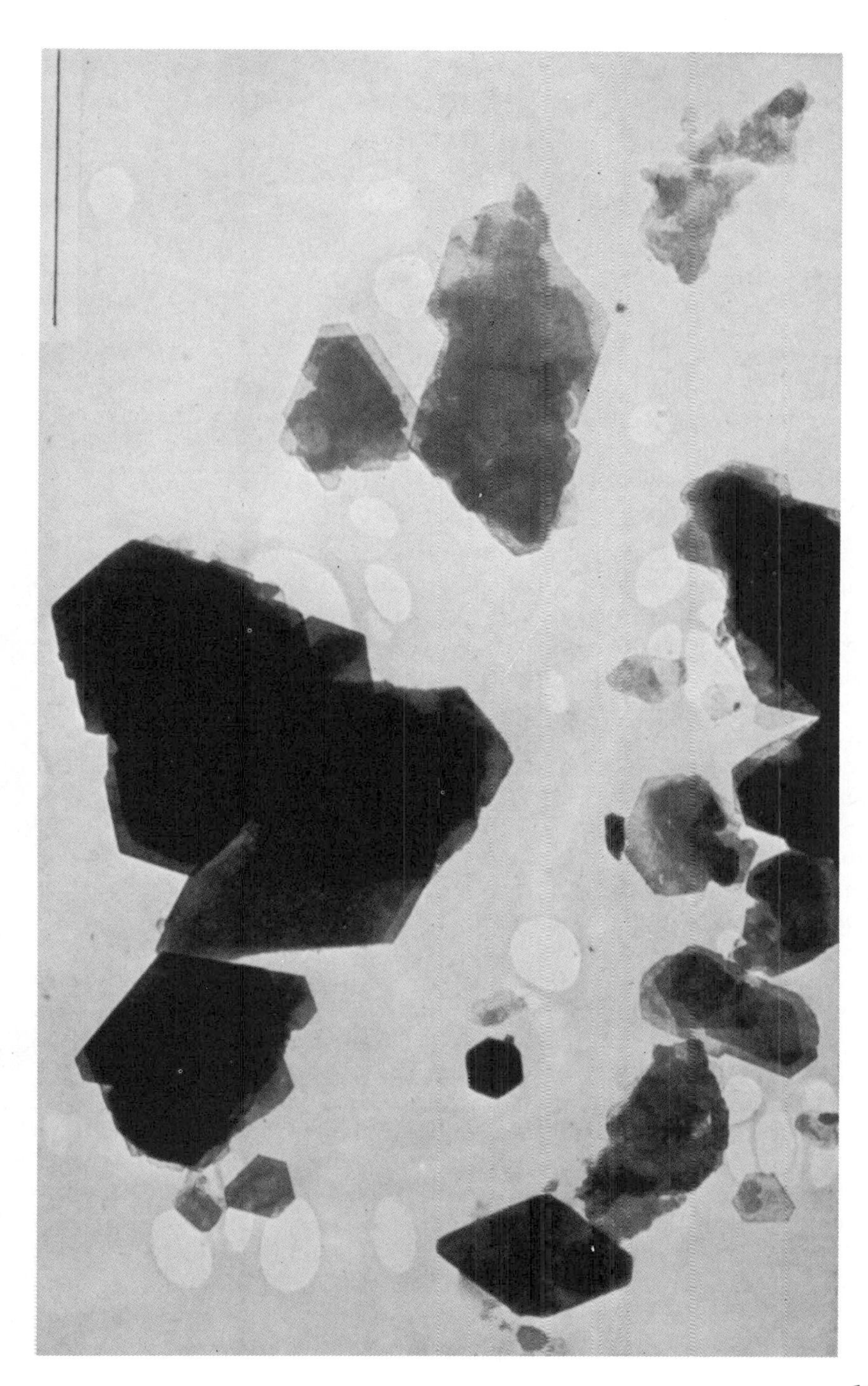

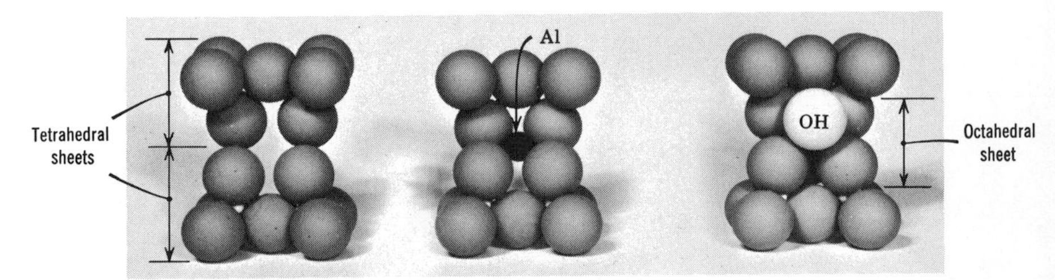


**Fig. 7–11** Models illustrating a 2:1 structure consisting of 2 tetrahedral sheets and 1 octahedral sheet. On the left are 2 silicon-oxygen tetrahedral "sheets" facing each other. The center model shows the addition of Al that is shared by 4 apical oxygens. The right model shows the addition of 2 hydroxyls (one in rear is invisible) to place the Al in 6 coordination in an octahedral arrangement. The Al shares half of a valance bond from each of the 6 surrounding anions (4 oxygen and 2 hydroxyl).

During the formation of montmorillonite some of the aluminum atoms in the interstices of the octahedral layer are replaced by magnesium. It has been mentioned that an abundance of magnesium in the weathering environment is a prerequisite for montmorillonite formation. The substitution of aluminum by magnesium can occur because the two atoms are sufficiently similar in size so that the replacement of only one-sixth of the aluminum atoms does not cause excessive strain in the lattice. This substitution is called *isomorphous substitution* (see Fig. 7–12).

Since the valence of the original lattice aluminum is three and the valence of the substituting magnesium is two, each substitution leaves the lattice with one unsatisfied negative charge or valency bond. In montmorillonite about one-sixth of the aluminum ions have been replaced and the result is a highly, negatively charged lattice or layer. This negative charge is permanent; it originates within the lattice (octahedral sheet) and is satisfied by hydrated cations which remain on the exterior of the particle or in close proximity of the surface. Thus a most important function of the clays is evident—that of holding hundreds or thousands of pounds of nutrient ions per acre furrow slice. These ions are adsorbed strongly enough to retard their movement out of the soil by leaching but weakly enough to be readily used by plants. In fact, one theory holds that plant roots can exchange hydrogen ions directly with cations on the colloidal surfaces.

Illite also has a 2:1 lattice. Some of the tetrahedral $Si^{+4}$ has been replaced by $Al^{+3}$, as indicated in Fig. 7–13. As in montmorillonite a negatively charged lattice results. In illite, however, most of the lattice charge is neutralized by potassium ions that sit in the "hexag-

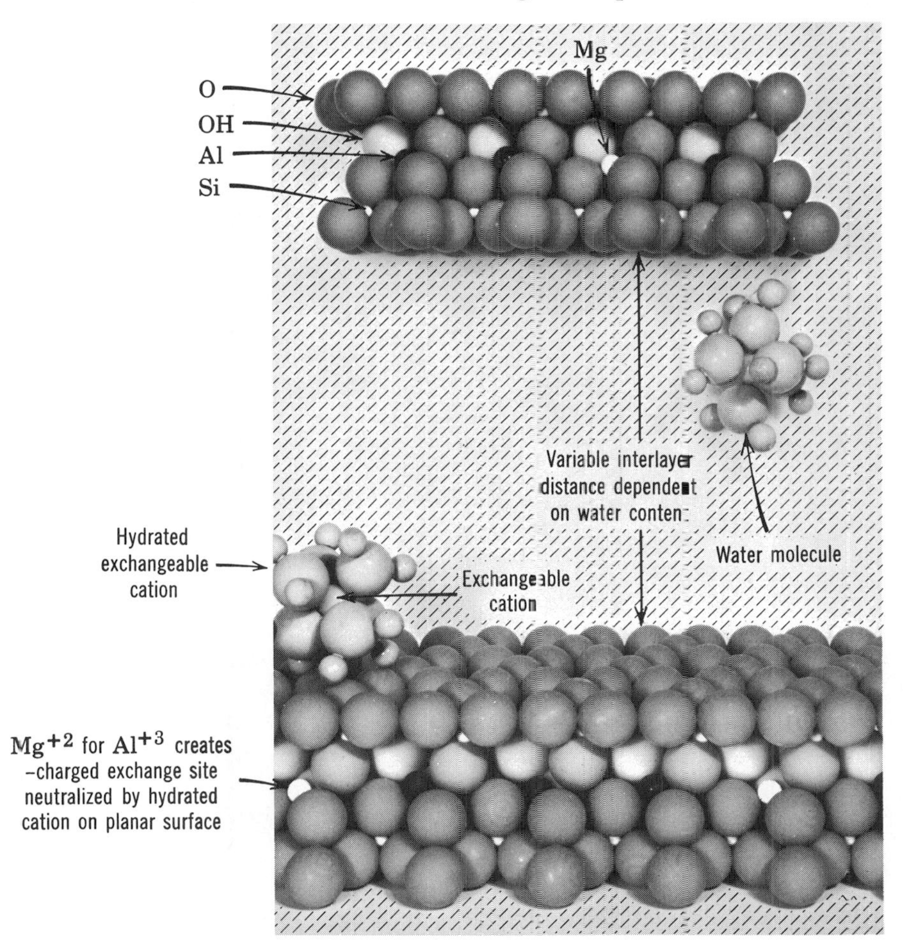


**Fig. 7–12** Models of 2:1 clay mineral montmorillonite showing (1) expanding lattice and variable water content, (2) cation-exchange site originating from isomorphous substitution of $Mg^{+2}$ for $Al^{+3}$, and (3) adsorption of 2 hydrated cations on the planar surface cation-exchange sites. In a moist soil the interlayer water is believed to be all or primarily water associated with exchangeable cations.

onal holes" of the planar surface of adjacent units. The potassium forms a bonding mechanism called a "potassium bridge" (O − K − O), which holds adjacent layers together and produces a nonexpanding lattice (Fig. 7–13). Some of the tetrahedral Al around the edges of the particles are not involved in forming potassium bridges and serve as sites for cation exchange. The lattice charge for cation exchange is

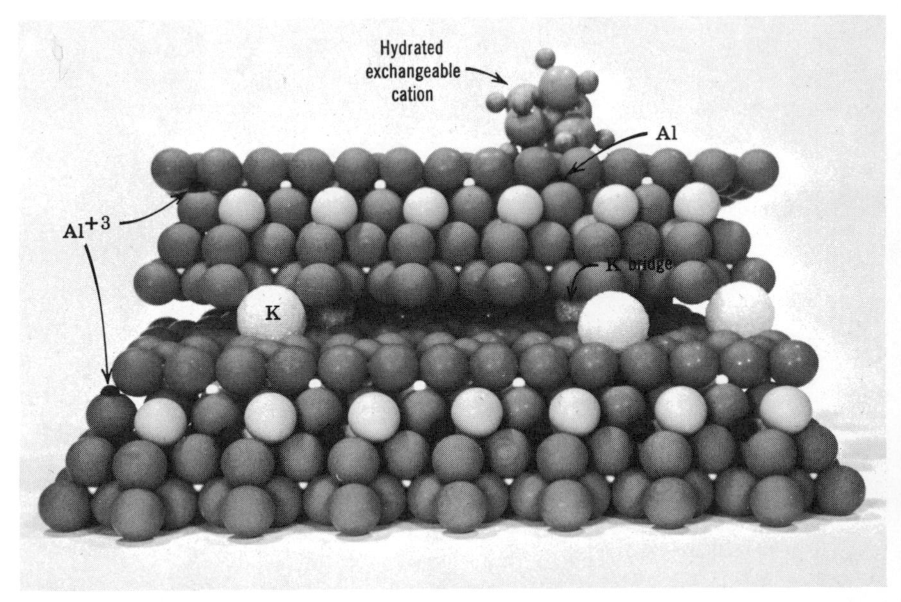


**Fig. 7–13** Model of the 2:1 clay mineral illite showing (1) some tetrahedral $Al^{+3}$ substituted for $Si^{+4}$, which creates a negatively charged lattice, (2) "potassium bridges" that neutralize most of the lattice charge and hold the layers together to form a nonexpanding lattice, and (3) adsorption of a hydrated cation on an exchange site created by Al substitution.

about one-third as great as it is for montmorillonite. During weathering, if all or most of the potassium is removed from the interlayer positions, the clay becomes an expanding lattice type of clay called *vermiculite.* The cation-exchange capacity of the clay is thereby greatly increased since none of the lattice charge is neutralized by K bridges.

In all clays cation-exchange sites originate from the dissociation of H from exposed OH along the edges of particles. Cation-exchange sites also exist where exposed oxygens are bonded to silicon and have one "free" bond. Isomorphous substitution in montmorillonite accounts for about 80 percent of the cation-exchange sites, and exposed edge bonds account for about 20 percent, whereas, in kaolinite most or all of the sites result from exposed edge bonds. A summary of cation-exchange capacity and other properties of layered-silicate clays is given in Table 7–4.

Some expanding clays, like montmorillonite, can hold water many times the volume of the clay itself. The water films between the layers and between the clay particles give soil its shrink and swell char-

**Table 7–4** Some Generalized Properties of Clay Minerals

| Clay Mineral | Ratio of Tetra-hedra to Octa-hedra Sheets | Relative Particle Size | Cation-Exchange Capacity, me. per 100 grams | Isomorphous Substitution | Structural Formula |
|---|---|---|---|---|---|
| Kaolinite | 1:1 | Large | 8 | Little or none | $(OH)_8Si_4Al_4O_{10}$ |
| Montmorillonite | 2:1 | Small | 100 | Mg for Al in octahedral layer | $(OH)_4Si_8(Al_{3.34}Mg_{.66})O_{20}$ |
| Illite | 2:1 | Intermediate | 30 | Al for Si in tetrahedral layer | $(OH)_4K_2(Si_6Al_2)Al_4O_{20}$ |

acteristics, its plasticity, and its cohesion (Fig. 7–14). Expanding clays can push in basement walls. Some scientists at Iowa State University are experimenting with the use of expanding clays as jacks. For example, dry clay is placed under road pavements that settle and when the clay is wetted the expansion raises the pavement. The leaning of the Tower of Pisa and the settling of buildings in Mexico City are associated with changes in the water content of clays.

### Soil Water and the Exchangeable Cations

The water existing in the immediate vicinity of clay particles is considered to be basically water associated with exchangeable cations. This is explained on the basis that exchangeable cations attract water molecules more strongly than the exposed oxygen atoms held in clay particles. In the case of sand and silt particles, with essentially none or very low cation-exchange capacity, the water in the immediate vicinity of the particle surface is mostly hydrogen bonded, as shown in Fig. 7–15. These innermost water molecules probably exist in a "quasi" crystalline state.

The innermost adsorbed water of both clay and silt and sand particles is unavailable to plants and exists in soils in the air dry state. Further out from the soil particle surfaces, cations are much less abundant than near the clay surfaces, and water exists in a "free" state. This water contains hydrated cations in fairly low concentration and can be considered equivalent to the soil solution.

### Oxide Clays

All or most soils contain at least a small amount of colloidal-sized particles composed of oxides of iron and aluminum. Representative oxide clays include gibbsite, $Al(OH)_3$, of weathering stage 11, and

**Fig. 7–14** Close-up of large cracks formed by drying in a field where the soil had a high content of expanding clay. The tendency for thin flaky pieces to form on the very surface of the soil is an expression of the plate-shaped nature of clay particles. (Photo USDA.)

hematite, $Fe_2O_3$, of weathering stage 12. Soils with properties dominated by oxide clays and kaolinite (weathering stage 10) are in the advanced weathering stages and are usually found in the humid tropics.

Soils dominated by oxide and kaolinite clays are characterized by very stable soil aggregates and they exhibit a low degree of plasticity. Large amounts of oxide and kaolinite clays in a soil contribute to the formation of extremely stable soil aggregates because the clays tend to neutralize each other. Kaolinite, as we have seen, has a net negative charge, while the oxide clays have a net positive charge (in acid soils).

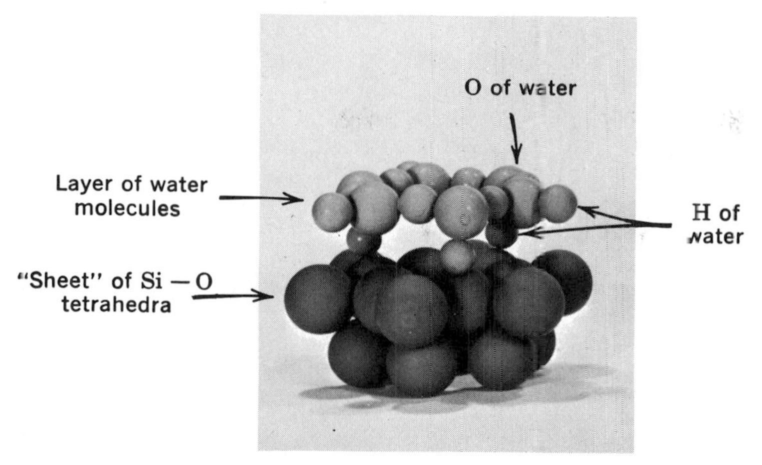


**Fig. 7–15** Model to show H-bonding of water molecules to exposed oxygen atoms of a mineral surface where the oxygen are coordinated with silicon in tetrahedrons.

This promotes flocculation. Oxide gels also probably contribute to the hardness of aggregates that resist crushing when moistened and rubbed in the palm of the hand. Such soils act as "sands" even though they contain 100 percent clay as is common for soils in the Hawaiian Islands that have weathered from basalt. Infiltration of water is rapid and the soil resists erosion. Plasticity is low when wet and soils are easily tilled. Adsorption and release of water to plants occurs largely at low soil moisture tensions. Soil textures by feel are frequently silt loam and silty clay loam or, in some cases, sandy clay loam.

## CATION AND ANION EXCHANGE IN SOILS

Clay and humus are of utmost importance in soils. Because they are in a colloidal state, they expose a relatively large amount of surface area for adsorption of water and ions. Nutrients set free in solution during weathering tend to be adsorbed on the humus and clay surfaces. In this section we shall emphasize the adsorption and storage of nutrients in exchangeable form on colloidal surfaces.

### The Nature of Cation Exchange in Soils

The adsorption of a cation by a colloidal nucleus or micelle and the accompanying release of one or more ions held by the micelle is termed *cation exchange*. For example, assume that the micelle has one-half of its capacity satisfied with Ca ions, one-quarter with K ions, and one-quarter with H ions. The situation would be as shown in the diagram below. Now suppose that the colloidal material is treated with a solution of strong KCL. In time the K ions from the KCL will replace virtually all the cations on the micelle creating a micelle that is entirely potassium saturated, and the adsorbed calcium and hydrogen will exist in solution as chlorides.

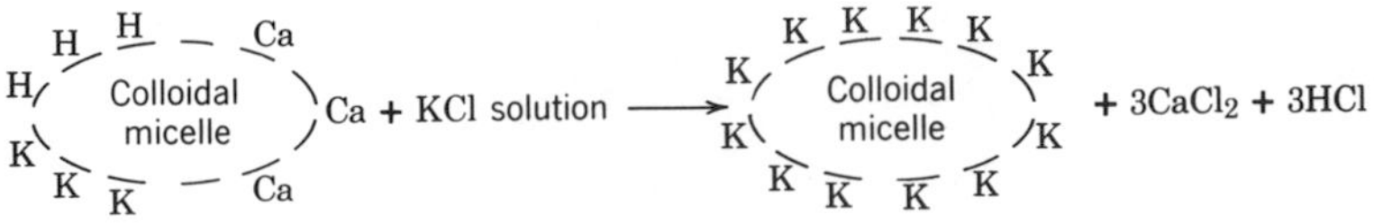

H H Ca H K K K Ca Colloidal micelle Ca + KCl solution ⟶ K K K K K K K K K K Colloidal micelle + $3CaCl_2$ + 3HCl

The efficiency with which ions will replace each other is determined by such factors as (*a*) relative concentration or numbers of the ions, (*b*) the number of charges on the ions, and (*c*) the speed of movement or activity of the different ions. The first factor is an application of the well-known chemical law of mass action. The greater the number of charges carried by the ion the greater is its efficiency, other factors being the same. The speed or activity of an ion is primarily a function of its size, but the degree of hydration must also be considered. If we consider the ions Li, Na, K, and Rb, we find that they are listed in Table 7–5 in the ascending order of size, and hence we would expect their efficiency of replacement to be Li Na K Rb. During hydration, however, the Li ion associates itself with so many water molecules that its speed is much reduced. Because of its large hydrated radius Li can not get as close to the micelle as other ions. Likewise, the Na ion is more highly hydrated than the K ion. The result is that the order of replacement is reversed: Rb K Na Li (see Table 7–5).

**Table 7–5** Ionic Radii, Hydration, and Exchange Efficiency of Several Monovalent Ions

| Ion | Radii of Ions Angstroms, ($10^{-8}$ cm) | | Order of Cation-Exchange Efficiency |
|---|---|---|---|
| | Dehydrated | Hydrated | |
| Li | 0.78 | 10.03 | 4th |
| Na | 0.98 | 7.90 | 3rd |
| K | 1.33 | 5.32 | 2nd |
| Rb | 1.49 | 5.09 | 1st |

Considering some of the most common cations in soils, the replaceability series is usually Al > Ca > Mg > K > Na. Exchangeable H is difficult to put in the series because of the uncertainty of its hydration properties. It is usually considered quite easily replaced as are the other monovalent ions.

## Cation Exchange Capacity of Soils

The colloidal fraction carries a positive as well as a negative charge. The negative charge, however, is of much greater magnitude and of greater significance for plant growth in most soils. The cation exchange capacity (CEC) is an expression of the *number* of cation adsorption sites per unit weight of soil. It is defined as the sum total of exchangeable cations adsorbed, expressed in *milliequivalents per 100 grams of oven-dry soil.* An equivalent is that quantity which is chemically equal to one gram of hydrogen. The number of hydrogen atoms in an equivalent is equal to Avagadro's number ($6.02 \times 10^{23}$). A milliequivalent is equal to 0.001 gram of hydrogen. If there was one milliequivalent of cation exchange capacity in a teaspoon of soil, it would contain $6.02 \times 10^{20}$ negatively charged adsorption sites. Such large numbers are difficult to comprehend but will be better understood when these numbers are converted into weight of adsorbed cations later in this chapter. It suffices at this time to mention that on an acre furrow slice basis this may amount to several thousands of pounds of exchangeable cations.

The total cation exchange capacity of the soil is the total number of exchange sites of both the organic and mineral colloids. For our purposes it can be assumed that the organic matter of a mineral soil will have a cation-exchange capacity of 200 milliequivalents per 100 grams. Each constituent in the clay fraction of the soil has an average cation-exchange capacity that is rather characteristic. The approximate cation-exchange capacities of some of the common clay minerals are: montmorillonite, 100; illite, 30; and kaolinite, 8 (Table 7–4). The exchange capacity of the hydrous oxides of iron and aluminum is much less than that of the other clay minerals. For many soils of the temperate zone, the average cation-exchange capacity of the clay fraction as a whole might be in the vicinity of 50 milliequivalents per 100 grams because the clay fraction is a mixture of clay minerals. Under these conditions, the cation-exchange capacity of 100 grams of soil is 2 milliequivalents for each percent of organic matter and ½ milliequivalent for each percent of clay. A soil with 3 percent organic matter and 24 percent clay would have a cation-exchange capacity of approximately 18 milliequivalents per 100 grams $[2(3) + \frac{1}{2}(24) = 18]$.

**Table 7–6** Cation-Exchange Capacity, Exchangeable Cations, Percent Base Saturation, and pH of the Horizons of the Tama Silt Loam

| Depth, inches | Horizon | CEC | Exchangeable Cations, me/100 gm | | | | | Percent Base Saturation | pH |
|---|---|---|---|---|---|---|---|---|---|
| | | | Ca | Mg | K | Na | H | | |
| 0–6 | A1 | 24.4 | 11.3 | 4.1 | 0.8 | 0.2 | 8.0 | 67 | 5.1 |
| 6–12 | A1 | 24.7 | 13.2 | 4.7 | 0.6 | 0.2 | 6.0 | 75 | 5.3 |
| 12–18 | AB | 23.0 | 13.0 | 5.0 | 0.5 | 0.2 | 4.3 | 81 | 5.5 |
| 18–36 | B2 | 25.5 | 15.0 | 6.5 | 0.5 | 0.2 | 3.3 | 87 | 5.2 |
| 36–48 | B3 | 25.7 | 15.1 | 7.6 | 0.4 | 0.2 | 2.4 | 91 | 5.5 |
| 48+ | C | 22.1 | 13.4 | 7.2[a] | 0.3 | 0.2 | 1.0 | 95 | 5.8 |

Data from "Prairie Soils of the Upper Mississippi Valley," G. D. Smith, W. H. Allaway, and F. F. Riecken, *Advan. Agron.*, **2**:157–205, 1950. Used by permission.

[a] Estimated by authors. Exchangeable sodium was not determined and the values given are also estimates.

The cation-exchange capacity of the sand and silt fractions is omitted in the approximation of exchange capacity since it is so small. An accurate determination can be made by saturating all exchange positions with a single cation, such as ammonium, and then determining the total amount of ammonium adsorbed.

The cation-exchange capacity ranges from less than 5 for soils containing very little clay or organic matter to about 200 for a muck soil. The cation-exchange capacity of the horizons of a typical Prairie (Mollisol) soil are given in Table 7–6.

## Kinds and Amounts of Exchangeable Cations

The amounts of the exchangeable cations for the horizons of a typical Prairie soil are also given in Table 7–6. For practical purposes, the sum of the cations shown in the table is considered synonymous with cation-exchange capacity, recognizing that very small but important amounts of exchangeable iron, copper, manganese, and other cations are present.

The upper 6 inches of the Prairie (Brunizem) soil cited in Table 7–6 has 11.3 me of exchangeable calcium per 100 grams of soil. To calculate the pounds of exchangeable calcium per acre furrow slice requires that the milliequivalence of calcium be converted to a weight unit. The equivalent weight of calcium is 20, making the milliequivalent weight equal to 0.02 gram. Thus the A1 horizon contains 0.226 gram ($11.3 \times 0.02$) of exchangeable calcium per 100 grams of soil. Conversion to pounds can be made by simply stating that there would be 0.226 pound of exchangeable calcium per 100 pounds of soil. Assume this 6-inch layer weighs two million pounds or 20,000 times more than 100 pounds. Multiplying 0.226 by 20,000 produces 4520, which is the pounds of exchangeable calcium. The formula for calculating the weight of any cation per acre furrow slice is:

pounds per acre furrow slice

= (20,000) (milliequivalents/100 g) (milliequivalent weight in grams)

The total quantity of exchangeable calcium in the soil ramified by the root system of plants is many times the amount in this particular soil. The importance and magnitude of the nutrients available to plants as exchangeable cations in soils is readily apparent. Taking into consideration the differences in the milliequivalent weight of the various cations, there would be 934 pounds of exchangeable magnesium and 604 pounds of exchangeable potassium in the plow layer. Knowledge such as this is useful in predicting the response, if any, of crops to these cations applied as fertilizers.

## Percent Base and Hydrogen Saturation

The exchangeable bases include Ca, Mg, K, and Na.[1] Note that they are found in this order of decreasing abundance in Table 7–6. As exchangeable cations they act in the formation of hydroxyls as follows:

$$\boxed{\text{micelle}}\text{—K} + H_2O \rightleftharpoons \boxed{\text{micelle}}\text{—H} + K^+ + OH^- \text{ (in solution)}$$

The percentage base saturation is the percentage of the cation-exchange capacity saturated by these cations. Thus, in Table 7–6 it can be seen that all the horizons have well over 50 percent base saturation; a common characteristic of Prairie soils. The AB horizon has a base saturation of 81 percent calculated as follows:

$$\%\text{ base saturation} = \frac{\text{me exchangeable bases}}{\text{cation exchange capacity}} \times 100$$

$$= \frac{18.7}{23.0} \times 100 = 81$$

The percentage of hydrogen saturation is calculated by substituting milliequivalent hydrogen in the place of the milliequivalence of bases in the formula for determining base saturation. Its importance will

[1] Technically, a base is a proton acceptor like the OH ion while an acid is a proton donor like the H ion. However, the exchangeable cations Ca, Mg, K, and Na are all associated with compounds in the soil like $CaCO_3$, $MgCO_3$, $K_2CO_3$, and $Na_2CO_3$, which are more basic than acidic in reaction. For this reason Ca, Mg, Na, and K are commonly referred to as *exchangeable bases*, while H is commonly called an *exchangeable acid*.

be brought out in Chapter 8 when the relationship of percentage of hydrogen saturation and pH will be discussed.

## Cation Exchange and Plant Growth Relationships

There are two important points about cation exchange and plant growth. One point concerns the total amount of nutrients available to plants as exchangeable cations and the other point concerns the degree to which the exchange is saturated with bases as contrasted to hydrogen. The exchangeable hydrogen contributes to soil acidity, so it produces a special effect, which we shall consider in detail in the next chapter. For our purposes, two specific cases will be cited to show the importance of the kinds and amounts of cations in the exchange.

Over a period of some years researchers had collected soil samples under good productive stands of many tree species. Wilde reasoned that the soil properties found where trees naturally grow well in the forest would be desirable properties to have in nursery soils for growing the seedlings. On this basis he formulated the nursery standards given in Table 7–7. The most demanding tree species require both a greater total amount of exchangeable Ca, Mg, and K, but also a higher percentage base saturation (and pH) than the less demanding species that do well on soils of moderate or low fertility.

**Table 7–7** Standards for Raising Nursery Stock

| Fertility level | Species | CEC, me/100 grams | Exchangeable Cations me/100 grams | | | % base saturation | pH range |
|---|---|---|---|---|---|---|---|
| | | | Ca | Mg | K | | |
| High | White oak, pecan, hard maple, basswood, white cedar | 10 | 5.0 | 2.0 | .3 | 73 | 5.5–7.3 |
| Moderate | White spruce, white pine, Douglas fir, yellow birch | 7 | 2.5 | 1.0 | .2 | 53 | 5.0–6.0 |
| Low | Jack pine, Scotch pine, Virginia scrub pine | 4 | 1.5 | 0.5 | .1 | 52 | 4.8–5.5 |

Adapted from "Forest Soils," S. A. Wilde, The Ronald Press, New York, 1958.

Horticulturists have studied highbush blueberry plantations to determine the soil conditions best suited for blueberries. Soil samples were collected and analyzed from 55 locations where the growth of the blueberries had been judged to be either good or poor. Soils were analyzed for the major exchangeable cations, cation exchange capacity, and pH. The only factor that consistently correlated with the growth

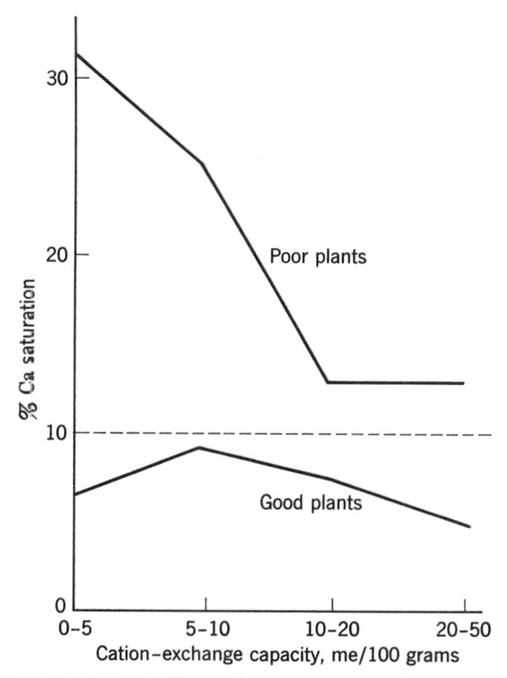


**Fig. 7–16** Effect of percentage of the soil cation-exchange capacity occupied by calcium and the general vigor of blueberry plants. Good vigor occurred when exchangeable calcium was less than 10 percent over a wide range in cation-exchange capacity. (Adapted from "Production in Michigan Blueberry Plantations in Relation to Nutrient-Element Content of the Fruiting-Shoot Leaves and Soil," W. E. Ballinger, et al., *Mich. State Agr. Exp. Sta. Quart.* **40**:2–11, 1958)

of the blueberries was exchangeable calcium. The results plotted in Fig. 7–16 show that the best growth of blueberries occurred on soils with less than 10 percent exchangeable calcium. This is consistent with the very low base saturation and highly acid soil conditions required.

Most garden and agricultural crops grow best when the base saturation is 80 percent or more and the pH is 6 or above. These examples serve to illustrate the great importance of cation exchange relationships for plant growth. The entire next chapter on pH will use these ideas to explain the nature, importance, and control of pH.

## Anion Exchange in Soils

Although cation exchange in soils has been studied extensively and its importance in supplying nutrients to plants is recognized, the phenomenon of the exchange of anions in the colloidal complex has received less attention. It has been established that such exchange does occur, though apparently to a much lesser extent in most soils than that of cations. Anions may replace the OH groups in the clay minerals, and, as these groups are much more plentiful in the kaolinite minerals than in the montmorillonite minerals, the kaolinite minerals are considered the seat of most anion exchange in temperate or arid region soils.

Although the areas are not extensive, some of the most weathered soils in the humid tropics have little or no kaolinite and the clays are almost entirely oxides of iron and aluminum (to a lesser extent also manganese and titanium). Such soils may have a *net positive* charge where the charge due to the clay over balances the negative charge of the organic matter. Imagine what this means. Instead of having calcium adsorbed and nitrate very mobile and subject to leaching, the reverse occurs. We need to consider the reasons for such an important reversal in soil properties.

Gibbsite is an example of an important oxide clay. Gibbsite consists of aluminum in six coordination (aluminum in octahedral position) surrounded by six hydroxyls. In the normally acid environment of highly weathered tropical soils, the hydroxyls take on hydrogen atoms (protons) or are protonated as follows:

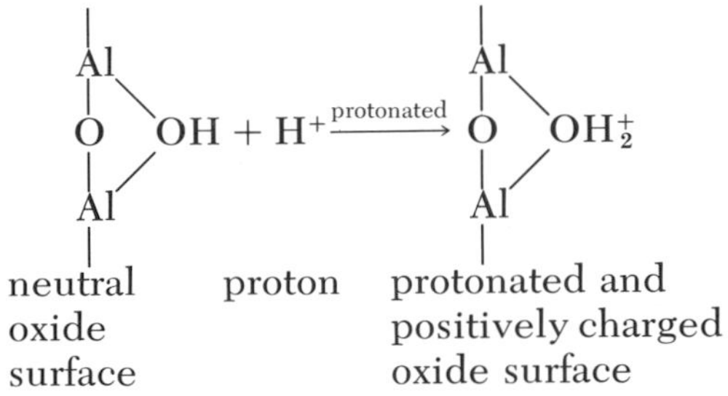

$$\begin{matrix} | \\ Al \\ | \\ O \\ | \\ Al \\ | \end{matrix} \!\!\begin{matrix} \diagdown \\ \\ \diagup \end{matrix} OH + H^+ \xrightarrow{\text{protonated}} \begin{matrix} | \\ Al \\ | \\ O \\ | \\ Al \\ | \end{matrix} \!\!\begin{matrix} \diagdown \\ \\ \diagup \end{matrix} OH_2^+$$

| neutral oxide surface | proton | protonated and positively charged oxide surface |
|---|---|---|

The importance of the phenomenon of positively charged colloid fraction in soils can hardly be over estimated since most important soil-plant nutrition relationships are reversed. In summary soils with net positively charged colloids: (1) Adsorb anions like nitrate and chloride ions. (2) Cations like calcium, magnesium and potassium are repelled and remain in the soil solution very susceptible to leaching. In addition, the term percentage base saturation is meaningless. (3) Phosphate and sulfate ions are tightly fixed by replacement of hydroxyls and the soils have very high phosphorus fixing capacity and naturally a very low level of available phosphorus. One reason for some failures in the use of temperate region practices to crop production on some tropical soils has been due to the lack of understanding concerning positively charged soil colloids. Use your imagination to think of some specific reasons for such failures.

## pH Dependent Charge

The silicate clays carry a permanent negative charge that is unaffected by pH of the soil environment. A good example is the lattice charge of montmorillonite resulting from the isomorphous substitution of aluminum by magnesium. We noted in Chapter 6 that cation-exchange capacity or negative charged sites arise in humus by the dissociation of hydrogen ions from hydroxyls of carboxyl groups. A reduced concentration of hydrogen ions in solution, as occurs when the pH increases, encourages more dissociation of hydrogen from hydroxyls and thus produces more negatively charged sites or cation exchange capacity. To a limited extent the same occurs for the exposed hydroxyls of the silicate clay minerals.

The charge of oxide clays is entirely pH dependent. For example, consider what happens when an acid oxide clay soil is limed (pH increased):[2]

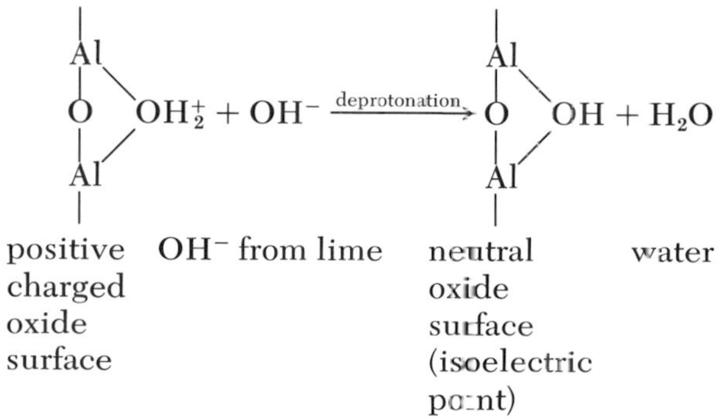

$$\text{(Al–O–Al)–}OH_2^+ + OH^- \xrightarrow{\text{deprotonation}} \text{(Al–O–Al)–}OH + H_2O$$

| positive charged oxide surface | $OH^-$ from lime | neutral oxide surface (isoelectric point) | water |
|---|---|---|---|

If more liming occurs, further deprotonation occurs.

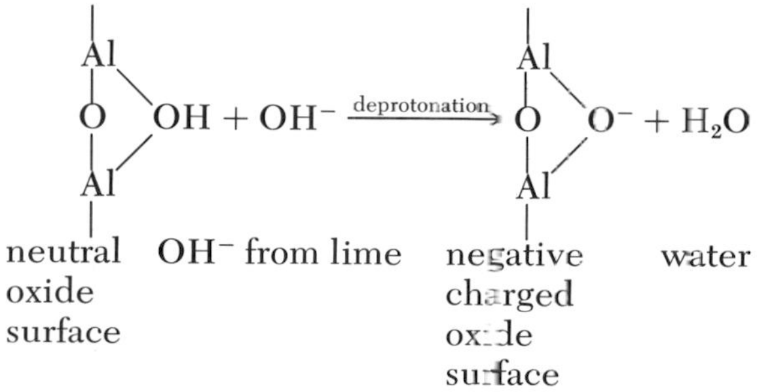

$$\text{(Al–O–Al)–}OH + OH^- \xrightarrow{\text{deprotonation}} \text{(Al–O–Al)–}O^- + H_2O$$

| neutral oxide surface | $OH^-$ from lime | negative charged oxide surface | water |
|---|---|---|---|

This effect of pH on the charge of soil colloids calls attention to the importance of soil pH—the topic of the next chapter.

[2] Schofield, R. K. "The Effect of pH on Electric Charges Carried by Clay Particles." *J. Soil Sci.* Vol. 1:1–8, 1949.

# 8

# Soil pH – Causes, Significance, and Alteration

The particular pH measured in a soil is caused by a particular set of chemical conditions. Therefore, a determination of soil pH is one of the most important tests that can be made to diagnose plant growth problems. For example, suppose some diseased plants have light green leaves that could be caused by several factors. If the pH of the soil is as low as 5.5 or less, the disease is probably not an iron deficiency, since iron compounds are soluble under acid conditions. If the soil pH is 8, one should seriously consider the possibility of iron deficiency because iron compounds are very insoluble in soils with a pH of 8. As an analogy, the pH of the soil is like the temperature of an animal. Both tests are easily made and provide basic information useful in diagnosing what is likely to be the disease or problem. Our attention in this chapter will be focused on the causes, significance, and alteration of soil pH.

## DEFINITION AND CAUSES OF SOIL pH

Usually, soils of humid regions are acid and soils of arid regions are alkaline. In acid soils the soil solution contains more hydrogen ions ($H^+$) than hydroxyl ($OH^-$) and in alkaline soils the soil solution contains more $OH^-$ than $H^+$. Our purpose is to examine the causes for various concentrations of $H^+$ and $OH^-$ that are responsible for the range of pH from about 4 to 10, which is normally found in soils.

## Soil pH Defined

Water is neutral because the concentration of $H^+$ and $OH^-$ are equal. At neutrality the pH is 7. For our purposes it is sufficient to consider water as composed of water molecules, hydrogen ions, and hydroxyl ions. Water dissociates or ionizes as follows:

$$HOH \rightleftharpoons H^+ + OH^- \quad (1)$$

At equilibrium the reaction is strongly to the left producing the following composition of water expressed as moles per liter: (at 25°C a liter of water weighs 997 grams and 1 mole of water weighs 18 grams resulting in 55.4 moles of water per liter):

| | |
|---|---|
| HOH | 55.399,999,8 moles per liter |
| $H^+$ | 0.000,000,1 moles per liter |
| $OH^-$ | 0.000,000,1 moles per liter |

Two points should be noted. First, only one water molecule in 554 million is ionized. Second, the number, concentration, or moles per liter of $H^+$ is equal to $OH^-$.

The pH scale has been devised for conveniently expressing the extremely small concentrations of $H^+$ found in water and in many important biological systems. The pH is defined as:

$$\mathrm{pH} = \log \frac{1}{[H^+]} \text{ (where } [H^+] \text{ equals moles of } H^+ \text{ per liter)} \quad (2)$$

The pH of pure water is calculated as follows:

$$\mathrm{pH} = \log \frac{1}{0.000{,}000{,}1} = \log 1{,}000{,}000 = 7 \quad (3)$$

Each unit change in pH is associated with a tenfold change in the concentration of $H^+$ and $OH^-$ as shown in Fig. 8–1.

## Exchangeable Cations as Sources of $H^+$ and $OH^-$

From the previous chapter we know that the exchangeable cations are adsorbed with sufficient energy to retard their leaching from the soil, but a significant number dissociate from the cation exchange surfaces and exist in the solution where they are readily utilized by plants. On dissociation the exchangeable bases cause hydrolysis and OH ions are produced, as shown in Fig. 8–2.

Exchangeable H dissociates and contributes $H^+$ to the soil solution according to Equation 4.

$$\underset{\text{(adsorbed)}}{\text{micelle-H}} \rightleftharpoons \underset{\text{(in solution)}}{H^+} \quad (4)$$

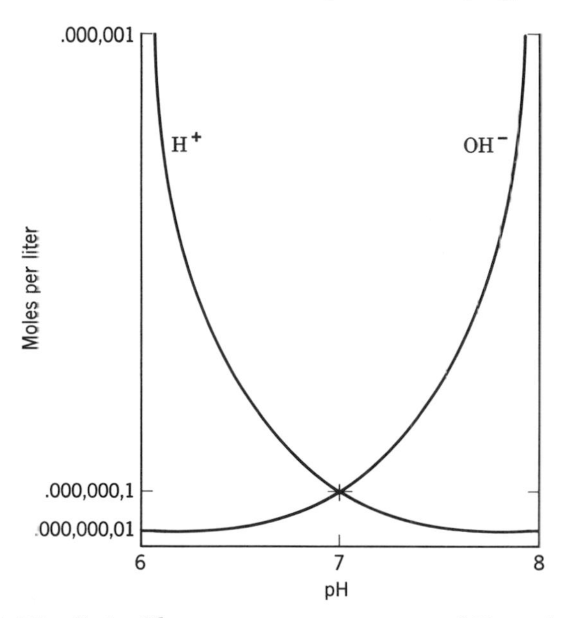


**Fig. 8–1** Changes in concentration of $H^-$ and $OH^-$ with changes in pH of solutions.

Exchangeable H is the principal source of $H^+$ until the pH of the soil goes below 6 when Al in the Al octahedral sheet of clays becomes unstable and is adsorbed as exchangeable Al. The exchangeable Al is a source of $H^+$ according to Equation 5.

$$\text{micelle-Al} + 3H_2O \rightleftharpoons \underset{\text{(insoluble)}}{Al(OH)_3} + \text{micelle}\begin{matrix} \diagup H \\ -H \\ \diagdown H \end{matrix} \rightleftharpoons H^+ \qquad (5)$$

The net effect of the hydrolysis by exchangeable Al is an increase in the $H^+$ concentration of the soil solution resulting from the dis-

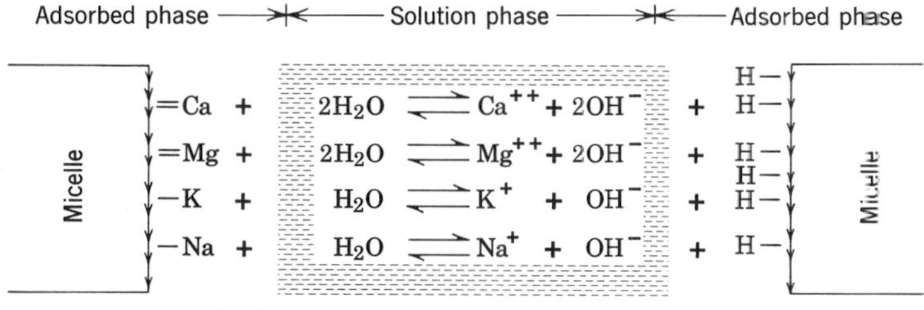


**Fig. 8–2** Hydrolysis of exchangeable bases as a source of $OH^-$ for the soil solution. (The exchangeable bases are hydrated, but have been shown to be not hydrated for simplicity of illustration.)

sociation of the exchangeable (micelle) H that is produced (same as in Equation 4). For all practical purposes, exchangeable H and Al are both sources of $H^+$ and, for simplicity, the term exchangeable H will be used with no attempt to indicate whether Al is involved.

## Relationship Between Soil pH and Percent Base and Hydrogen Saturation

One of the most logical questions at this point is, "What conditions produce equal concentrations of $H^+$ and $OH^-$ and a pH of 7?" In every soil there is a relationship between the percentage of base (or H saturation) and pH. For broad geographic areas where soils have similar mineralogy, a general relationship exists. In southern Michigan the data from thousands of soil samples were analyzed and for mineral soils the relationship shown in Equation 6 was found.

$$\text{pH} = \frac{187 - 0.3(\text{me CEC}) - \%\ \text{H saturation}}{24} \tag{6}$$

Using Equation 6, Fig. 8–3 was constructed. From the diagram it can be seen that a pH of 7 was associated with a 15 percent H saturation and 85 percent base saturation (assuming a CEC of 13). Furthermore, for the soils studied, the minimum and maximum pH that could be developed by the dissociation of cations from the exchange was 3.5 and 7.6, respectively.

At 50 percent H and base saturation the pH was 5.5. This is because the exchangeable H is less tightly adsorbed to the micelles than are the exchangeable bases which are predominately divalent calcium and magnesium. In kaolinitic soils, where the exchange sites are mainly from dissociation of H from OH on the exposed edges of the clay particles, a pH between 6 and 7 may be associated with 50 percent base saturation because OH bound H is much more tightly adsorbed than is H adsorbed on an exchange site resulting from isomorphous substitution.

Earlier it was mentioned that the pH of some soils was as high as 10. The exchangeable cations in a "leached" humid region soil can account for a pH of 7 and slightly more, but to account for pH values of as high as 9 or 10 one must examine some other soil conditions. The first to be examined will be the conditions found in calcareous soils.

## pH of Calcareous Soils

A *calcareous* soil contains calcium carbonate ($CaCO_3$) and when treated with hydrochloric acid (HCl) a bubbling can be observed representing the evolution of carbon dioxide. Calcium carbonate

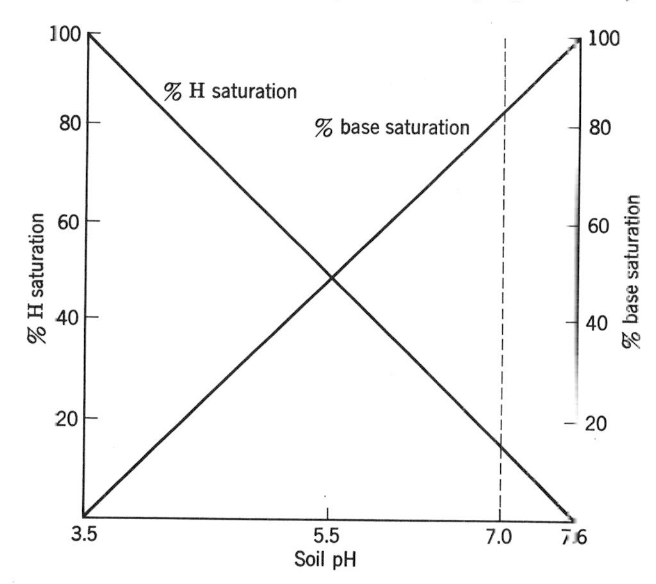


**Fig. 8–3** Relationship between soil pH and percentage base and H saturation for loamy textured soils of southern Michigan using equation 6 and assuming a cation-exchange capacity of 13. (Courtesy John Shickluna.)

is relatively insoluble, but when present in soils it creates a constant pressure to saturate the exchange with calcium as follows.

$$CaCO_3 + H_2\text{-micelle} \rightarrow \text{Ca-micelle} + H_2O + CO_2 \tag{7}$$

For this reason calcareous soils are 100 percent base saturated and the pH is mainly controlled by the hydrolysis of calcium carbonate as follows.

$$CaCO_3 + 2H_2O \rightarrow Ca(OH)_2 + H_2CO_3 \tag{8}$$

The greater dissociation of the calcium hydroxide and production of $OH^-$, as compared to the production of $H^+$ from the weak carbonic acid, creates an alkaline effect. As a result the pH of calcareous soils usually ranges from $7^+$ to a maximum of 8.3.

## Effect of Sodium on Soil pH

Sodium is released from the weathering of minerals. In the humid regions leaching readily removes the sodium because of the weak attraction of sodium for cation exchange sites. In the arid regions the sodium may accumulate as sodium carbonate and the sodium will tend to occupy some of the exchange positions. The hydrolysis of sodium carbonate and exchangeable sodium produces a very strong base, NaOH. When the soil is 15 percent or more sodium saturated or a significant amount of sodium carbonate exists in the soil, the pH value may be between 8.5 and 10.

### pH of Saline Soils

Saline soils have sufficient soluble salts to impair plant growth, mainly by increasing the osmotic pressure of the soil solution and restricting water uptake. Soluble salts may naturally accumulate in soils in arid regions or as a result of the addition of irrigation water. Excessive use of soluble fertilizer salts produces saline soils and is a particular problem in managing greenhouses. The major potassium fertilizer is KCl. Its hydrolysis produces both a strong base, KOH, and a strong acid, HCl, which are about equal in their ability to produce $H^+$ and $OH^-$. Saline soils tend to have a pH at or near 7 because of the hydrolysis by soluble salt.

### Other Factors Affecting Soil pH

Carbon dioxide released from respiration dissolves in the soil solution and forms weak carbonic acid. This influence has existed in calcareous and other alkaline soils for thousands of years indicating that the formation of carbonic acid in soils plays a very minor role in causing a particular soil pH. Carbonic acid in soil does facilitate weathering and plays a role in leaching (see Fig. 8–4).

A few other factors influencing soil pH are worthy of mention. Sulfur is a by-product in industrial gases and is sometimes responsible for soil acidity in nearby soils as a result of the formation of sulfuric acid. Some unusual soils that contain a significant amount of iron sulfide (FeS) also contain significant amounts of acid. Plants are killed when the sulfur is oxidized and converted to sulfuric acid and the soil pH becomes very low. A small amount of nitric acid is a natural component of rain, but its effect appears to be insignificant.

## SIGNIFICANCE OF SOIL pH

The major effects of soil pH are biological. Some organisms have rather small tolerances to variations in pH, but other organisms can tolerate a wide pH range. Studies have shown that the actual concentrations of $H^+$ or $OH^-$ are not very important except under the most extreme circumstances. It is the associated conditions of a certain pH value that is most important.

### Nutrient Availability and pH Relationships

Perhaps the greatest general influence of pH on plant growth is the effect of pH on the availability of nutrients. In Fig. 8–3 the pH was

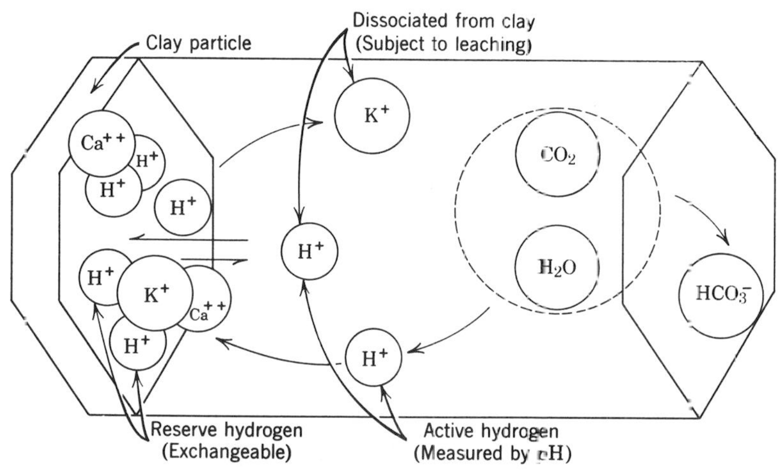


**Fig. 8–4** Schematic diagram showing the relationship between reserve and active acidity. Carbon dioxide and water form carbonic acid which provides hydrogen ions that replace exchangeable bases. In the humid regions the exchanged bases are subject to leaching and removal from the soil as soluble carbonates. In this way carbonic acid contributes to the development of soil acidity. (From *Laboratory Manual for Introductory Soil Science*, 3rd Ed., W. C. Brown and Co., Dubuque, Iowa. Used with permission.)

seen to be related to percentage base saturation. When the base saturation is less than 100 percent, an increase in pH is associated with an increase in the amount of calcium and magnesium in the soil solution, since calcium and magnesium are usually the dominate exchangeable bases. Many studies have been conducted that relate increases in plant growth with increases in the percentage of calcium in plants and with increasing pH or percentage base saturation. The general relationship between pH and availability of calcium and magnesium are shown in Fig. 8–5.

Another nutrient whose availability is increased as the pH is increased on the low end of the pH range in soils is molybdenum. At low pH molybdenum forms insoluble compounds with iron and is rendered unavailable. Under these conditions plants susceptible to molybdenum deficiency, such as cauliflower, clover, and citrus, will show a growth response to an increase in pH. The fact that increased availability of molybdenum occurs as the pH increases is substantiated by data in Fig. 8–6, which shows that molybdenum concentration in cauliflower leaves increase with increasing pH.

Potassium availability is usually good in alkaline soils that reflect the limited leaching and removal of exchangeable potassium (Fig. 8–5).

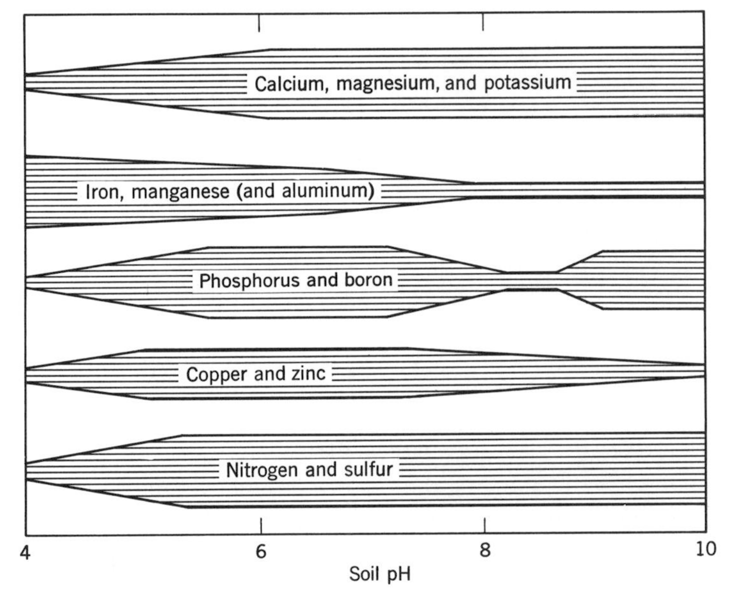


**Fig. 8–5** General relationship between soil pH and availability of plant nutrients: the wider the bar, the more available the nutrient.

The availability or solubility of some plant nutrients decreases with an increase in pH. Iron and manganese are two good examples (see Fig. 8–7). Iron and manganese are commonly deficient in calcareous soils. Phosphorus and boron also tend to be unavailable in calcareous soils resulting from reactions with calcium. Phosphorus and boron also tend to be unavailable in very acid soils. Copper and zinc have reduced availability in both highly acid and alkaline soils. For the plant nutrients as a whole, good overall nutrient availability is found near pH 6.5.

## Effect of pH on Soil Organisms

The greater capacity of fungi over bacteria to thrive in highly acid soils was mentioned in Chapter 5. The pH requirements of some disease organisms can be used by soil management practices as a means to control diseases. One of the best known cases is that of the maintenance of acid soils to control potato scab. Damping-off disease in nurseries is controlled by maintaining the pH at 5.5 or less. Nitrifying organisms also become inhibited when the pH is less than 5.5. The availability of nitrogen in soils is related mainly to the effect of pH on decomposition of organic matter (see Fig. 8–5). High soil acidity has also been shown to inhibit earthworms in soils. Peter

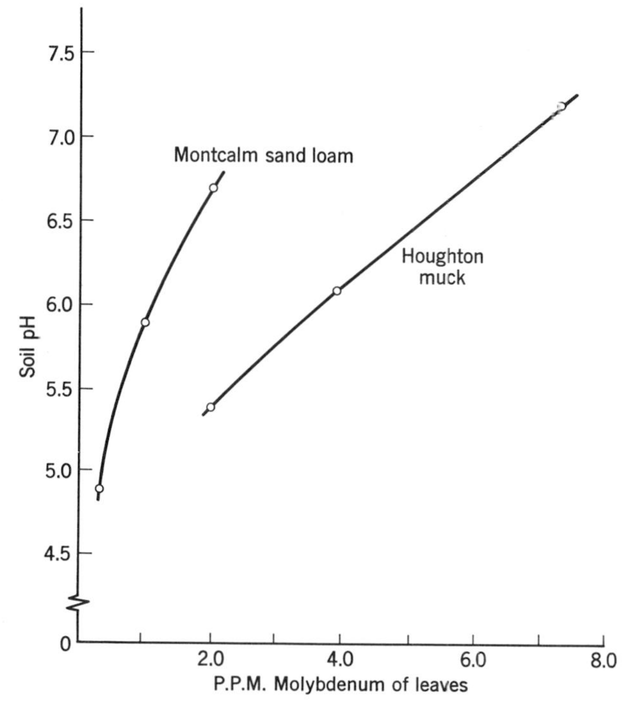


**Fig. 8–6** Correlation between molybdenum content of cauliflower leaves and soil pH in a greenhouse experiment. (From "Studies on Crop Response to Molybdenum and Lime in Michigan," F. Turner and W. W. McCall, *Mich. Agr. Exp. Sta. Quart. Bull.*, **40**:268–281, 1957.)

Farb relates an interesting case where soil pH influenced both earthworm and mole activity.

On a certain tennis court in England, each spring after the rains and snow had washed away the court marking, the marking lines could still be located because moles had unerringly followed them. This intriguing mystery was investigated and it produced the following results. The tennis court had been built on acid soil and chalk (lime) was used to mark the lines. As a result, the pH in the soil below the marking lines was increased. Earthworms then inhabited the soil under the markings and the remainder of the soil on the tennis court remained uninhabitated. Since earthworms are a primary food of moles, the moles restricted their activity to the soil only under the marking lines.[1]

[1] From "The Living Earth," Peter Farb, New York: Harper Brothers, 1959, p. 44.

**Fig. 8–7** Pin oak is a popular ornamental tree that is susceptible to iron deficiency when grown on neutral or alkaline soil. Iron deficient leaves have dark green veins and yellow intervein areas. This tree is seriously deficient in iron as indicated by the death and defoliation of many leaves on the upper branches.

## Aluminum, Iron, and Manganese Toxicity in Acid Soils

The data in Fig. 8–5 show that iron (aluminum) and manganese are most available in highly acid soils. The relationship between manganese in corn leaves and soil pH is shown in Fig. 12–13. Manganese toxicity occurs when the pH is about 4.5 or less. Evidence indicates that the high exchangeable aluminum in many acid soils of southeastern United States restricts root growth in many subsoils. Aluminum toxicity was probably a factor in an unsuccessful attempt to grow Gaines wheat in Maryland. Gaines is a short-strawed variety that was developed on high base-saturated soils of the state of Washington. In Washington yields as high as 209 bushels per acre were attained. When Gaines wheat was grown in Maryland it yielded only 17 bushels per acre, although locally adapted wheat varieties yielded 60 bushels per acre. Plants, even varieties of the same species, exhibit differences in tolerance to high levels of aluminum, iron, or manganese as well as other soil conditions associated with soil pH. This gives rise to pH preferences of plants—our next topic.

### pH Preferences of Plants

Blueberries are well known for their highly acid soil requirement. In the previous chapter it was shown in one study that poor blueberry growth occurred when the calcium saturation exceeded 10 percent. Normally this would mean that the soil had a very low base saturation and pH. In Table 8–1 the optimum range for highbush blueberries is given as 4.0–5.0. Other plants preferring a pH of 5 or less include azaleas, orchids, sphagnum moss, jack pine, black spruce, and cranberry. Those plants listed in Table 8–1 that prefer a pH of 6 or less are italicized. Most field and vegetable crops prefer a pH of about 6 or higher.

The fact that plants have specific soil pH requirements gives rise to the need to alter soil pH for successful growth of many plants. We shall examine this topic next.

## ALTERATION OF SOIL pH

There are two approaches to assure that plants will grow without serious inhibition from unfavorable soil pH: (1) Plants can be selected that will grow well with the existing soil pH or (2) the pH of the soil can be altered to suit the preference of the plants. Considerations for altering soil pH will be examined first.

### Theory of Changing Soil pH

In leached soils where there is no significant amount of salts, such as calcium or sodium carbonate, the soil pH is determined mainly by the base or H saturation, as shown in Fig. 8–3. In theory, if a soil represented by the data plotted in Fig. 8–3 had a pH of 5.5, the base saturation would be 50 percent. If the base saturation of such a soil was increased to 85 percent, the soil pH would be 7. Conversely, if the base saturation was lowered from 85 to 50 percent, the pH would be lowered from 7 to 5.5. Thus, alteration of soil pH under these conditions is dependent on changes in base (or H) saturation. Increases in base saturation produce increases in pH and decreases in base saturation produce decreases in pH.

**Table 8–1** Optimum pH Ranges of Selected Plants

| Plant | pH Range |
|---|---|
| **Field Crops** | |
| Alfalfa | 6.2–7.8 |
| Barley | 6.5–7.8 |
| Bean, field | 6.0–7.5 |
| Beets, sugar | 6.5–8.0 |
| Bluegrass, Ky. | 5.5–7.5 |
| Clover, red | 6.0–7.5 |
| Clover, sweet | 6.5–7.5 |
| Clover, white | 5.6–7.0 |
| Corn | 5.5–7.5 |
| Flax | 5.0–7.0 |
| Oats | 5.0–7.5 |
| Pea, field | 6.0–7.5 |
| Peanut | 5.3–6.6 |
| Rice | 5.0–6.5 |
| Rye | 5.0–7.0 |
| Sorghum | 5.5–7.5 |
| Soybean | 6.0–7.0 |
| Sugar Cane | 6.0–8.0 |
| Tobacco | 5.5–7.5 |
| Wheat | 5.5–7.5 |
| **Vegetable Crops** | |
| Asparagus | 6.0–8.0 |
| Beets, table | 6.0–7.5 |
| Broccoli | 6.0–7.0 |
| Cabbage | 6.0–7.5 |
| Carrot | 5.5–7.0 |
| Cauliflower | 5.5–7.5 |
| Celery | 5.8–7.0 |
| Cucumber | 5.5–7.0 |
| Lettuce | 6.0–7.0 |
| Muskmellon | 6.0–7.0 |
| Onion | 5.8–7.0 |
| Potato | 4.8–6.5 |
| Rhubarb | 5.5–7.0 |
| Spinach | 6.0–7.5 |
| Tomato | 5.5–7.5 |
| **Flowers & Shrubs** | |
| African violet | 6.0–7.0 |
| Almond, flowering | 6.0–7.0 |
| Alyssum | 6.0–7.5 |
| *Azalea* | *4.5–5.0* |
| Barberry, Japanese | 6.0–7.5 |
| Begonia | 5.5–7.0 |
| Burning bush | 5.5–7.5 |
| Calendula | 5.5–7.0 |
| Carnation | 6.0–7.5 |
| **Flowers & Shrubs** (cont.) | |
| Chrysanthemum | 6.0–7.5 |
| *Gardenia* | *5.0–6.0* |
| Geranium | 6.0–8.0 |
| *Holly, American* | *5.0–6.0* |
| Ivy, Boston | 6.0–8.0 |
| Lilac | 6.0–7.5 |
| Lily, Easter | 6.0–7.0 |
| *Magnolia* | *5.0–6.0* |
| *Orchid* | *4.0–5.0* |
| *Phlox* | *5.0–6.0* |
| Poinsettia | 6.0–7.0 |
| Quince, flowering | 6.0–7.0 |
| *Rhododendron* | *4.5–6.0* |
| Rose, hybrid tea | 5.5–7.0 |
| Snapdragon | 6.0–7.5 |
| Snowball | 6.5–7.5 |
| Sweet William | 6.0–7.5 |
| Zinnia | 5.5–7.5 |
| **Forest Plants** | |
| Ash, White | 6.0–7.5 |
| *Aspen, American* | *3.8–5.5* |
| Beech | 5.0–6.7 |
| *Birch, European (white)* | *4.5–6.0* |
| *Cedar, White* | *4.5–5.0* |
| *Club Moss* | *4.5–5.0* |
| *Fir, balsam* | *5.0–6.0* |
| *Fir, Douglas* | *6.0–7.0* |
| *Heather* | *4.5–6.0* |
| *Hemlock* | *5.0–6.0* |
| Larch, European | 5.0–6.5 |
| Maple, Sugar | 6.0–7.5 |
| *Moss, Sphagnum* | *3.5–5.0* |
| Oak, Black | 6.0–7.0 |
| Oak, Pin. | 5.0–6.5 |
| Oak, White | 5.0–6.5 |
| *Pine, Jack* | *4.5–5.0* |
| *Pine, Loblolly* | *5.0–6.0* |
| *Pine, Red* | *5.0–6.0* |
| *Pine, White* | *4.5–6.0* |
| *Spruce, Black* | *4.0–5.0* |
| Spruce, Colorado | 6.0–7.0 |
| *Spruce, White* | *5.0–6.0* |
| *Sycamore* | *6.0–7.5* |
| Tamarack | 5.0–6.5 |
| Walnut, Black | 6.0–8.0 |
| Yew, Japanese | 6.0–7.0 |
| **Weeds** | |
| Dandelion | 5.5–7.0 |
| Dodder | 5.5–7.0 |
| Foxtail | 6.0–7.5 |
| Goldenrod | 5.0–7.5 |
| Grass, Crab | 6.0–7.0 |
| Grass, Quack | 5.5–6.5 |
| *Horse Tail* | *4.5–6.0* |
| *Milkweed* | *4.0–5.0* |
| Mustard, Wild | 6.0–8.0 |
| Thistle, Canada | 5.0–7.5 |
| **Fruits** | |
| Apple | 5.0–6.5 |
| Apricot | 6.0–7.0 |
| Arbor Vitae | 6.0–7.5 |
| Blueberry, High Bush | 4.0–5.0 |
| Cherry, sour | 6.0–7.0 |
| Cherry, sweet | 6.0–7.5 |
| Crab apple | 6.0–7.5 |
| *Cranberry, large* | *4.2–5.0* |
| Peach | 6.0–7.5 |
| *Pineapple* | *5.0–6.0* |
| Raspberry, Red | 5.5–7.0 |
| Strawberry | 5.0–6.5 |

Data from "Soil Reaction (pH) Preferences of Plants," C. H. Spurway, *Mich. Agr. Exp. Sta. Special Bul.*, **306**, 1941. Plants with an optimum pH range of 6.0 and below are italicized.

The case depicted in Fig. 8–3 can be used for illustration. The cation exchange capacity of the soil was 13 milliequivalents per 100 grams and, as we have already indicated, at 50 percent base saturation the pH is 5.5. To increase the soil pH to 7 would require increasing the base saturation 35 percent (from 50 to 85 percent). The quantity of bases needed is calculated as follows:

$$13 \text{ me} \times 35\% = 4.55 \text{ me bases needed per 100 grams of soil}$$

The amount of lime required is usually expressed on the basis of some convenient volume of soil such as an acre furrow slice, commonly considered to weigh 2 million pounds. In our example if $CaCO_3$ were used as source of base (Ca), the 4.55 me of base needed per 100 grams of soil would require 0.2275 grams of $CaCO_3$ per 100 grams of soil calculated as follows.

$$\text{molecular weight of } CaCO_3 = 100 \text{ grams} = 2 \text{ equivalents} = 2000 \text{ me}$$

$$\frac{100 \text{ grams } CaCO_3}{2000 \text{ me}} = 0.05 \text{ grams of } CaCO_3 \text{ per me}$$

$$4.55 \text{ me} \times 0.05 \text{ grams} = 0.2275 \text{ grams of } CaCO_3 \text{ per 100 gram soil}$$

We need not convert grams to pounds, but we can just state that 0.2275 *grams* of $CaCO_3$ per 100 *grams* of soil is equal to 0.2275 *pounds* of $CaCO_3$ per 100 *pounds* of soil. To put the theoretical lime requirement on an acre furrow slice basis we use a simple proportion:

$$\frac{0.2275 \text{ pounds } CaCO_3}{100 \text{ pounds soil}} = \frac{x \text{ pounds } CaCO_3}{2 \text{ million pounds soil}} = 4550 \text{ pounds } CaCO_3$$

For each milliequivalent of base needed per 100 grams of soil, the theoretical lime requirement is equal to 1000 pounds of pure calcium carbonate per acre furrow slice.

Turning the situation around, suppose the pH were 7 and a pH of 5.5 was desired. Now, 4.55 me of acid-producing material per 100 grams of soil would be required to furnish the H needed to increase H saturation from 15 to 50 percent. Sulfur is commonly used to increase soil acidity and has a milliequivalent weight of 0.16 grams ($32 \div 2000$) because sulfur is oxidized in the soil to sulfuric acid and each sulfur atom is the source of two $H^+$ (see Equation 3 of Chapter 5). Substituting 0.0728 for 0.2275 in the above equation results in a theoretical sulfur requirement of 1456 pounds of sulfur per acre furrow slice. For each milliequivalent of exchangeable acidity needed per 100 grams of soil, the theoretical sulfur requirement is 320 pounds per acre furrow slice.

## Role of Cation-Exchange Capacity in Altering Soil pH

Suppose the soil just used to illustrate the theoretical lime requirement had a CEC twice as large, 26 instead of 13. Using Equation 6, one finds that at pH 5.5 the soil with twice as much CEC has a 53 percent base saturation compared to 50 for the soil with a CEC of 13. Notice this in Table 8–2. This means that the same concentration of $H^+$ in the soil solution (same pH) is produced with only a slight difference in percentage H or base saturation. The soil with a CEC of 26 has nearly twice the exchangeable H as the soil with a CEC of 13 when both have a pH of 5.5. The data in Table 8–2 also show that similar percentage base saturation increases are needed to increase pH from 5.5 to 7. This results in the need for almost two times more lime for the soil with twice the CEC (Table 8–2) 4.55 me versus 9.36 me per 100 grams of soil.

**Table 8–2** Comparisons of Base Saturation Relationships of Soils with Varying Cation Exchange Capacity as Calculated Using Equation 6

| | Cation Exchange Capacity of Soil, me/100 grams | |
|---|---|---|
| | 13 | 26 |
| Base saturation at pH 5.5 | 50% | 53% |
| Exchangeable H at pH 5.5 | 6.5 me | 12.2 me |
| Base saturation at pH 7 | 85% | 89% |
| Base saturation change needed to increase pH from 5.5 to 7 | 35% | 36% |
| Base needed per 100 grams soil | 4.55 me | 9.36 me |

The greater capacity of soils with greater exchange capacities to adsorb bases for a given pH change is illustrated in Fig. 8–8.

Soil acidity has two components: (1) the active or solution H and (2) the exchangeable or reserve H. These two forms tend toward equilibrium so that a change in one produces a change in the other. When a base is added to an acid soil, the solution H is neutralized and some exchangeable H ionizes to reestablish the equilibrium. The amount of exchangeable H is slowly decreased, the solution H is decreased, and the pH slowly increases. The resistance of the soil pH to change gives rise to the buffering phenomenon in soils. Soils with the largest CEC capacities offer the greatest resistance to change in pH and are the most strongly buffered.

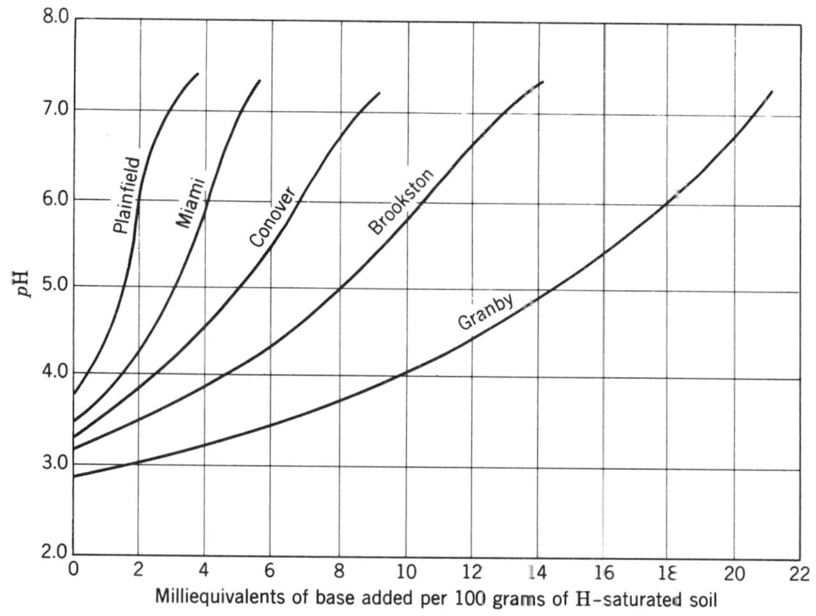


**Fig. 8–8** Titration curves showing the different amounts of base needed to increase the pH of soils having increasing cation-exchange capacities from left to right (different degrees of buffering). All soils were H-saturated when titration was started.

## Methods Used to Determine the Lime Requirement

In practice, the first decision in arriving at a recommendation of lime is deciding on the pH value desired. The desired pH depends in part on the type of soil and the kind of plant to be grown. Identification of the factors limiting the growth of alfalfa on 13 highly acid soils of southeastern United States showed that manganese toxicity was the most limiting in 8 soils, aluminum toxicity in 1 soil and calcium deficiency in 1 soil. The results of the study are presented in Table 8–3. At present there are two schools of thought on how high the pH should be raised by liming. Some feel that liming to remove only the limiting factor, such as the toxicity of manganese or aluminum, is sufficient. Others feel that alfalfa should be limed to a pH of 6.8. Future research will probably lead to a more precise definition of the most desirable pH to which acid soils should be limed. Many soil-testing laboratories are currently recommending that acid soils be limed to a pH of about 6.5 for most crop and garden situations.

It is common practice in soil-testing laboratories to make a direct measurement of the lime requirement by using a buffer solution. One

**Table 8–3** Identification of Factors Limiting the Growth of Alfalfa on Acid Surface Soils

| Soil | pH | Region | Sample Site | Growth Limiting Factor |
|---|---|---|---|---|
| Zanesville | 4.9 | Ozarks | Fayetteville, Ark. | Mn toxicity |
| Johnsburg | 5.1 | Ozarks | Fayetteville, Ark. | Mn toxicity |
| Taloka-complex | 5.2 | Ozarks | Fayetteville, Ark. | Mn toxicity |
| Waynesboro | 5.2 | Ozarks | Fayetteville, Ark. | Mn toxicity |
| Richland | 5.1 | Loessial Terraces | Marianna, Ark. | Mn toxicity |
| Loring | 5.3 | Loessial Hills | Colt, Ark. | Mn toxicity |
| Dundee | 4.8 | Mississippi Bottoms | Earle, Ark. | Mn toxicity |
| Cecil | 5.2 | Piedmont | Watkinsville, Ga. | Mn toxicity |
| Bladen | 4.8 | Lower Coastal Plains | Fleming, Ga. | Al toxicity |
| Leon | 4.2 | Lower Coastal Plains | Fleming, Ga. | Ca deficiency |
| Lakeland | 5.1 | Lower Coastal Plains | Live Oak, Fla. | Complex of factors not separated |
| Tifton | 5.6 | Middle Coastal Plains | Tifton, Ga. | Complex of factors not separated |
| Rains | 4.7 | Middle Coastal Plains | Tifton, Ga. | Complex of factors not separated |

From "Toxic Factors in Acid Soils," C. D. Foy and G. B. Burns, *Plant Food Review*, **10**, p. 2, 1964.

such test was developed at Ohio State University and is widely used. A buffer solution with pH of 7.5 is mixed with a known quantity of soil. The exchange acidity is replaced from the exchange sites and the depression in pH from 7.5 (buffer solution pH) to the pH of the mixture of buffer solution and soil is a measure of the total acidity. The lime requirement is obtained from tables that have been developed which relate depression of buffer pH to tons of limestone required to raise the pH of the soil to a desired value.

A less accurate but useful method, especially for the home owner, is to determine the pH of the soil with inexpensive indicator dyes (see Fig. 8–9) and to obtain the lime requirement from a table that relates soil pH and texture (used as an indication of cation-exchange capacity) to lime requirement. Such a table used for liming a garden is presented in Table 8–4. The lower lime requirement values for the soils of the southern states is a reflection of the lower cation exchange

**Table 8–4** Suggested Applications of Finely Ground Limestone to Raise the pH of a 7-Inch Layer of Several Textural Classes of Acid Soils, in Pounds per 1000 Square Feet

| | pH 4.5 to 5.5 | | pH 5.5 to 6.5 | |
|---|---|---|---|---|
| Textural Class | Northern and Central States | Southern Coastal States | Northern and Central States | Southern Coastal States |
| Sands and loamy sands | 25 | 15 | 30 | 20 |
| Sandy loams | 45 | 25 | 55 | 35 |
| Loams | 60 | 40 | 85 | 50 |
| Silt loams | 80 | 60 | 105 | 75 |
| Clay loams | 100 | 80 | 120 | 100 |
| Muck | 200 | 175 | 225 | 200 |

From U.S.D.A. Yearbook "Soil," 1957, p. 678.

capacity resulting from kaolinitic clays. Caution must be observed to prevent overliming because some of the micronutrients may become unavailable resulting in nutrient deficiencies.

## Forms of Lime

Chemically, lime is CaO, but an extension of the meaning of the word lime is now used to include all limestone products used to neutralize soil acidity. Limestone deposits are widely distributed and constitute the most important source of lime (Fig. 8–10). Limestone is a carbonate form of lime with $CaCO_3$ and $MgCO_3$ as the major components. The oxide form, CaO, is produced by heating calcium carbonate and driving off carbon dioxide. Hydroxide forms of lime are produced by "slaking" or adding water to the oxide forms. In the soil all forms act similarly in that the exchangeable H is replaced and converted to water and the base saturation of the soil is correspondingly increased. The reaction of the carbonate and hydroxide forms with exchangeable hydrogen is as follows:

$$CaCO_3 + {}^{H}_{H}\!\!-\text{micelle} \rightarrow \text{Ca–micelle} + H_2O + CO_2 \qquad (9)$$

$$Ca(OH)_2 + {}^{H}_{H}\!\!-\text{micelle} \rightarrow \text{Ca–micelle} + H_2O \qquad (10)$$

The oxide form of lime reacts with water in the soil to form calcium hydroxide and then neutralizes soil acidity according to equation 10.

**Fig. 8–9** Determination of soil pH using pH indicator dye solution and pH color chart.

Replacement of exchangeable hydrogen tends to drive the reactions to the right and the reactions usually occur until the lime is consumed. The greater solubility of the oxide and hydroxide forms as compared to $CaCO_3$ results in more rapid reaction with exchangeable hydrogen.

On a weight basis the three forms of lime have different neutralization capacity. The neutralizing value of liming materials is calculated on the basis of pure calcium carbonate as 100 percent. It has been pointed out that when pure calcium carbonate is burned in a kiln,

**Fig. 8–10** Many areas have limestone near the surface that can readily be mined, ground, and applied to acid soils as an inexpensive corrective for excess soil acidity. (Monroe County, Wisconsin.)

100 pounds of the dry material give off 44 pounds of carbon dioxide gas, leaving 56 pounds of calcium oxide. When the 56 pounds of calcium oxide are moistened, they react with 18 pounds of water to form 74 pounds of hydrated lime. It is obvious, then, that 100 pounds of pure calcium carbonate, 74 pounds of pure calcium hydroxide, and 56 pounds of pure calcium oxide all contain the same amount of calcium and all have the same power to neutralize soil acidity because it is the calcium ions that displace the hydrogen ions in decreasing acidity.

The relative ability of different liming materials to neutralize acidity is frequently expressed on a percentage basis. Thus, the neutralizing power of the different forms of lime in the pure state is determined by their molecular weights. The molecular weight of calcium carbonate is 100; of calcium hydroxide, 74; and of calcium oxide, 56. By dividing 100 by 74 the figure 1.35 is obtained, which means that 1 pound of calcium hydroxide supplies the same amount of calcium as 1.35 pounds of calcium carbonate. In other words, if expressed on a percentage basis ($1.35 \times 100$), pure calcium hydroxide has a neutralizing value of 135 percent relative to calcium carbonate. Likewise, calcium oxide ($100/56 \times 100 = 178$) has a neutralizing power of 178 percent. Pure magnesium carbonate with a molecular weight of 84 has a neutralizing value of 119 percent (100/84 × 100 = 119). A limestone containing 80 percent calcium carbonate and 20 percent magnesium carbonate would have a neutralizing value of 103.8 percent (80 + 20 × 1.19 = 103.8). Frequently magnesium limestones have a neutralizing power of 107 or 108 percent. The molecular weights, neutralizing values, and calcium carbonate equivalents for the common chemical forms of liming materials in a pure state are given in Table 8–5. Locally, many other materials are used for liming. These include marl, by-product lime from manufacturing, and oyster shells.

**Table 8–5** Relative Neutralizing Power of Different Forms of Lime

| Form of Lime | Molecular Weight | Neutralizing Value, percent | Pounds Equivalent to 1 Ton of Pure $CaCO_3$ |
|---|---|---|---|
| Calcium carbonate | 100 | 100 | 2000 |
| Magnesium carbonate | 84 | 119 | 1680 |
| Calcium hydroxide | 74 | 135 | 1480 |
| Magnesium hydroxide | 58 | 172 | 1160 |
| Calcium oxide | 56 | 178 | 1120 |
| Magnesium oxide | 40 | 250 | 800 |

## Importance of Limestone Particle Size

With all factors being equal, the finer a limestone is ground, the more rapidly it will dissolve and the more thoroughly it can be mixed with the soil. The effectiveness of a ground substance of a given neutralizing power is determined not only by its rate of solubility but also by its contact with the colloidal particles. However, the finer the stone is ground, the greater will be its cost and the less will be its lasting qualities. Furthermore, a very finely ground limestone is not only difficult to handle but also unpleasant to distribute. Therefore, it is generally recommended that a ground limestone of medium fineness be purchased. Such a grade can be ground rather cheaply, and it contains a sufficient quantity of fine material to give immediate effects and a sufficient amount of coarse material to give it lasting qualities. Such a ground limestone would be one which would pass an 8-mesh screen and 25 to 50 percent would pass a 100-mesh screen. It can be seen from Fig. 8–11 that the fractions 8 to 20 mesh and larger are not very effective in increasing the pH of acid soils.

## Method and Time of Applying Lime

The principal requirement of any method of applying lime is that it should be distributed evenly and, except when applied to grass-

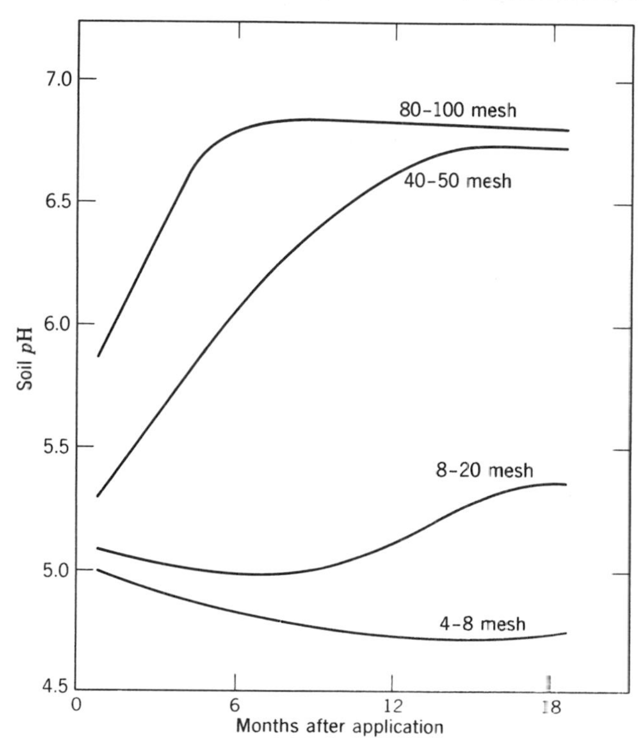


**Fig. 8-11** Effect of dolomitic lime of different particle size on the pH of the soil at various times after application. (From "Effect of Particle Size of Limestone on Soil Reaction, Exchangeable Cations, and Plant Growth," T. A. Meyer and G. W. Volk, *Soil Sci.*, Vol. **73**, pp. 37–52, 1952. Used by permission.)

lands, it should be thoroughly mixed with the soil. Lime even as it dissolves moves to no appreciable extent horizontally and only to a limited extent vertically. Movement is not sufficient to distribute the lime evenly over the field or to mix it thoroughly with the soil. Since soil acidity is due largely to the colloidal clay acids, it is essential that lime come in contact with all the soil particles so far as possible. This requires a thorough and even mixing of the lime with the soil. An even distribution may be accomplished by the use of most of the standard lime spreaders on the market. The only way of mixing the lime thoroughly with the soil is by tillage operations.

Generally, lime may be applied any time during the year when it is most convenient. Often, however, the type of the rotation, the system of farming, and the form of lime used will be the deciding factors.

If the caustic forms of lime like $CaO$ or $Ca(OH)_2$ are to be used for spring-seeded crops, it is usually best to apply the lime in the fall. These forms will change over to the carbonate form by seeding time and will lose their caustic properties and have no injurious effect on seed germination. Lime in the carbonate form may be applied any time without danger of injury.

It is advisable of course, to apply lime where it can be used to the best advantage in the rotation, for example, preceding the legume crop or in connection with a green-manure crop. It is usually best to apply lime considerably in advance of seeding legumes, in order that the lime will have time to correct the acid condition of the soil. To insure best results, lime should be in the soil at least six months prior to seeding legumes, although successful legume seedings sometimes are obtained by applying lime immediately preceding or with the legume seeding.

The data from lysimeter studies conducted in Ohio and cited in Table 4–1 showed that the annual loss of calcium in the drainage water could be equal to or greater than the calcium removed in some harvested crops. This shows that in the humid regions leaching may be as important as is crop production in the removal of exchangeable bases and development of soil acidity. Factors which affect leaching, such as soil permeability and slope, will influence the frequency of liming. In general it can be stated that several hundred pounds of lime per acre will be needed annually in the humid regions. Weathering can be expected to supply part of the need, but it is obvious that the maintenance of a pH near 6.5 will require liming once every five to ten years.

## Acidulation of Soils

The addition of acid sphagnum peat to soil may have some acidifying effect. However, significant and dependable increases in soil acidity are perhaps best achieved through the use of sulfur. The sulfur is slowly converted to sulfuric acid so the change in soil pH is gradually decreased over several months or a year. Earlier it was noted that 320 pounds of sulfur per acre furrow slice would theoretically increase exchangeable acidity or exchangeable hydrogen equal to one milliequivalent per 100 grams of soil. In a forest nursery near Orono, Canada, the use of sulfur resulted in significant increases in soil acidity and the growth of red pine seedlings. About 900 pounds of sulfur per acre lowered soil pH one unit in the experiment.[2]

As with lime, the amount of sulfur required varies with the cation

[2] For additional information see "Soil Acidulation with Sulfur in a Forest Tree Nursery," R. E. Mullin, *Sulfur Journal*, Vol. **5**:2–3, 1969.

exchange capacity of the soil. Recommendations for using sulfur on small areas are given in Table 8–6. The effect of nitrogen fertilizer on soil acidity is discussed in Chapter 12.

**Table 8–6** Suggested Applications of Ordinary Powdered Sulfur to Reduce the pH of 100 Square Feet of an 8-inch Layer of Sand or Loam Soil

| | Pints of Powdered Sulfur, for Desired pH | | | | | | | | | |
|---|---|---|---|---|---|---|---|---|---|---|
| | 4.5 | | 5.0 | | 5.5 | | 6.0 | | 6.5 | |
| Original pH | Sand | Loam | Sand | Loam | Sand | Loam | Sand | Loam | Sand | Loam |
| 5.0 | 2/3 | 2 | | | | | | | | |
| 5.5 | 1 1/3 | 4 | 2/3 | 2 | | | | | | |
| 6.0 | 2 | 5 1/2 | 1 1/3 | 4 | 2/3 | 2 | | | | |
| 6.5 | 2 1/2 | 8 | 2 | 5 1/2 | 1 1/3 | 4 | 2/3 | 2 | | |
| 7.0 | 3 | 10 | 2 1/2 | 8 | 2 | 5 1/2 | 1 1/3 | 4 | 2/3 | 2 |

Taken from *Soil*, the 1957 Yearbook of Agriculture, p. 678, "Home Gardens and Lawns" by C. E. Kellogg.

## Managing the pH of Calcareous Soils

Many millions of acres of soil are calcareous in arid regions, on flood plains, and on recently drained lake plains. Some of these soils are excellent for agriculture like those located in the Palouse Country in Washington, the flood plains of the Red, Platte and Mississippi Rivers, and on the lake plains around the Great Lakes. Plants growing on calcareous soils are sometimes deficient in iron, manganese, zinc, copper, or boron. To appreciably lower the pH of calcareous soils, the calcium carbonate must be leached out and this is impractical. As a result, crops are fertilized with appropriate nutrients that are deficient or crops are selected that are adapted to the alkaline soils.

Plants show considerable differences in their ability to utilize certain nutrients in calcareous soils. Sorghum may be iron deficient, but alfalfa will not be iron deficient when grown on the same calcareous soil. Varieties of the same species may exhibit remarkable differences to tolerate toxicity concentrations of some nutrients or to remove certain nutrients from the soil. For example, the Hawkeye soybean is not susceptible to iron deficiency, but the variety PL-54619-5-1 variety is susceptible to iron deficiency. Since the pH of calcareous soils cannot be practically altered, crop selection and fertilization with deficient nutrients are commonly the only practical solutions for growing many crops on calcareous soils.

# 9

# Soil Genesis and the Soil Survey

About 8000 soil series have been described and mapped in the United States. On a particular farm there may be more than half a dozen kinds of soil series. Even though we have many different soils, a few basic processes occur in the development of all soils. These processes proceed at different rates and in different ways to produce the various types. The five groups of factors responsible for the kind, rate, and extent of soil development are: *climate, vegetation, parent material, topography,* and *time.* We need to know how these factors influence soil development in order to understand why soils differ, why soils vary in their productivity, and ultimately how they may be properly used. This understanding is also necessary for the efficient mapping of soils showing their geographic distribution. In studying the influence of variations in each factor on soil properties we shall need to assume that the other factors are constant in the situations cited.

## HORIZON DIFFERENTIATION AND SOIL GENESIS

Soil as defined in Chapter 1 included mixtures of organic and mineral matter which were used for growing plants in the greenhouse as well as the soil body on the landscape which has a profile. The discussion in this chapter is restricted to soils that have experienced some degree of genetic evolution and, consequently, have horizons. The presence of horizons in all genetically developed soils suggests that certain processes are common to the development of all soils and therefore each kind of soil is not the product of a distinctly different set of processes. We will next consider the major kinds of horizons found in soils, and will follow this with a discussion of the processes responsible for horizon differentiation.

## Master Horizons or Layers

Broadly speaking, all soil profiles contain two or more master horizons. Short descriptions of these horizons follow:[1]

O—Organic horizons of mineral soils. Horizons: (1) formed or forming in the upper part of mineral soils above the mineral part; (2) dominated by fresh or partly decomposed organic material; and (3) containing more than 30 percent organic matter if the mineral fraction is more than 50 percent clay, or more than 20 percent organic matter if the mineral fraction has no clay. Intermediate clay content requires proportional organic-matter content.

A—Mineral horizons consisting of: (1) horizons where organic matter is accumulating or forming at or adjacent to the surface; (2) horizons that have lost clay, iron, or aluminum with resultant concentration of quartz or other resistant minerals of sand or silt size; or (3) horizons dominated by 1 or 2 above but transitional to an underlying B or C.

B—Horizons in which the dominant feature or features is one or more of the following: (1) an illuvial concentration (moved in from some other horizon such as A horizon) of silicate clay, iron, aluminum, or humus, alone or in combination; (2) a residual concentration of sesquioxides or silicate clays, alone or mixed, that has formed by means other than solution and removal of carbonates or more soluble salts; (3) coatings of sesquioxides adequate to give conspicuously darker, stronger, or redder colors than overlying and underlying horizons; or (4) an alteration of material from its original condition that obliterates original rock structure, that forms silicate clays, liberates oxides, or both, and that forms granular, blocky, or prismatic structure if textures are such that volume changes accompany changes in moisture.

C—A mineral horizon or layer, excluding bedrock, that is either like or unlike the material from which the solum is presumed to have formed, only slightly affected by pedogenic processes, and lacking properties diagnostic of A or B.

[1] From the *U.S. Department of Agriculture Handbook* 18, 1951. Supplement issued September 1962.

R—Underlying consolidated bedrock, such as granite, sandstone, or limestone.

## Master Horizon Subdivisions

The master horizons usually consist of one or more subdivisions which are indicated by numbers. For example, the O1 refers to the undecomposed litter of twigs and plant leaves. In the O2 only partially decomposed organic matter is found (Fig. 9–1).

That layer in which mineral particles predominate and the dark

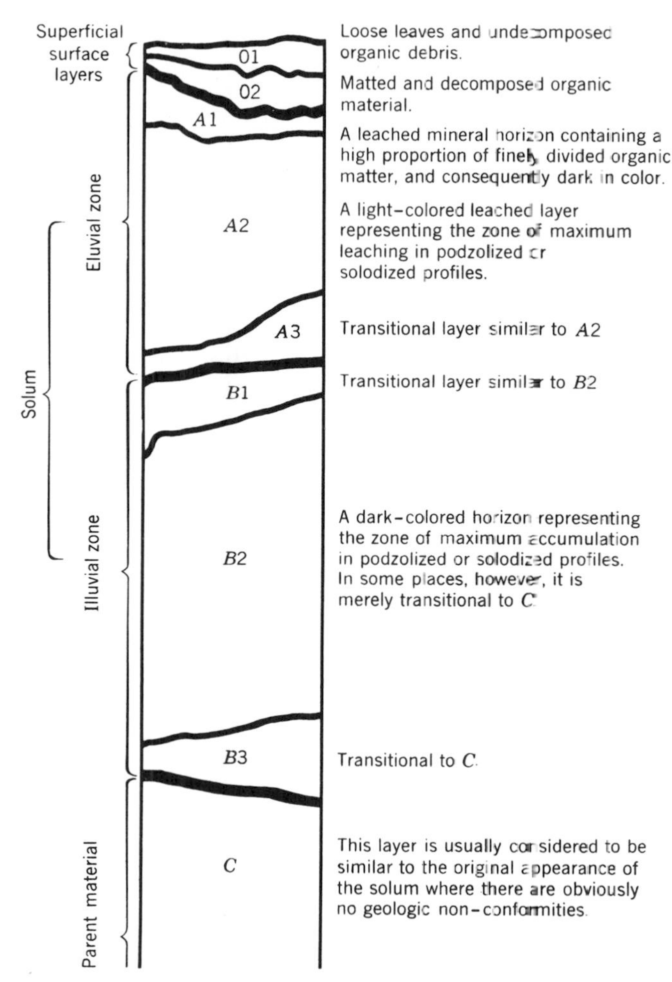


**Fig. 9–1** A generalized profile of a timbered soil developed in a humid climate with moderate temperature. The thickness of the various horizons varies as indicated.

color of organic residues persists is designated the A1 horizon. The A2 layer, on the contrary, is comparatively light in color and shows the maximum effects of leaching or *eluviation* (washing out).

The horizon of accumulation or *illuviation* (washing in) is technically designated the B horizon. It is often subdivided into the B1, B2, and B3 sections, depending on the degree of accumulation in evidence. These terms and their relationships are presented in Fig. 9–1. Notice that horizons A and B together constitute the *solum*.

As previously stated, all horizons are not present in every soil. Figure 9–2*b* shows a profile in which the light-colored A2 horizon is not evident. Limited leaching of the soil because of limited rainfall and grass vegetation are two factors that contribute to the development of this type of profile.

Sometimes neither the A2 nor the B horizon is discernible, and a profile similar to the one in Fig. 9–2*c* is produced. A high water table which has limited the activity of weathering agencies, limited rainfall, and a comparatively short period of activity of soil-developing processes are some of the conditions that give rise to this type of profile.

Additional features of the horizons are indicated with the use of uncapitalized letters. For example, cultivated fields may no longer have the original upper layer intact, and possibly the O horizons as well as the A1 and part of the A2 have been mixed together to form a plow layer. This layer is designated Ap to indicate disturbance by cultivation or pasturing. An accumulation zone in the B horizon may be high in clay or iron oxide. The zone of maximum clay in a soil is commonly labeled a B2t, the t indicating illuvial clay. Spodosols (Podzols) commonly have a Bir horizon indicating illuvial iron. Where both iron oxide and humus have accumulated in the same horizon, as in some Spodosols (Podzols), it is called a Bhir horizon. Illuvial humus is indicated by the "symbol" h. Other characteristics are indicated as follows:

b—Buried soil horizon
ca—An accumulation of carbonates of alkaline earths, commonly of calcium
cs—An accumulation of calcium sulfate
cn—An accumulation of concretions
f—Frozen soil
g—Strong gleying
m—Strong cementation, induration
sa—An accumulation of salts more soluble than calcium sulfate
si—Cementation by siliceous material, soluble in alkali
x—Fragipan character (firmness, brittleness, and high density layer)

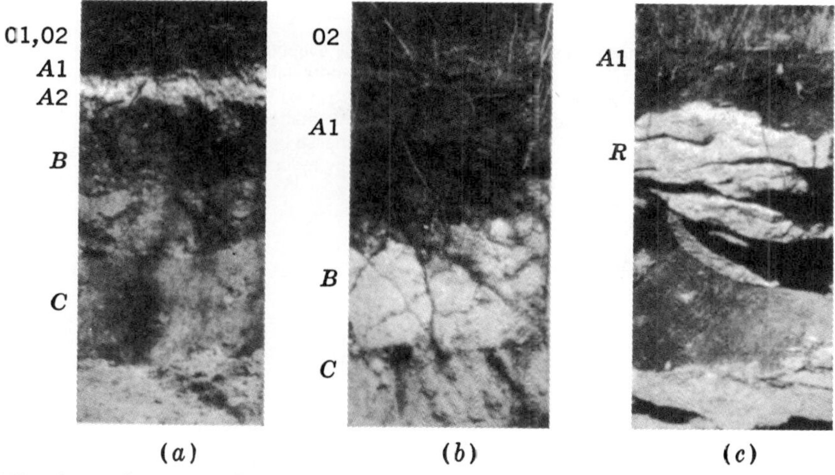


**Fig. 9–2** Compare the thin A1 and the strongly developed A2 horizons of the Spodosol (Podzol) profile on the left with the deep A1 and the absence of an A2 horizon in the Mollisol (Chernozem) profile in the center. Profile C shows the profile of a soil developing from limestone. Note the absence of a B horizon. *(Profiles B and C by courtesy of the late Dr. Harper of the Oklahoma Experiment Station.)*

The O1 horizon in forest soils is often referred to as the L layer or litter layer and the O2 horizon as the F layer if the structure of the organic matter is evident, or the H layer if the organic matter is amorphous.

## Processes of Horizon Differentiation

Soil genesis or horizon differentiation includes processes that can be viewed as *additions, losses, transformations,* or *translocations.* Plants and animals find a habitat in all soils and become a part of the organic fraction. Carbon in organic matter is lost from the soil as carbon dioxide resulting from microbial decomposition. Nitrogen is transformed from the organic to inorganic forms. Further, organic matter is subject to translocation from place to place in the soil by means of water and animal activity.

Mineral constituents undergo changes that can be similarly considered. In all soils minerals weather with the simultaneous formation of secondary minerals and other compounds of varying solubility which may be moved from one horizon to another. A summary of these processes is presented in Fig. 9–3. The great diversity of soils in the world results not from the operation of many distinctly different processes, but rather from variations in the intensity and length of time the processes have operated.

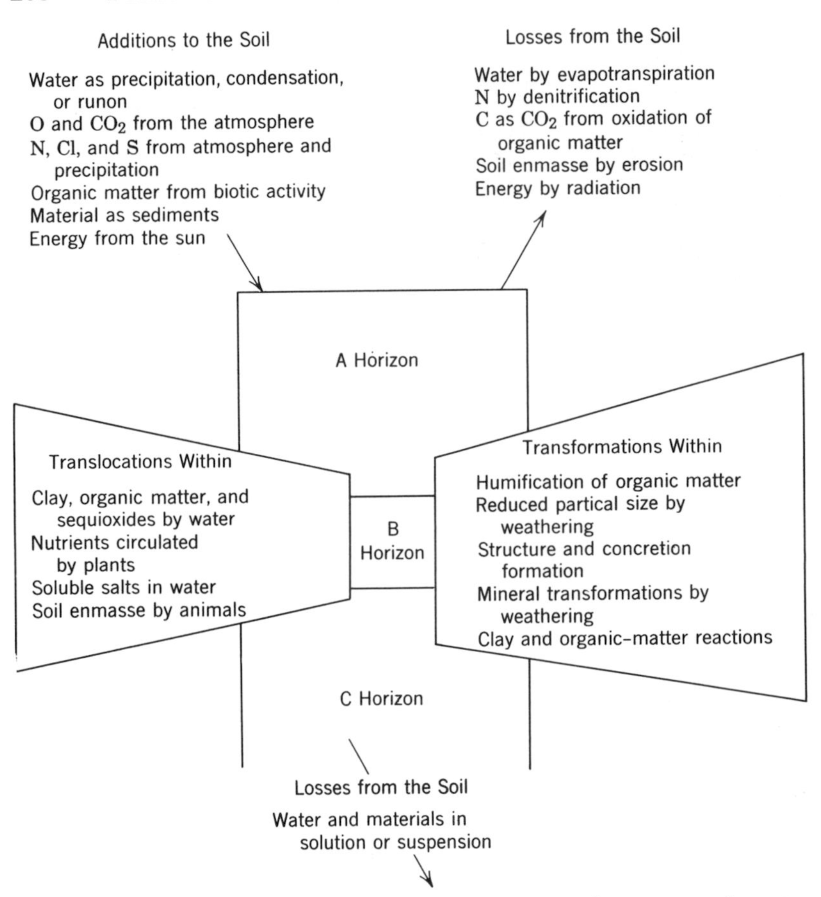


**Fig. 9–3** Diagrammatic presentation of additions, losses, translocations, and transformations involved in horizon differentiation.

## SOIL DEVELOPMENT IN RELATION TO TIME

Soils are constantly undergoing change. The changes take place slowly, and many people hastily conclude that none occur. In Chapter 1 reference was made to the life cycle of soils. The life cycle includes the stages of parent material, immature soil, mature soil, and old soil. A discussion of these stages and the amount of time required for soil development follow.

### Major Stages in Soil Development

The parent material may be transformed into an *immature* or young soil in a relatively short period of time if conditions are favorable.

This stage is characterized by organic-matter accumulation in the surface soil and by little weathering, leaching, or translocation of colloids. Only the A and C horizons are present and soil properties to a large extent have been inherited from the parent material. The mature stage is attained with the development of the B horizon. Eventually, if sufficient time has elapsed, the mature soil may become highly differentiated so that large differences exist in the properties of the A and B horizons. This is the *old-age stage*. Many clay-pan soils are characteristic of those in the old-age group and they have low fertility and productivity. Highest natural productivity is found in the mature and immature soils.

A summary of the stages in the development of soils in central United States in unconsolidated, medium-texture material and under the influence of prairie vegetation is shown in Fig. 9–4. Under these conditions, development proceeds from parent material, to Entisol (Regosol) (immature), to Mollisol (Brunizem) (mature), to Alfisol (Planosol) (old age).

Mohr and van Baren[2] have recognized five stages in the development of tropical soils.

1. Initial stage—the unweathered parent material.
2. Juvenile stage—weathering has started, but much of the original material is still unweathered.
3. Virile stage—easily weatherable minerals have largely decomposed; clay content has increased and a certain mellowness is discernible.
4. Senile stage—decomposition arrives at a final stage, and only the most resistant minerals have survived.
5. Final stage—soil development has been completed and the soil is weathered out under the prevailing conditions.

The names used to refer to the stages are very descriptive. For instance, virile, referring to the stage at which the capacity of the soil to support vegetation is at a maximum.

### Amount of Time Required for Soil Development

A question which has many interesting aspects is, "How much time is required to form an inch of soil or for a soil to develop?" For development from hard rock, the time may be very great. On the other hand, development can proceed rapidly in permeable, unconsolidated material in a warm and humid climate. Plant growth can occur on

[2] For a more detailed discussion see E. C. J. Mohr, and F. A. van Baren, *Tropical Soils*, Interscience Publishers, New York, 1954.

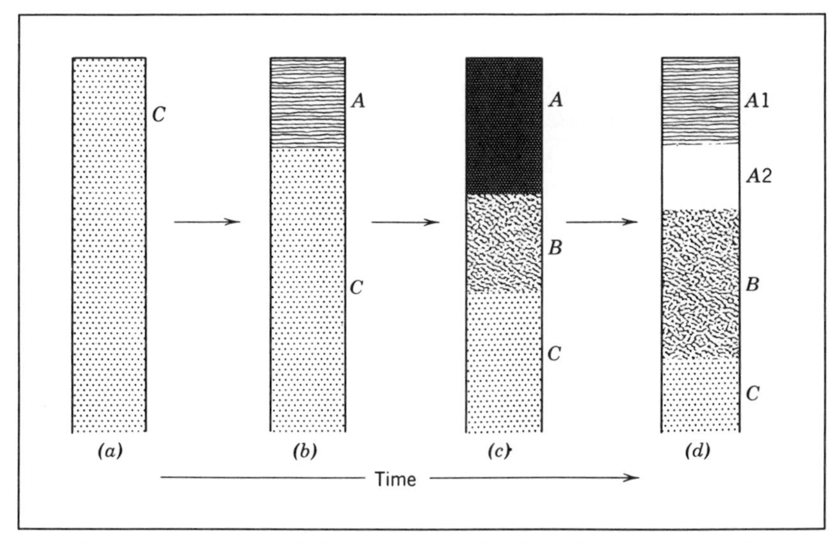


**Fig. 9–4** A summary of the stages in the development of soils in central United States under tall-grass vegetation.

| (*a*)<br>Parent Material | (*b*)<br>Young Soil<br>(Regosol)<br>Entisol | (*c*)<br>Mature Soil<br>(Brunizem)<br>Mollisol | (*d*)<br>Old Soil<br>(Planosol)<br>Alfisol |
|---|---|---|---|
| Original material before soil development begins. | Thin, solum, organic-matter accumulation in A horizon, from which carbonates have been leached. Minimal weathering and eluviation. | Organic-matter content is at a maximum. Has moderate clay accumulation in the B horizon and the solum is acid. Stage of maximum productivity for corn. | Very acid in reaction, severely weathered and has less organic matter than mature stage. Clay accumulation in B horizon has formed a clay pan. An A2 horizon exists. |

freshly exposed parent material, so soil development need not precede plant growth. This is readily seen in areas where a plant cover is established on freshly exposed road cuts along the highway. The answer to the question, therefore, lies partly in knowledge of the nature of the material from which the soil develops.

From loessial, glacial, volcanic, and other unconsolidated deposits,

Entisols (Regosols) can develop in less than 100 years. Mature Spodosol soils (Podzols) which are about 1000 years old have been found in Alaska. On the Kamenetz fortress in the Ukraine, the modern soil is 4 to 16 inches thick and has developed from limestone slabs since the fortress was abandoned in 1699. On the late Wisconsin glacial materials, which are about 10,000 years old, most of the soils are in the mature stage and clay-pan soils (Planosols) are rarely found.

Aridity and the rapid removal of soil by erosion on steep slopes can delay or prevent the development of mature soils. It becomes clear then that the rate of development varies greatly from one soil to another. A given period of time may produce much change in one soil and little in another. For this reason the maturity of the soil is expressed in the degree of horizon development rather than number of years. Conditions that hasten the rate of soil development are: warm, humid climate; forest vegetation; permeable, unconsolidated material low in lime content; and flat or depressional topography with good drainage. Factors which tend to retard development are cold, dry climate; grass vegetation; impermeable, consolidated material high in lime; and steeply sloping topography.

## Rate of Soil Development

As is typical of many processes in nature, the rate of soil development as a whole varies over time as do many of the individual processes. First, it can be mentioned that the characteristics of a soil change most rapidly when the soil is young and that detectable changes occur more slowly with age. As stated earlier, Spodosols (Podzols) are mature soils and were observed on material deposited about 1000 years ago in Alaska. These soils are very similar to Spodosols (Podzols) in other parts of the world that are many times older in terms of years but are not older in terms of degree of development.

Secondly, the individual processes vary in intensity over time. Changes in the organic-matter content of a soil can be separated into three phases. In young soils, the organic matter content is increasing rapidly because the rate of addition exceeds the rate of decomposition. Maturity is characterized by a constant organic-matter content as additions are counterbalanced by losses. Old age is characterized by a declining organic-matter content indicating that the rate of addition is waning as the soil becomes more weathered. The fertility declines and the reduced rate of organic-matter production allows decomposition to exceed the rate of addition.

Another illustration of the rate of soil development is silicate clay formation. A youthful soil which has a low clay content and a high content of primary minerals might be characterized by a high rate of clay formation. In a mature or old soil in which most of the primary minerals have already been weathered, silicate clay formation will necessarily be low. The high clay content, however, encourages a relatively high rate of clay decomposition. Thus, it is seen that some of the processes are more operative in youthful soils whereas others are more operative in old soils.

## SOIL DEVELOPMENT IN RELATION TO CLIMATE

Important climatic influences that affect soil development are precipitation and temperature. The climate also influences soil development indirectly in determining the natural vegetation. It is not surprising that there are many parallels in the distribution of climate, vegetation, and soil on the earth's surface.

### Climate a Factor in the Organic-Matter Content of Soils

The quantity of organic matter in a soil represents the balance between addition and decomposition. Accordingly, climatic factors which affect the quantity of organic material developed in or returned to the soil and the activity of decay organisms have a bearing on the amount accumulated. Studies have shown that when average annual temperatures increase and the moisture and other relations remain constant, the quantity of organic matter decreases in soils of similar characteristics and covered by the same type of vegetation (Fig. 9–5). The decrease is somewhat greater in grassland soils than in forest soils. Although this relationship exists in the continental United States, it cannot be validly extrapolated to the equatorial regions. Many soils in the humid tropics contain abundant quantities of organic matter.

On the contrary, an increase in moisture supply with temperatures remaining constant results in an increase in organic content in soils of similar characteristics and vegetative cover. Again the change is more pronounced in grassland than in timbered soils (Fig. 9–6).

In view of the long growing season and high rainfall of the southeastern states, a combination which results in a large amount of plant growth, one might expect the organic content of the soils to be correspondingly high. Such is not the case, however, because the long, warm, moist seasons are also favorable for decomposition, with the result that corresponding soils in the North are higher in organic matter than in the South (see also Fig. 6–2).

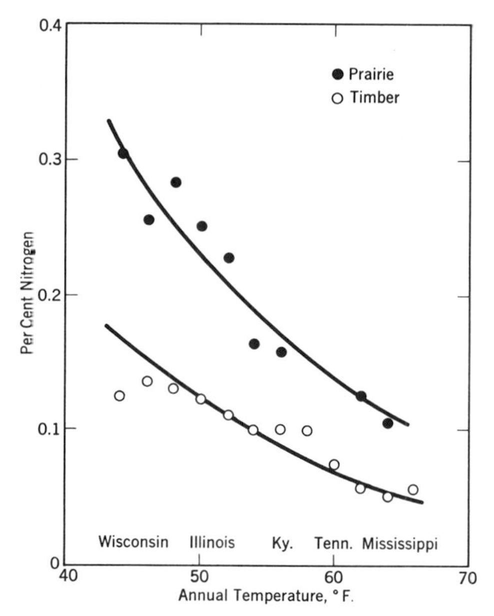


**Fig. 9–5** Nitrogen-temperature relation in humid prairie (upper curve) and humid timber soils (lower curve) for silt loams. (*Missouri Res. Bull.*, **152**.) These curves show that nitrogen and organic-matter content of soils under similar moisture conditions decrease from north to south, and that under similar conditions the prairie soils are considerably higher in nitrogen and organic matter than the timber soils.

Indirectly, precipitation has another influence – if through leaching it causes the pH of the surface layers to become about 4.5 or less. Under these conditions, microbial decomposition of organic matter may become so restricted that the litter added to the surface of the soil accumulates rather than decomposes. This accounts in large part for the existence of prominent O horizons in some forest soils.

## Weathering and Clay Formation as Influenced by Climate

Mineral weathering occurs through physical and chemical reactions whose rates are influenced by temperature. All other things being equal, an increase in temperature causes an increased rate of weathering and clay formation. The rate of weathering is also related to precipitation as the presence of water enhances the weathering reaction. High average temperature and precipitation tend to encourage rapid weathering and clay formation. The conditions that result in a minimum degree of weathering are found where the climate is warm and dry, cold and dry, or cold and moist.

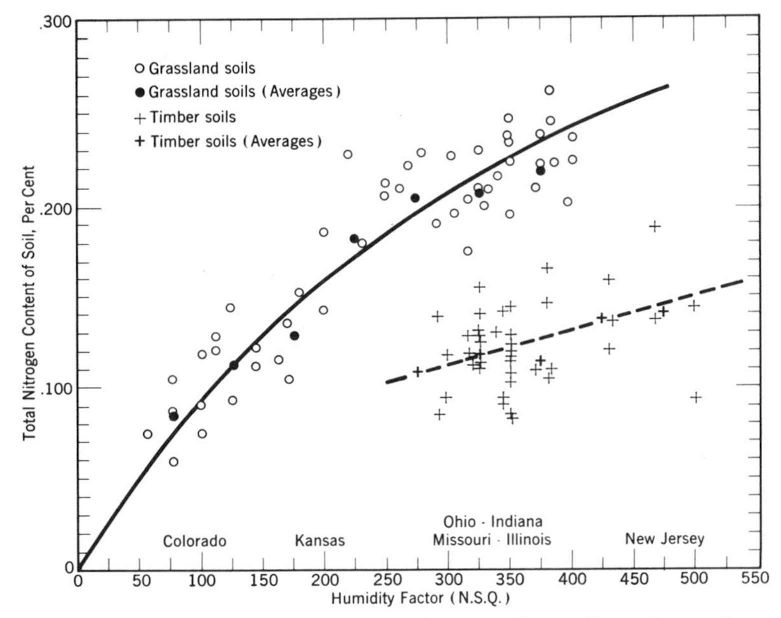


**Fig. 9–6** Soil nitrogen-humidity factor relationship along the annual isotherm of 11° C. (*Missouri Res. Bull.*, **152.**) It is observed that with an increase in humidity, with all other factors constant, including the mean annual temperature, there is an increase in content of soil nitrogen in both the grassland and timber soils. N.S.Q. is the ratio of precipitation to the absolute saturation deficit of the air. These values are used instead of precipitation values as such since rainfall alone is not a satisfactory index of soil-moisture conditions because of the great variations in evaporation. The N.S. quotients include therefore the effect not only of temperature but also of air humidity on evaporation.

## Chemical Properties

It has been pointed out that an increase in precipitation is commonly associated with an increase in organic matter and clay content. Since the cation-exchange capacity is directly related to the amount of these two fractions, it also increases with precipitation.

Where the annual precipitation is small and evaporation is rapid, insufficient water moves through the soil to leach away the exchangeable bases. This is the general case in central United States, where the annual precipitation is less than 26 inches and the soils are about 100 per cent saturated. In the more humid areas, the downward movement of water leaches away exchangeable bases. The bases are replaced by hydrogen from the water, and this results in a decrease in the percentage of base saturation. In the humid regions, soils in

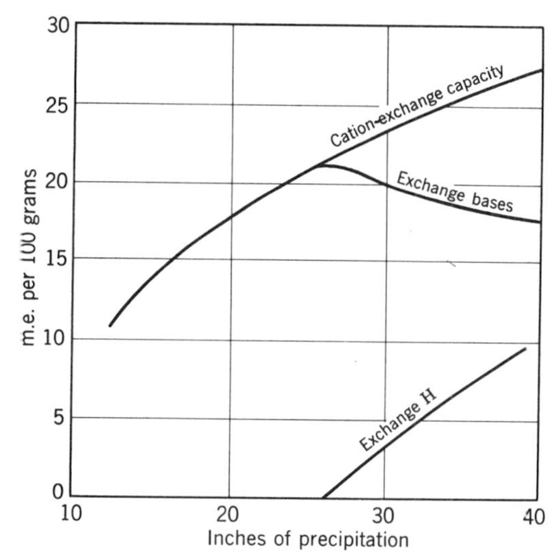


**Fig. 9–7** The relationships between annual precipitation and cation-exchange capacity, exchangeable bases and hydrogen in soils of central United States. (From "Functional Relationships between Soil Properties and Rainfall," Hans Jenny and C. D. Leonard, *Soil Sci.*, Vol. **38**, p. 375, 1934. Courtesy of Williams and Wilkins Co.)

their natural state tend to be acid in reaction, and the extent of acidity is related to the amount of effective precipitation. The relationships between annual precipitation and cation-exchange capacity, exchangeable bases and exchangeable hydrogen are shown in Fig. 9–7.

### Type of Clay Minerals

Several generalizations concerning the type of clay in soils and climate can be made. In the soils of the northern part of the United States, the clay fraction is generally dominated by 2:1 silicate clay minerals (illite, montmorillonite). Kaolinite or 1:1 lattice silicate clays and oxides of iron or aluminum are more common in the soils of southeastern United States and in many tropical areas. In the humid tropics, intense weathering can result in an almost complete loss of silica, and the clay fraction of the soil will then be high in iron and aluminum oxides. There are many exceptions to these generalizations as they only apply to the clay formed in the soil.

## SOIL DEVELOPMENT IN RELATION TO VEGETATION

Natural vegetation may be divided, very broadly speaking, into the two general classes of trees and grass, and the soils supporting them are termed forest soils and grassland soils, respectively. There are several characteristics in soils developed in association with grass which are of considerable agricultural significance. The different effect of each kind of vegetation on the soil supporting it is brought out in the following discussion.

### Amount and Distribution of Organic Matter in Soil Profile

In Chapter 6 the differences in the amount and distribution of organic matter in grassland and forest ecosystems were compared (see Fig. 6–3). Considering the discussion in Chapter 6, it is sufficient now to restate that under comparable environmental conditions that grassland soil profiles contain more organic matter which is more uniformily distributed with depth than do forest soil profiles as shown in Fig. 9–8.

### Differences in Nutrient Cycle

Plants absorb nutrients from the soil and transport the nutrients to the tops of the plants. When the tops die and fall onto the soil surface, decomposition of the organic matter releases the nutrients in a self-fertilizing "do-it-yourself manner." Bases returned to the soil surface in this manner retard the loss of exchangeable bases by leaching and retard the development of soil acidity. Wide differences in the uptake of ions and consequently in the chemical composition of plant tissues

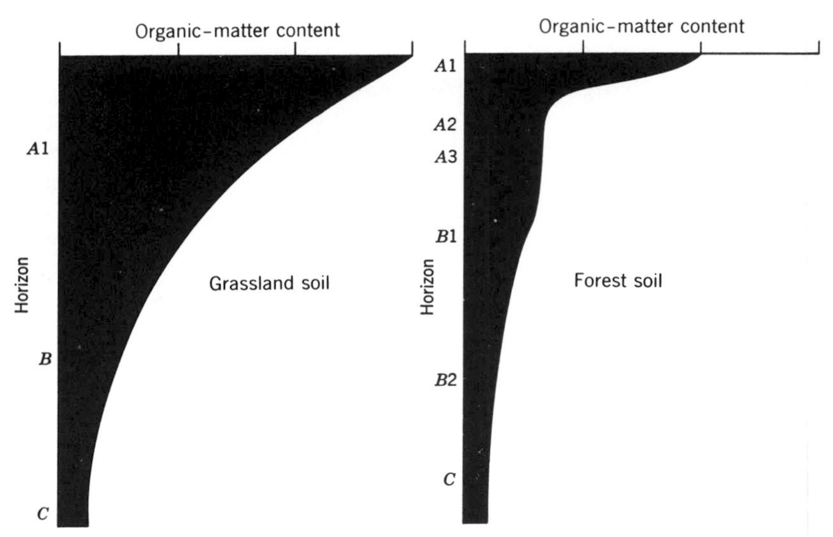


**Fig. 9–8** Grassland soil profiles contain about twice as much organic matter which is more uniformly distributed through the profile than do forest soils under similar environmental conditions.

have been well substantiated. Even between tree species there are large differences and this plays a role in soil development. Species that normally absorb large quantities of alkaline earths and alkali metals will delay the development of soil acidity because of the large amount of bases returned to the surface of the soil in vegetative residues. The data in Table 9–1 confirm the fact that hardwoods maintain a higher pH and percentage base saturation than spruce when grown on parent material with the same mineralogical composition.

**Table 9–1** Effect of Tree Species on Soil pH and Base Saturation

| Forest Type | Horizons | pH | Percentage Base Saturation |
|---|---|---|---|
| Spruce | O2 | 3.45 | 13 |
| | A2 | 4.60 | 20 |
| | B1 | 4.75 | 27 |
| | B2 | 4.95 | 27 |
| | C1 | 5.05 | 23 |
| Hardwood | O2 | 5.56 | 72 |
| | A1 | 5.05 | 47 |
| | B1 | 5.14 | 36 |
| | B2 | 5.24 | 34 |
| | C1 | 5.32 | 34 |

Adapted from "Soil Classification and The Genetic Factors of Soil Formation," R. J. Muckenhirn, et al., *Soil Sci.*, **67**:100, 1949.

## Rate of Eluviation and Leaching

Under the same climatic conditions, where both forest and grasslands exist side by side and have comparable parent material and slope, the forest soils will show evidence of greater eluviation and leaching. Three possible causes for this difference have been offered.[3] First, the forest vegetation returns fewer alkaline earths and alkali metals to the surface in vegetation each year. Second, water is intercepted for transpiration at a greater depth by trees so the water is more effective in leaching before it is absorbed by roots. Third, water entering the soil is more acid. Hydrogen ions dissolved from the organic acids in the O horizon, which is more prominently developed under trees, cause greater replacement and leaching of exchangeable bases.

Closely associated with the leaching of bases is the translocation

[3] See "Prairie Soils of the Upper Mississippi Valley," G. D. Smith, W. H. Allaway, and F. F. Riecken, *Advan. Agron.*, **2**:157–205, 1950.

of clay. That clay eluviates downward is evident from the higher clay content of the B horizon and from the occurrence of more pronounced clay coatings on the peds of the B horizon. Greater movement of clay in the forested soil is based on the higher clay content of the B horizon and the lower clay content of the A horizon of the forested soil as compared to the grassland soil. Thus, the permeability and other physical properties of the subsoil also exhibit a degree of difference.

Two important points stand out in summarizing the differences between forest and grassland soils. The forest soil has about half as much organic matter in the solum as does grassland soil, and it is less uniformly distributed vertically. The forest soil shows evidence of greater age or development. The horizons of the solum are more acid and have a lower percentage base saturation. Relatively more clay has been translocated from the A to B horizon. It can be seen that the differences are one of degree and not kind. This supports the view that the same basic processes have been operative in both. Eventually both kinds of soils can evolve into clay-pan soils and these modest but agriculturally important differences become less important. The soils may become strikingly similar in old age regardless of the vegetative cover under which they evolved.

## SOIL DEVELOPMENT IN RELATION TO PARENT MATERIAL

The nature of the parent material will have a decisive effect on the properties of young soils and may exert an influence on even the oldest soils. Where parent material is derived from consolidated rock, the formation of parent material and the soil may occur simultaneously. Properties of the parent material that exert a profound influence on soil development include texture, mineralogical composition, and degree of stratification.

### Consolidated Rock as a Source for Parent Material

Consolidated rock is not parent material, strictly speaking, but serves as a source for parent material. Soil formation may begin immediately after the deposition of volcanic ash, but must await the physical disintegration of hard rock where granite is exposed. During the early stages of soil formation, rock distintegration may limit the rate and depth of soil development. Where the rate of rock distintegration exceeds the rate of removal of material by erosion, productive soils with thick solums may develop from bedrock. This is the case in the Blue Grass Region of Kentucky where soils developed from limestone.

## Water-Deposited Sediments

Alluvial deposits are scattered in narrow, irregular strips bordering streams and rivers. A common characteristic of this material is its stratification, layers of different-sized particles overlying each other. Mineralogically, alluvium is related to the soils which served as a source of the material.

Most alluvium is carried and deposited during floods because it is at this period that erosion is most active and the carrying capacity of streams is at a maximum. When a flooding stream overflows its banks, its carrying power is suddenly reduced as the flow area increases and velocity decreases. This causes the coarse sands and gravels to settle along the bank, where they sometimes form conspicuous ridges called *natural levees*. As the water reaches the *flood plains* of the valley, the rate of flow is slow enough to permit the silt to settle. Finally the water is left in quiet pools, from which it seeps away or evaporates leaving the fine clay. Levees are characterized by good internal drainage during periods of low water, whereas flood plains exhibit poor internal drainage.

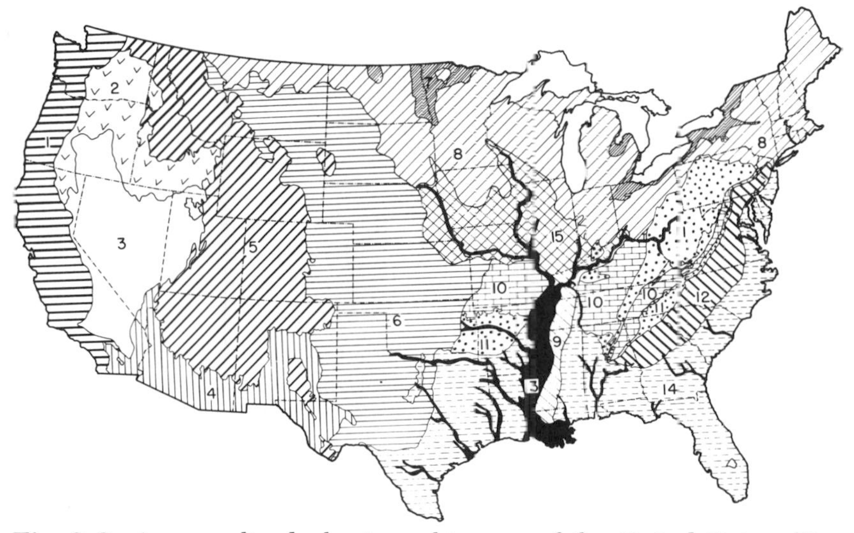


**Fig. 9–9** A generalized physiographic map of the United States. (Drawn from a map prepared by the Division of Soil Survey and presented in *USDA Bull.*, **96**.) Legend of Areas: 1. Pacific Coast region. 2. Northwest intermountain region. 3. Great Basin region. 4. Southwest arid region. 5. Rocky Mountain region. 6. Great Plains region. 7. Glacial lake and river terraces. 8. Glacial region. 9. Loessial deposits. 10. Limestone Valleys and Uplands. 11. Appalachian Mountains and Plateaus. 12. Piedmont Plateaus. 13. River flood plains. 14. Atlantic and Gulf coastal plains. 15. Loessial deposits over glacial material.

*Terraces* are developed from flood plains as streams cut deeper channels because of lowered outlets. Several terraces may be found along a stream or a lake which has undergone repeated changes in level. In glaciated areas extensive terraces were formed as the glaciers receded and their outwash plains were no longer covered with water. Terraces usually are quite well drained and may be droughty. They exhibit stratification.

Streams flowing from hills or mountains into dry valleys or basins drop their sediments in a fan-like deposit as the water spreads out. These *alluvial fans* are usually coarse-textured, being composed of sands and gravels, and are well drained.

Sediments not deposited as flood plains are carried to the lake, gulf, or other body of water into which a stream empties. The decrease in velocity at the stream's mouth together with the coagulating effect of the salt content of the receiving water body results in the deposition of much of the suspended material, thus producing a delta. These deposits are poorly drained, but, where drainage is provided, they constitute important crop-producing areas, as is evidenced by the deltas of the Nile, Po, Tigris, Euphrates, and the Mississippi.

Flood plains as well as deltas are in general rich in plant nutrients and comparatively high in organic-matter content. Terraces and alluvial fans, on the other hand, are more likely to be less fertile. Special crops such as vegetables and fruits frequently are grown on the latter formations because the soil warms up quickly and their good drainage and coarse texture permit free root development.

It is common to find *marine* deposits along the coastlines. This material was derived from sediments carried by streams and deposited in the ocean and gulf through decreased current velocity and chemical coagulation. Much of it is sandy but is interspersed with beds of silt and clay which were deposited in estuaries or other sheltered bodies of water or farther out in the ocean. When raised above water level these deposits were subjected to soil-forming processes.

## Glacial Materials

Ice was the transporting agent for much of the mantle of northern Europe, Asia, and North America. In the United States the Ohio and Missouri rivers form a general southern boundary for ice-carried material. As the great continental ice-sheets moved southward from their accumulation centers, they first followed and filled the great drainage valleys and then gradually spread out over the intervening upland and divides. As the ponderous ice-mass moved forward, it pushed before it and gathered within itself a large part of the unconsolidated surface layer. It also scooped up great rock fragments, which scraped at the rock floor over which they passed. Sharp corners and edges of even the hardest rocks were ground smooth by this abrasive action to form the rounded rocks and boulders that are characteristics of glaciated landscapes. Large quantities of weathered and unweathered rocks, varying in size from the fine rock powder to massive boulders, were thus incorporated into the ice and carried along in the glacier.

The movements of this continental ice-sheet depended on the changes in climatic conditions which took place during the glacial age. During mild periods the ice melted rapidly. In cold seasons melting ceased and the ice front would creep southward. Sometimes during extremely mild periods the ice would melt faster than it was pushed forward. This would lead to a rapid recession of the ice front, and all debris carried in the ice was, of course, dropped. Generally, after this type of recession, the land surface appeared as a rolling plain, called a *till plain* or *ground moraine.*

At certain times climatic conditions allowed the glacier to melt back just as fast as its rate of advance, and this process resulted in the front of the ice remaining at one place for some time. All debris carried by the ice was brought to the line of the stationary ice-front and there dumped as melting proceeded. This process resulted in the formation of ridges or a series of hills, called *terminal* and *recessional moraines.* Lateral moraines formed along the sides of the ice sheets (Fig. 9–10). Moraines are usually composed of an unassorted, heterogeneous mass

**Fig. 9–10** The Bierstadt moraine is a lateral moraine in the Rocky Mountain National Park. Valley glaciers from the mountains to the left deposited the debris along their margins as the ice melted.

of boulders, rocks, sand, silt, and clay, briefly called *till,* but in places a water-sorting also occurred. As would be expected, the proportions of these materials vary greatly, and hence some moraines are relatively high in sand content whereas others contain a large proportion of fine particles. Not only does the material vary in texture, but the shape of the surface is also variable.

As the ice melted, giving rise to moraines, great volumes of water rushed away. These waters carried quantities of sediment, the coarser of which was deposited as the current diminished. These coarse-textured, comparatively level deposits are known as *outwash plains* (Fig. 9–11). Most of the finer silt and clay were carried into slowly moving water or lake basins, where they settled out to form lake bed or *lacustrine* plains. In Michigan there is evidence of as many lakes which have become extinct as there are lakes in existence, whereas along the Great Lakes extensive areas have been exposed through the disappearance of ice barriers, lowering of the outlet at Niagara Falls, and tilting of the land surface. The disappearance of Lake Agassiz has laid bare a great land surface in Minnesota and the Dakotas, and the basin of old Lake Bonneville occupies an immense area in Utah.

The glaciers also produced *kames* and *eskers,* but these formations are only of local significance. Minor readvances of the ice-sheet resulted in the modification of these glacial formations in many places. Sometimes till was relaid over a lake plain, or sandy and gravelly

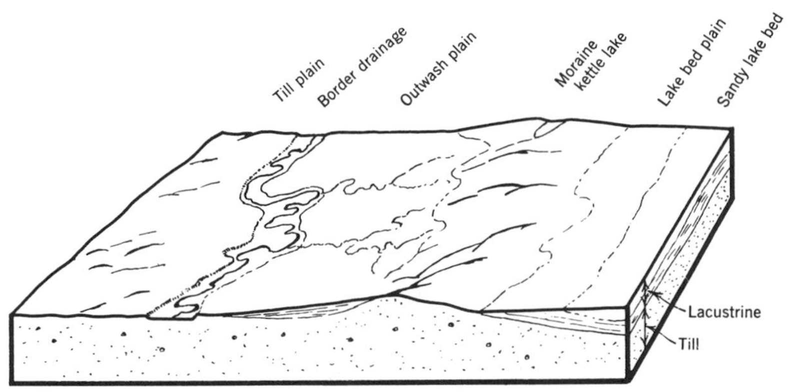


**Fig. 9–11** A generalized view of physiographic features in a glaciated area. On the left is a till plain separated from the moraine by an outwash plain and a border-drainage way developed while the ice front was at the moraine line and melting as rapidly as it advanced to deposit the morainic material. To the right a glacial lake has receded to leave a plain of heavy sediment partially covered with lacustrine sands (deposit nearer the water's edge).

outwash was pushed up to form gravelly moraines. Major readvances "streamlined" glacial material in some parts of the country to form *drumlins*. It is customary to designate all the material deposited by glaciers and their melt waters as *glacial drift*. In general, the effect of glaciation was to scrape off and smooth the tops of high land forms while filling valleys and depressions. Thus continental glaciation decreased the local relief intervals and flattened topography.

Glaciation results not only from continental glaciers, but also from numerous piedmont and valley glaciers (Fig. 9–12) which are currently active thoughout the world today.

## Wind-Transported or Aeolian Material

There are three classes of wind-moved soil material: (1) sand of variable fineness which may be collected into low swells or steep ridges, like the dunes occurring on the leeward side of large bodies of water and on sandy deserts; (2) volcanic ash; and (3) silt-like material, called *loess*, which occupies large areas in the United States, Europe, and Asia.

*Dune sand* is of little agricultural value, although crops are pro-

**Fig. 9–12** The origin of a glacier. Notice the parent snow fields in the background and the small tributary glaciers. (Photograph by Bradford Washburn.)

duced on it to a limited extent in humid regions. At times dunes are a hazard to agriculture, for in their movement they sometimes cover good land.

*Loess* was deposited in central United States after the recession of the ice-sheet. This material was derived in part from sediments deposited by huge rivers which were fed by the melting continental glaciers in a broad belt, even beyond the southern limits of glaciation. A period of aridity after the recession of the glaciers with strong westerly winds set the stage for the transportation of this wind-blown material to its present resting place, as shown in Fig. 9–13. The great thickness of the deposits of loess on the east and northeast banks of the Mississippi and Missouri rivers is one of the facts which have led to this explanation of the accumulation of the material. Glaciers from the Rockies probably supplied the sediments making up the western part of the loessial deposits. Deserts are also sources of loess. Extensive deposits of loess are also found along the Rhine and its tributaries and over a large part of the immense valley of Hwang Ho. Other deposits occur in southern Russia, several Balkan countries, and northern France, Belgium, and Poland and the Argentine pampa.

Loess is composed largely of silt and is commonly grayish yellow or buff. This wind-blown material stands in almost vertical walls so that gullies and streams cutting through it have very steep banks. Its content of mineral plant nutrients was originally high, as was the

**Fig. 9–13** Loess was being deposited when this picture was taken. During this Colorado dust storm total darkness lasted for half an hour. (Courtesy of Soil Conservation Service.)

quantity of calcium compounds. Loess is one of the most uniform of soil parent materials, but even it varies considerably in texture and mineralogical composition. Soils developed from these deposits are frequently referred to as fertile; however, no single material always gives rise to productive soil, as the parent material is only one of the factors involved in soil formation.

### Nature and Source of Organic Soil Parent Material

In locations where considerable quantities of plant material grow and decay is limited because of much water or low temperatures, a large accumulation of partially decayed vegetable matter gradually develops (Fig. 9–14). Such deposits are of wide occurrence and are not restricted to any given climatic zone. They are found in Europe, Asia, Africa, Canada, South America, and the United States, and various other places including the tropics. It may be said, however, that accumulations of this nature are more common in northern latitudes and occupy a larger percentage of the land surface in Norway, Sweden, Ireland, Scotland, northern Germany, Russia, and Holland than in countries lying farther south. In the Tundra region organic deposits are of frequent and extensive occurrence.

Even when a rank growth of vegetation occurs each year and remains on the soil, peat does not accumulate unless decay processes are very slow. The most common factor which limits decay is an excess of water. Accordingly in moderate to warm climates peat ac-

**Fig. 9–14** A lake which is rapidly being filled with vegetation and will soon become a peat swamp.

cumulates in shallow lakes and swamps. The topography, resulting from glaciation, in Minnesota, Wisconsin, and Michigan, and to a lesser extent in Maine, New York, and New Jersey, has given rise to innumerable small lakes and swamps. Thus conditions have been suitable for peat accumulation. Along the Atlantic and Gulf coasts many swamps have developed because of the slight elevation above sea level. Peat has accumulated in many of these areas.

### Texture of Parent Material and Soil Properties

The textures of transported materials are related to their origin, and they may have great variability. Glacial and water-laid deposits range from sands to silty clays. Loessial deposits are high in silt and many soils developed in loess have silt loam A horizons. When the parent material is consolidated rock, the texture (size of mineral grains) of the rock becomes an important factor.

Granite and rhyolite are igneous rocks that have the same chemical composition. Rhyolite has the finer texture, or smaller mineral-grain size, because during formation it was subjected to more rapid cooling. This causes the rhyolite to weather more slowly and results in a finer-textured soil than that developed from granite. A similar comparison can be made between basalt and gabbro in young soils, but the textures of the old soils developed from these two materials are similar because all the minerals are weatherable.

Since the minerals in basalt weather more easily than those in granite, the finer-textured soil will develop from the basalt. Large areas of deep, clay-textured soils have developed from basalt in India and Australia. The complete weathering of minerals in basalt in humid tropic regions produces soils composed only of clay-sized mineral particles.

For some of the sedimentary rocks, a few generalizations can be drawn. Sandstones high in quartz weather to produce sandy soils. Soils developed from limestone and shale are usually fine-textured. Some cherty limestones, however, result in the formation of stony soils.

It is logical that the texture of the parent material will have a direct influence on the texture of the soil horizons in immature soils. The texture of resistant minerals will have a direct influence on the texture of even mature or old soils. Three additional ways in which the texture of the parent material influences soil development will be discussed in the following paragraphs. These are organic matter content, soil permeability (or the downward movement of water), and solum thickness.

Soils developed from fine-textured materials usually have a higher organic-matter content than those formed from coarser-textured materials. The finer texture may enhance plant growth by providing a greater water and nutrient supply. This results in a greater annual addition of organic matter to the soil. Fine-textured soils also tend to be less well aerated and have slightly lower average temperatures. This has the effect of retarding the rate of organic-matter decomposition and thereby aiding its accumulation. In addition, certain organic compounds may combine with the clay to render soil organic matter resistant to decomposition, as discussed in Chapter 6.

The permeability will determine to a certain extent the quantity of precipitation that will run off and that which will infiltrate into the soil. In humid regions the development of acidity can readily occur in soils developing in calcareous materials, if they are permeable. The more water that moves through the soil, the more rapidly acidity develops, weathering proceeds, and colloidal materials are translocated. Certain soils of the coastal plains of southeastern United States have developed in clay-textured marine sediments which are many thousands of years old. Where these parent materials are impermeable to water, some of the soils are still alkaline even though the average annual precipitation exceeds 40 to 50 inches.

If the parent material is very coarse-textured or gravelly, little surface is exposed to weathering and little water is retained for weathering and plant growth. In this case, very rapid permeability is associated with slow soil development.

It has been shown that fine-textured parent materials tend to retard leaching and the translocation of colloids. This in effect contributes to the development of soils with thin solums. On sloping lands the fine-textured soils have greater runoff and, consequently, less water available for leaching. There is also more water active in erosion, which contributes to the development of a thin solum. Soils that develop in the coarser or permeable parent materials have the thicker solums. The relationships between texture and solum thickness are shown in Fig. 9–15.

### Composition of Parent Material Is Important

The mineral composition of the parent material has much to do with the characteristics of the profile developed, at least until the soil becomes very old. If the material contains a large portion of aluminosilicate minerals which decompose with relative ease, there will be much clay produced. Under suitable conditions, some of the clay will accumulate in the B horizon, thus making a finer-textured subsoil. On the other hand, if the parent material is composed almost entirely of minerals which weather slowly, there will be very little clay formation

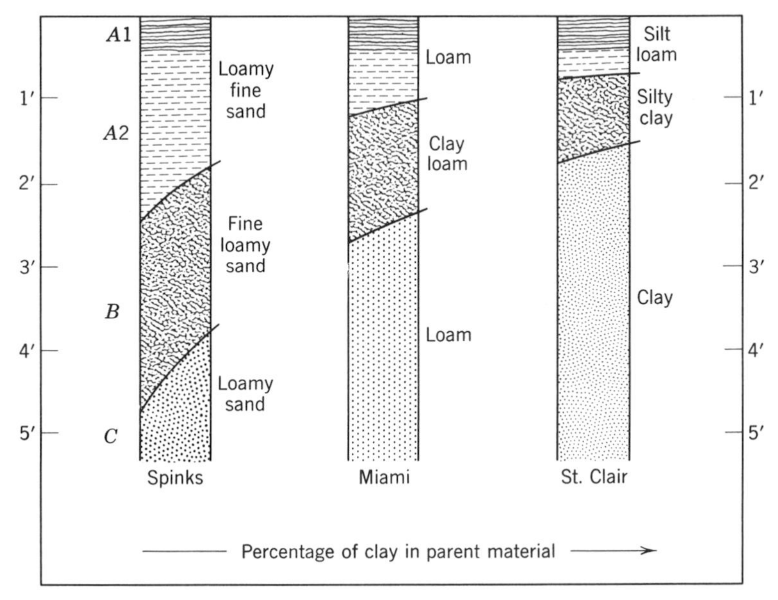


**Fig. 9–15** Relationship between the texture of the parent material and the thickness and texture of the horizons of three forest soils of north central United States (Gray-Brown Podzolic soils or Alfisols).

or clay accumulation in the B (illuviated) horizon. In nature we find all variations between these extremes.

It has been pointed out that soil acidity encourages mineral decomposition, translocation of colloids, and the overall development of the soil profile. Therefore, where parent materials are rich in lime, development of the soil is delayed, and it remains in the immature stage for a longer period of time.

## Stratification of Parent Material Is Important

The discussion thus far has assumed that the parent material was uniform throughout its depth. Soils that develop in water-laid sediments, however, usually develop in stratified parent material. Stratification could cause horizons of the same soil profile to develop in layers which have different textures and other characteristics. Where there is evidence of a lithological discontinuity, Roman numerals are used in horizon designation. The horizon sequence, A1-A2-B1-IIB2-IIB3-IIIC, would indicate that three different materials served as parent material for the horizons of the profile. Such a horizon sequence *could* indicate a situation where the upper three horizons developed from loess, the B2 and B3 developed from till, and where the C horizon was disintegrated bedrock.

In central United States there are large areas where a thin layer of loess was deposited over glacial till. Soils in such areas frequently have A and B horizons that have developed in loess. In cases where the loess is very thin, only the A horizon may have developed in loess whereas the B and C horizons have developed from glacial till. The horizons formed in loess, compared to those developed in glacial till, frequently have more rapid permeability. This makes the thickness of the loess an important factor in the design of terraces and the use of lister furrows for water-erosion control.

Where streams dissect an area underlain by strata of varying composition, various parent materials will be exposed along a traverse from the base to the top of a hill. This results in the development of a sequence of soils whose differences are due to differences in parent material. Such a sequence of soils is a *lithosequence* (Fig. 9–16).

## SOIL DEVELOPMENT IN RELATION TO TOPOGRAPHY

Topography modifies soil-profile development in three ways: (1) by influencing the quantity of precipitation absorbed and retained in the soil, thus affecting moisture relations; (2) by influencing the rate of removal of the soil by erosion; and (3) by directing movement of materials in suspension or solution from one area to another. As moisture is essential for the action of the chemical and biological processes of

**Fig. 9–16** A roadcut showing outwash overlying sandstone. Strata of different materials can produce a lithosequence of soils on a hillside where the materials are exposed, as in the situation on the left side of the photograph.

weathering and effectively acts in conjunction with some of the physical forces, it is evident that a modification of moisture relationships within a soil will materially influence profile development. In a humid region it is noteworthy that, in similar parent material of intermediate or fine texture, the soil of steep ridges or hills differs from the soil on gentle slopes or on a level to undulating topography. In arid climates these soil differences associated with differences in slope are much less pronounced because of the absence of water tables near the surface in the more level areas.

### Slope as a Factor in Soil Development

On steep slopes the continuous removal of surface soil by erosion keeps exposing the lower horizons and so modifies the profile. Consequently the soils on steep slopes have thinner solums, less organic matter, and less conspicuous horizons than soils on level or undulating topography when the water table is well below the solum. These profile differences due to slope are least pronounced in soils developed in coarse-textured parent material in which internal drainage is very rapid.

Topography indirectly plays another part in profile development by influencing the supply of moisture available for plant growth. It also has a bearing on the agricultural value of the land because it is related not only to both external and internal drainage conditions but also to the ease of performing tillage operations.

### Effect of Drainage Conditions on Soil Development

As pointed out before, drainage materially influences soil-forming processes. The accumulation of organic matter is usually facilitated because it is preserved by water (Fig. 9–17). Also, because of their low topographic position, poorly drained soils generally receive both organic and mineral matter from the adjacent slopes. In arid regions soluble salts also accumulate in areas receiving drainage water from surrounding soils. Colors of the soil material at shallow depths are changed from yellows, reds, and browns, denoting good aeration and oxidizing conditions, to the drabs, grays, and mottled yellows resulting from chemical reduction where drainage is poor. The horizon of eluviation is modified or may not be evident at all because of the slow or infrequent downward movement of water, and the B horizon is often replaced by a gray or bluish gray horizon known as a Bg or gleyed layer. In this layer iron is reduced to the ferrous form in the presence of organic matter to produce gray colors.

As a result of differences in topography or drainage or both, the soil profiles developed in similar parent material of a like age and within

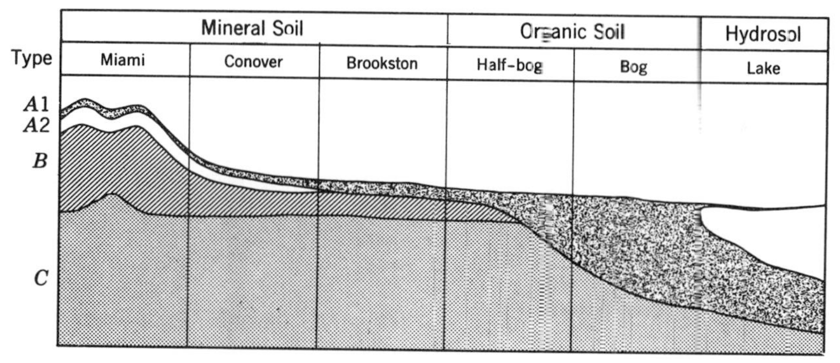


**Fig. 9–17** Topography, through its influence on drainage, is an effective factor in soil development in humid regions. The moderate humus accumulation, well-defined horizon of eluviation, and strongly developed B horizon of the zonal profiles are superseded by an increase in humus, diminution in eluviation, and accumulation in the B horizon as poorer drainage limits the illuvial processes and gleying becomes prominent. Finally, the bog is encountered and then the hydrosol or body of water itself. In the diagram the thickness of the A1 and A2 horizons is intentionally exaggerated to illustrate the effect of topography and drainage on them.

a single zonal region vary appreciably. A group of soils developed under such conditions and showing such variations in profile characteristics is designated as a *catena* or *toposequence of soils*. The Miami, Conover, and Brookston soils shown in Fig. 9–17 comprise a catena.

Locally, topography is perhaps the factor that most frequently causes soil differences. Parent material and to a lesser extent the vegetation also cause local soil differences as shown in Fig. 9–18. The landscape in Fig. 9–18 is near the prairie-forest transition and trees occur mostly in the protected steep-sided valleys. Erosion has removed loess from the steeper slopes so the soils are developing from the underlying glacial till rather than loess (Shelby soil). The Edina soil is developing in a depression where runoff water from adjacent areas accumulates and is the most leached soil because the water table is far below the surface. The gumbotil represents an ancient soil (paleosol) that existed before the loess was deposited.

## THE SOIL SURVEY

At the turn of the century there was an increasing awareness of the bonds between land and society. In an attempt to find the underlying causes of some agricultural problems and in an effort to build

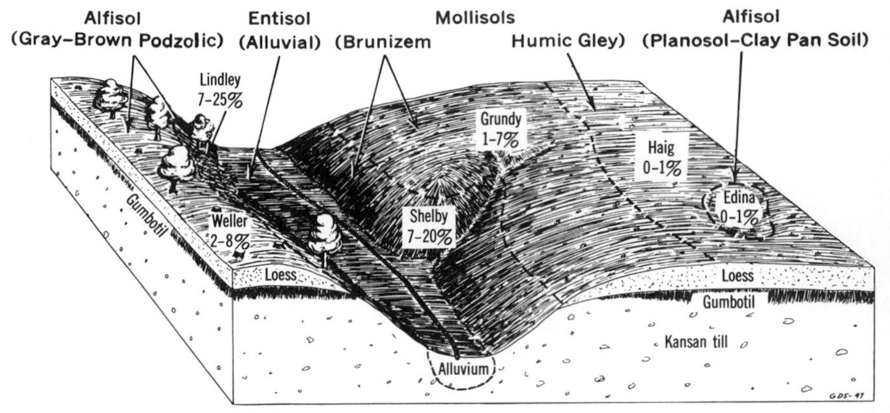


**Fig. 9–18** A block diagram that illustrates a soil pattern as related to topography, parent material, and vegetation. This landscape is located in south-central Iowa. (Adapted from *Understanding Iowa Soils*, R. W. Simonson et al., Brown Co., Dubuque, Iowa, 1952.)

a solid foundation for future research, the U.S. Department of Agriculture in cooperation with the various state experiment stations began at that time a systematic investigation of our soil resources. This investigation assumed the form of a national inventory and survey.

At the present time soil surveying is the process of studying and mapping the earth's surface in terms of units called soil types. A soil survey report thus consists of two parts: (1) the soil map, which is accompanied by (2) a description of the area shown on the map.

## Soil Mapping

The actual process of mapping or surveying consists of walking over the land at regular intervals, and taking notes on soil differences and all related surface features, such as slope gradients, evidence of erosion, land use, vegetative cover, and cultural features. Boundaries are drawn directly on aerial photographs representing in most places changes from one soil type to another (see Figs. 9–19 and 20).

Soil survey activities within a state are usually under the direction of the state agricultural experiment station in cooperation with the Soil Conservation Service of the United States Department of Agriculture. These organizations and other state and federal agencies share the expenses of mapping and work jointly from time to time in order not only to maintain a certain working standard but also to aid in correlating the mapping units in all sections of the country. After field work has been completed the data are assembled, printed, and made available to the public by the U.S. Department of Agriculture and the state agricultural experiment stations or other state agencies.

**Fig. 9–19** A soil mapper identifies the soil by inspecting all the soil horizons to a depth of about 5 feet. Lines are then drawn around areas of similar soil on an aerial photograph to produce a soil map. (USDA photo.)

## Types of Soil Surveys

Productive agricultural lands are inventoried in detail. A detailed map is one in which the scale is 2 or more inches to the mile. Four inches is the scale generally used at present, although occasionally an 8- or even a 12-inch scale is employed. In addition to soil-type boundaries, these maps show the location of gullies, railroads, houses, roads, streams, and other details needed in planning a soil-conserving practical farm- or ranch-management plan. This kind of work is expensive, costing from $300 to $400 for each square mile, that is, from about 50 to 60 cents per acre. Large expanses of land are too poor, however, to justify this cost. These areas are mapped on small scales, and the resulting study, called a reconnaissance survey, shows areas and regions that are dominated by soil associations in contrast to the detailed map where soil types are separated. Soil-association maps are useful in studying extensive land-use problems, and they possess the desirable characteristics of being made quickly at a comparatively low cost.

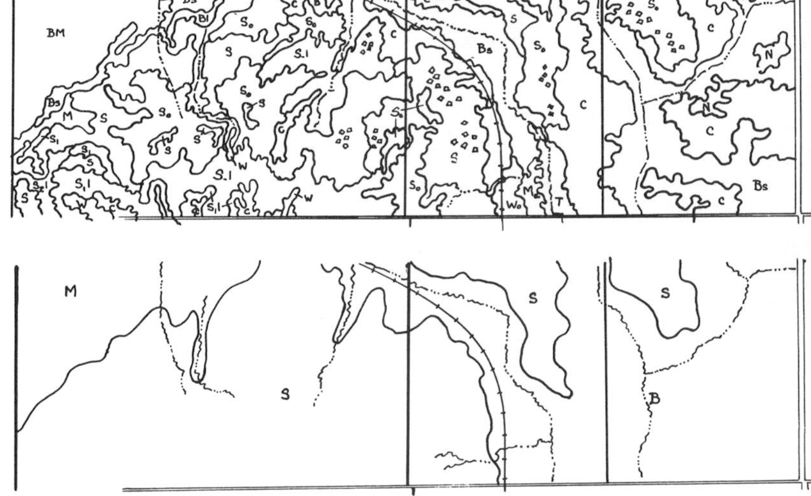


**Fig. 9–20** Top: Aerial photograph of good farming land in glaciated area. Such photographs are used extensively for a base map in making soil surveys. Center: A detailed soil survey map of the same area. The symbols indicate soil types. Bottom: A soil-association map of the area in which combinations of soil types occurring in close association are mapped as a unit.

## Purpose and Value of Soil Maps

Soil survey maps contain many types of information but perhaps that of greatest value is the soil type, slope, and degree of erosion that is recorded for each area delineated on the map. These maps form the basis to develop maps for a wide variety of uses. The areas can be grouped in land capability classes and a land capability map can be constructed like that shown in Fig. 14–6. The Soil Conservation Service personnel use these maps to develop conservation plans for farmers. The use of a soil survey map in wildlife management work is shown in Fig. 9–21.

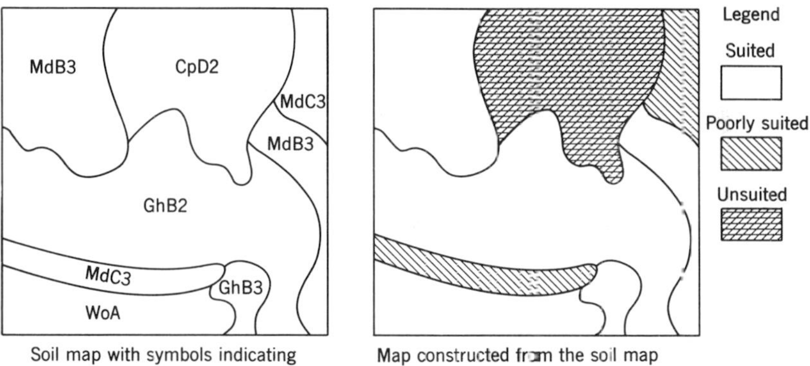


Soil map with symbols indicating soil type, slope, and erosion

Map constructed from the soil map showing suitability of land for grain and seed crops for wildlife management

**Fig. 9–21** Soil survey maps form the basis for the construction of many kinds of maps to serve a wide variety of needs: in this case for the wildlife management of a 40 acre tract. (Based on information from "Rating Northeastern Soils for Their Suitability for Wildlife Habitat," P. F. Allan, L. E. Garland, and R. F. Dugan, Trans. *28th North Am. Wildlife and Natural Resources Con.*, 1963.)

Life insurance companies, banks, and other money-lending agencies use soil surveys in determining security for loans. Real estate companies and individuals interested in buying or selling land make extensive use of soil surveys. They are also used by highway and drainage engineers, by various kinds of manufacturers in selecting suitable locations for factories, and by merchandising and advertising companies in selecting areas for intensive campaigns. County agricultural agents and other extension workers find soil surveys helpful in their work. And, last but by no means least, the farmers themselves are making increasing use of soil survey maps and reports in planning their management programs and in interpreting modern agricultural research in terms of their own farm conditions.

# 10

# Classification and Geography of the World's Soils

Classification schemes of natural objects seek to organize knowledge so that the properties and relationships of the objects may be most easily remembered and understood for some specific purpose. The ultimate purpose of soil classification is maximum satisfaction of human wants that depend on use of the soil. This requires grouping soils with similar properties so that lands can be efficiently managed for crop production. Furthermore, soils that are suitable or unsuitable for pipelines, roads, recreation, forestry, agriculture, wildlife, building sites, and so forth can be identified. This chapter is designed as an introduction to the classification, nature, distribution, and use of the major soils in the world.

## SOIL CLASSIFICATION

A genetic classification was suggested about 1880 by the Russian scientist Dokuchaev, and it has been further developed by European and American workers. This system is based on the theory that each soil has a definite morphology (form and structure) which is related to a particular combination of soil-forming factors. This system reached its maximum development in 1949 and was in primary use (especially, in the United States) until 1960. In 1960 the United States Department of Agriculture published *Soil Classification, A Compre-*

*hensive System.* This classification system places major emphasis on soil morphology and gives less emphasis to genesis or the soil forming factors as compared to previous systems. At present it is too early to totally give up the 1949 system and go entirely to the 1960 system. For this reason the major elements of both the 1949 and 1960 systems will be discussed. The 1960 system, however, will receive the most emphasis.[1]

## The Higher Categories of the 1949 Classification System

The highest category of the 1949 system is the *order.* There are three orders, which include *zonal, intrazonal,* and *azonal* (see Table 10–1). Soils with AC profiles belong to the azonal order. They are youthful soils. One or more of the following is responsible for their youthfulness: (1) hardness of parent material, (2) a rapid rate of erosion or deposition, or (3) an insufficient length of time. Under suitable conditions azonal soils develop B horizons and become either zonal or intrazonal soils.

**Table 10–1** A Classification of the Great Soil Groups in the 1949 System

| Order | Suborder | Great Soils Groups |
|---|---|---|
| Zonal soils | 1. Soils of the cold zone | Tundra soils |
| | 2. Light-colored soils of arid regions | Desert soils |
| | | Red desert soils |
| | | Sierozem |
| | | Brown soils |
| | | Reddish brown soils |
| | 3. Dark-colored soils of semi-arid, subhumid, and humid grasslands | Chestnut soils |
| | | Reddish Chestnut soils |
| | | Chernozem soils |
| | | Prairie soils |
| | | Reddish Prairie soils |
| | 4. Soils of the forest-grassland transition | Degraded Chernozem |
| | | Non-calcic Brown or Shantung Brown soils |
| | 5. Light-colored podzolized soils of the timbered regions | Podzol soils |
| | | Gray wooded or Gray Podzolic soils |
| | | Brown Podzolic soils |
| | | Gray-brown Podzolic soils |
| | | Red-yellow Podzolic soils |

[1] After a decade of revisions and improvements, the 1960 system will likely be re-published in 1972.

**Table 10–1** (Cont.)

| Order | Suborder | Great Soils Groups |
|---|---|---|
| | 6. Lateritic soils of forested warm-temperature and tropical regions | Reddish brown Lateritic soils<br>Yellowish brown Lateritic soils<br>Laterite soils |
| Intrazonal soils | 1. Halomorphic (saline and alkali) soils of imperfectly drained arid regions and littoral deposits | Solonchak or<br>Saline soils<br>Solonetz soils<br>Soloth soils |
| | 2. Hydromorphic soils of marshes, swamps, seep areas and flats | Humic-gley soils (includes *Wiesenboden*)<br>Alpine meadow soils<br>Bog soils<br>Half-bog soils<br>Low humic-gley soils<br>Planosols<br>Ground water Podzol soils<br>Ground water Laterite soils |
| | 3. Calcimorphic soils | Brown forest soils (*Braunerde*)<br>Rendzina soils |
| Azonal soils | | Lithosols<br>Regosols (includes dry sands)<br>Alluvial soils |

Prepared by James Thorp and Guy D. Smith, published in *Soil Science*, **67**:118, No. 2, 1949. Used through courtesy of Williams and Wilkins Co.

Included in the zonal order are those soils possessing well-developed profiles, that reflect the influence of the active factors of soil genesis, especially climate and vegetation. Topographically, zonal soils are situated on well-drained uplands and develop in parent material not extreme in texture or chemical composition.

Intrazonal soils have well-developed profile characteristics that reflect the dominating influence of some local factor, such as the nature of the parent material, topography, or drainage, over the zonal effect of climate and vegetation. They are found in more than one soil zone.

Suborders of soils are set apart on the basis of the factors which have been most instrumental in determining their characteristics or on the basis of the distinguishing characteristics of the soil group itself. For example, the cold and arid soils are direct products of climatic conditions. Grassland, transitional, and forest soils owe their properties to a combination of the influence of climate and vegetative cover. Excess moisture is responsible primarily for the nature of hydromorphic soils, and the accumulations of chlorides and of calcium salts are the distinguishing characteristics of halomorphic and calcimorphic soils, respectively.

The suborders of the 1949 system are divided into *great soil groups* (Table 10–1) and reference to some of them has been made. A time sequence of soils consisting of Regosol (azonal AC soils), Brunizem (zonal soils of the humid tall-grass prairie), and Planosol (intrazonal clay-pan soils) is presented in Fig. 9–4. The physical properties data in Fig. 3-5 is for a zonal Gray-Brown Podzolic soil developed under hardwood forest in a humid-temperate climate. An interesting parallel found in nature is the distribution pattern of climate, vegetation, and soil of the earth's surface. A diagrammatic visualization of the relation between climatic conditions and of occurrence of zonal great soil groups is given in Fig. 10–1. Further consideration of great soil groups will be included in discussions of the 1960 system.

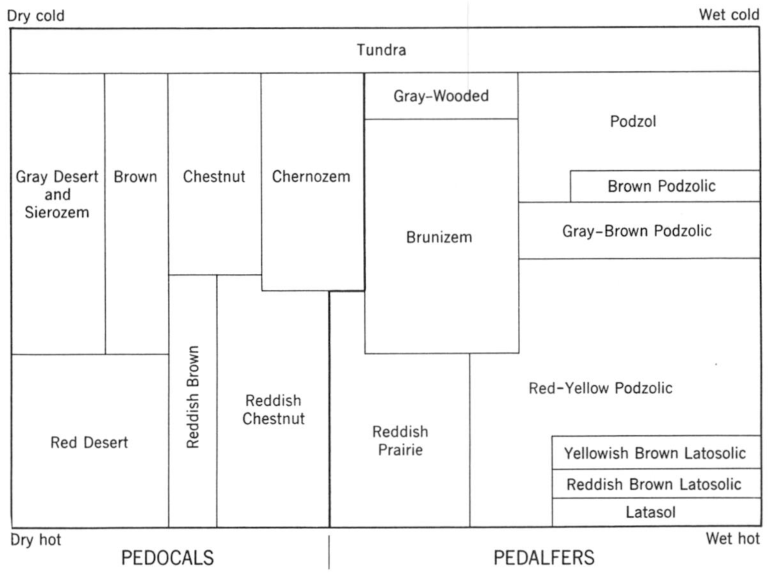


**Fig. 10–1** A diagrammatic visualization of the relation between climatic conditions and the occurrence of the great soil groups (1949 classification).

## Diagnostic Horizons Used with the 1960 System

The ABC nomenclature given in Chapter 9 is still used as the standard for describing and defining soil horizons. The 1960 comprehensive system had need for a more strict definition of soil horizons and diagnostic horizons were developed to be used in defining most of the orders. Two kinds of diagnostic horizons, surface and subsurface, are recognized. The surface diagnostic horizons are called epipedons (Greek *epi*, over; and *pedon*, soil). A brief description of the diagnostic horizons is given in Table 10–2.

**Table 10–2** Derivation and Major Features of Diagnostic Horizons

| Horizon | Derivation | Major Features |
|---|---|---|
| | *Surface diagnostic horizons-epipedons* | |
| Mollic | L. *mollis*, soft | Thick, dark colored, high base saturation and strong structure so that the soil is not massive or hard when dry. |
| Umbric | L. *umbra*, shade | Same as mollic but highly H saturated and may be hard or massive when dry. |
| Ochric | Gr. *ochros*, pale | Thin, light-colored and low in organic matter. |
| Histic | Gk. *histos*, tissue | Very high organic matter content and saturated with water at some time during the year unless artificially drained. |
| Anthropic | Gr. *anthropos*, man | Mollic-like horizon that has a very high phosphate content resulting from long time cultivation and fertilization. |
| Plaggen | Ger. *plaggen*, sod | Very thick, over 20 inches, produced by long continued manuring. |
| | *Subsurface diagnostic horizons* | |
| Argillic | L. *argilla*, white clay | Illuvial horizon of silicate clay accumulation. |
| Natric | *Natrium*, sodium | Illuvial horizon of silicate clay accumulation, over 15% exchangeable sodium and columnar or prismatic structure. |
| Spodic | Gk. *spodos*, wood | Illuvial accumulation of free iron and aluminum oxides and organic matter. |
| Oxic | L. *oxide*, oxide | Altered subsurface horizon consisting of a mixture of hydrated oxides of iron or aluminum and 1:1 clays. |
| Cambic | L. *cambiare*, to change | An altered horizon due to movement of soil particles by frost, roots and animals to such an extent to destroy original rock structure. |
| Agric | L. *ager*, field | An illuvial horizon of clay and organic matter accumulation just under the plow layer due to long-continued cultivation. |

## Soil Orders of the Comprehensive System (1960) or Soil Taxonomy (1970)

The order is the highest category and there are ten orders each ending in sol (L. *solum* meaning soil). The orders along with their meaning and approximate equivalents in great soil groups (1949) are given in Table 10–3. Entisols are very recent soils typified by azonal soils (see Table 10–3). Vertisols are soils high in clay that become "inverted" because of alternate swelling and shrinking (Grumusols). Inceptisols are young soils with just the beginning of genetic horizon development. Aridisols are soils of arid regions. Mollisols are the grassland soils with thick, "soft," dark-colored surface horizons (mollic epipedons). Spodsols have spodic horizons and are comparable to Podzols. Alfisol is derived from *pedalfer*—a word first used by Marbut to refer to humid region soils leached of lime and with a tendency for aluminum (Al) and iron (Fe) to accumulate in the subsoil. Ultisols are extremely leached soils, very low in bases. Oxisols are the red tropical soils rich in oxides of iron and aluminum and also 1:1 clays, that is, they have oxic horizons. The Histosols are bog soils composed mainly of plant tissue. A more detailed description of the orders along with their world distribution and use follows.

## SOIL ORDERS—PROPERTIES, DISTRIBUTION, AND USE

Most of the people of the world make their living by tilling the soil. The soil directly influences their lives everyday in that it determines how they build their houses and roads, and how they grow their crops. By affecting the amount and kinds of food they eat, soils affect their health. The order category is sufficiently general and yet sufficiently well defined to make a discussion of these and other aspects of the worlds' soils possible in one chapter. A broad schematic map of the soil orders of the world is presented in Fig. 10–2.

### Entisols

Entisols are soils that tend to be of recent origin. They are characterized by youthfulness and are without natural genetic horizons or have only the beginnings of horizons. The central concept of Entisols is soils in deep regolith or earth with no horizons except perhaps a plow layer. Some Entisols, however, have plaggen, agric, or A2 (albic) horizons and some have hard rock close to the surface. Soils classified in the 1949 system as Alluvial, Regosol, and Lithosol would now be classified mostly as Entisol.

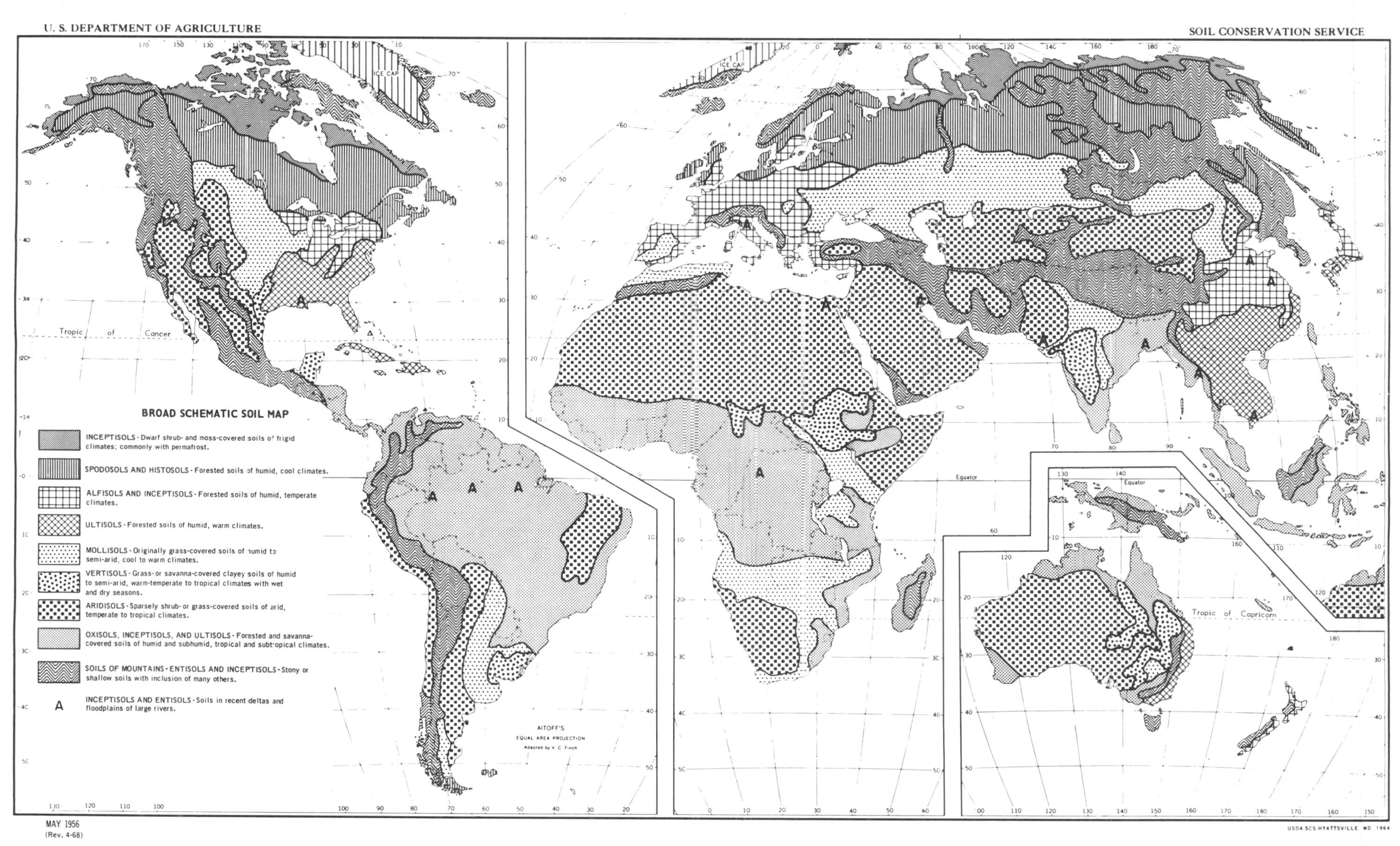


**Fig. 10–2** Broad schematic map of the soil orders of the world. (Map courtesy of Roy Simonson of the United States Department of Agriculture.)

**Table 10–3** New Soil Orders and Approximate Equivalents in Great Soil Groups

| Order | Formative Syllable | Derivation | Meaning | Approximate Equivalents |
|---|---|---|---|---|
| 1. Entisol | ent | Coined syllable | Recent soil | Azonal soils and some Low Humic Gley soils |
| 2. Vertisol | ert | L. verto, turn | Inverted soil | Grumusols |
| 3. Inceptisol | ept | L. inceptum, beginning | Inception, or young soil | Ando, Sol Brun Acide, some Brown Forest, Low Humic Gley, and Humic Gley soils |
| 4. Aridisol | id | L. aridus, dry | Arid soil | Desert, Reddish Desert, Sierozem, Solonchak, some Brown and Reddish Brown soils, and associated Solonetz |
| 5. Mollisol | oll | L. mollis, soft | Soft soil | Chestnut, Chernozem, Brunizem (Prairie), Rendzinas, some Brown, Brown Forest, and associated Solonetz and Humic Gley soils |
| 6. Spodosol | od | Gk. spodos, Wood ash | Ashy (podzol) soil | Podzols, Brown Podzolic soils, and Ground Water Podzols |
| 7. Alfisol | alf | Coined syllable | Pedalfer (Al-Fe) soil | Gray-Brown Podzolic, Gray Wooded, Noncalcic Brown, Degraded Chernozem, and associated Planosols and Half-Bog soils |
| 8. Ultisol | ult | L. ultimus, last | Ultimate (of leaching) | Red-Yellow Podzolic, Reddish-Brown Lateritic (of U.S.), and associated Planosols and Half-Bog soils |
| 9. Oxisol | ox | F. oxide, oxide | Oxide soils | Laterite soils, Latosols |
| 10. Histosol | ist | G. histos, tissue | Tissue (organic) soils | Bog soils |

From *The New Classification*, E. Joseph Larsen, *Soil Conservation* 30, No. 5, December 1964.

Alluvial soils develop on alluvium of recent origin and have very weakly developed profiles. In many of them the color change from the A to C horizon is hard to see or is nonexistent. They are, in large part, soils in which most of the properties have been inherited. They are usually characterized by stratification. The texture is related to the rate at which the water deposited the alluvium. For this reason they tend to be coarse-textured near the stream and finer-textured near the outer edges of the flood plain. Mineralogically, they are related to the soils which served as a source for the alluvium.

Periodic flooding brings fresh minerals to the soils and they tend to remain fertile. The soil remains youthful because it is buried before maturity is reached. Alluvial soils played an important role in the development of early agriculture before the development of fertilizers and manuring systems. Soils of the Nile Valley are a classic example. Most of the people in China live on floodplains. It has been estimated that one-third of the world's population obtains its food from them.[2] The major areas of Alluvial soils in the world are shown on the map in Fig. 10–2 with the symbol A.

Most soils in the world that developed from unconsolidated sediments were Regosols or Entisols when they were young. Steep slopes where erosion occurs rapidly, an insufficient length of time, or movement of the material, as in the case of sand dunes, are the major causes for their existence. The unstabilized sand dunes shown in Fig. 10–3 develop into Entisols or Regosols after vegetation has become established. These Entisols eventually develop into typical Spodosols (Podzols), as along the eastern shore of Lake Michigan, if the material remains stabilized for two to three thousand years.

Entisols developed from sand dunes have limited agricultural value. The moderating influence of the lake on the climate has made it profitable to raise fruit on some of them in Michigan. Small areas of Entisols frequently exist on the steepest parts of cultivated fields and are effectively used with the surrounding zonal soils that comprise the major portions of the fields. They are low in organic-matter content and are generally responsive to nitrogen fertilization. Many of them are neutral in reaction or calcareous at the surface.

Entisols in which the A horizon rests directly on hard rock were classified as Lithosols in the 1949 system. An AR type of Entisol is shown in Fig. 10–4. Two very important factors that contribute to their development are the hardness of the rock and the steepness of the slope. The rate of rock disintegration does little more than keep

[2] "Soil," by C. E. Kellogg, *Sci. Am.*, July 1950.

**Fig. 10–3** Sand dunes along the eastern shore of Lake Michigan are converted into Entisols or Regosols when stabilized by vegetation. Podzols form if the material remains stabilized 2000 or 3000 years or longer.

**Fig. 10–4** Entisol (Lithosol) developed from sandstone in the foothills of the Rocky Mountains in northcentral Colorado.

pace with removal of material by erosion. Cracks in the underlying rock may enable roots to penetrate much deeper than the A horizon. Where the A horizon is one to two feet thick; the land is profitably used for pasturing.

Lithosolic soils are common in mountainous areas and give evidence to the fact that deep soils did not cover the land everywhere before agriculture began. Many deep, productive soils were once Lithosols and, in a sense, these soils may be transitory in the development of well-differentiated profiles.

## Inceptisols

Inceptisol is derived from the Latin *inceptum*, meaning beginning. Development of genetic horizons is just beginning in Inceptisols, but they are still considered to be older than Entisols. Typically, Inceptisols have ochric epipedons and may have other diagnostic horizons, but show little evidence of eluviation or illuviation. Evidence of extreme weathering is lacking. They lack sufficient diagnostic features to be placed in any of the remaining eight soil orders.

Inceptisols occur in all climatic zones where there is some leaching in most years. On the soil order map of Fig. 10–2 two large areas are shown that include the Tundra of North America and Europe-Asia. Soils of the Tundra are characterized by high organic-matter content. The vegetation consists mostly of low growing mosses, lichens, and sedges. These plants grow slowly, but the low soil temperature inhibits organic matter decomposition resulting in soils with a high content of organic matter. They usually have permafrost, are slightly to strongly acid, and have a surface microrelief caused by freezing and thawing. Most of the soils on the Tundra show evidence of wetness or poor drainage.

Inceptisols of the Tundra support a sparse population of nomadic hunters that live almost entirely on the products of the caribou. In recent years some of these Eskimos have been in the news for having high radioactivity. Testing of nuclear bombs in the Arctic produced radioactive fallout that was absorbed by the lichens. The caribou ate the lichens and the radioactivity was transferred to man in the meat. Eskimos of the Brooks Range in Alaska were found to have about 100 times more radioactivity than persons in the "lower states."

Inceptisols are found widely distributed in the world and some make good agricultural and grazing lands. The data in Table 10–4 show that Inceptisols occupy 15.8 percent of the land surface of the world and rank second in order of abundance. These soils, however, play a minor part in the production of the world's food.

**Table 10–4** Area of Soils of the World by Soil Order

| Soil Order | Area in Thousands of Square Miles | Percentage of World Total | Rank Rank |
|---|---|---|---|
| Alfisols | 7,600 | 14.7 | 3 |
| Aridisols | 9,900 | 19.2 | 1 |
| Entisols | 6,500 | 12.5 | 4 |
| Histosols | 400 | 0.8 | 10 |
| Inceptisols | 8,100 | 15.8 | 2 |
| Mollicols | 4,600 | 9.0 | 6 |
| Oxisols | 4,800 | 9.2 | 5 |
| Spodosols | 2,800 | 5.4 | 8 |
| Ultisols | 4,400 | 8.5 | 7 |
| Vertisols | 1,100 | 2.1 | 9 |
| Ice Fields and Rugged Mountains | 1,200 | 2.4 | — |
| Islands, Unclassified | 200 | 0.4 | — |
| Grand Total | 51,600 | 100.0 | — |

Areas determined by John D. Rourke, Chief, World Soil Geography Unit, Soil Conservation Service, U.S.D.A., on "Soils of the World—Probable Occurrence of Orders and Suborders," May 1968.

### Aridisols

Aridisols are primarily the soils of dry places. They may have an ochric epipedon and usually some additional diagnostic horizons. Many deserts once had a more humid climate than the present and argillic horizons are common. Dust accumulation and transfer is a common phenomenon of the desert. There is almost always some carbonates in Aridisols, lending support to the theory that rain water leaches salts from transient calcareous dust and this becomes a source of salts, including lime. In most Aridisols there is a distinctive zone somewhere below the surface layer where lime has accumulated in large quantity, and in some Aridisols the lime has been cemented into a petrocalcic horizon (consolidated lime layer). Accumulations of calcium sulfate (gypsic horizon) and natric horizons also occur in some Aridisols.

The natural vegetation on Aridisols is sparse, consisting of desert shrubs and grasses. The organic matter content of the soil is low, but the soils are highly base saturated, many being calcareous. The primary use of Aridisols is grazing. When irrigated they make true the biblical verse, "and the deserts shall bloom as the rose" (Fig. 10–5). If water is available for irrigation the great potential of the deserts

**Fig. 10–5** Aridisol landscapes on the Sonoran Desert of southern Arizona. (Left) natural vegetation and the presence of desert pavement in the foreground. (Right) similar land used for irrigated agriculture.

can be seen from Table 10–4, which shows that Aridisols are the most abundant soils in the world and occupy 19.2 percent of the land surface. The world distribution of Aridisols is shown in Fig. 10–2. All Aridisols support vegetation, which means that large areas of blowing sand, etc. in the deserts are not areas of Aridisols but areas occupied by Entisols.

## Mollisols

Bordering many of the desert regions are areas of higher rainfall that support grasses that tend to cover the ground completely and produce an abundance of organic matter which decomposes within the soil. The rainfall, however, is sufficiently limited to prevent excessive leaching and base saturation remains high. Decomposition of abundant organic matter within the soil in the presence of calcium leads to the formation of mollic epipedons. The well-aggregated soil structure gives rise to the "softness" of the soil, which is neither massive nor very hard when dry. All Mollisols have mollic epipedons. Features of mollic horizons include (1) a thickness of 10 inches or greater, (2) dark color and at least one percent organic matter, and (3) over 50 percent base saturation. A soil with a mollic epipedon is shown in Fig. 10–6. In most Mollisols there has been sufficient clay migration to form a Bt or argillic horizon. As a group, Mollisols combine high soil fertility and fair to adequate rainfall so that they comprise perhaps the world's most productive agricultural soils.

**Fig. 10–6** A Mollisol with a mollic epipedon. The soil is the Aguilita clay loam from southwestern Puerto Rico and has developed from soft limestone. The soil was formerly called a Rendzina.

Geographically speaking, large areas of Mollisols and Aridisols usually share a common boundary. Illustrations shown in Fig. 10–2 include the great plains of North America, southern South America (Argentina), southern Soviet Union, northeastern China, and south Africa. In these areas increasing precipitation from Aridisols to the Mollisols results in gradual changes in soil properties, as illustrated in Fig. 10–7. With increasing precipitation there is a gradual increase in solum thickness, organic-matter content, development of Bt horizon, and depth to the Ca layer. The most humid Mollisols (Brunizems) lack a lime accumulation zone and typically have a pH less than

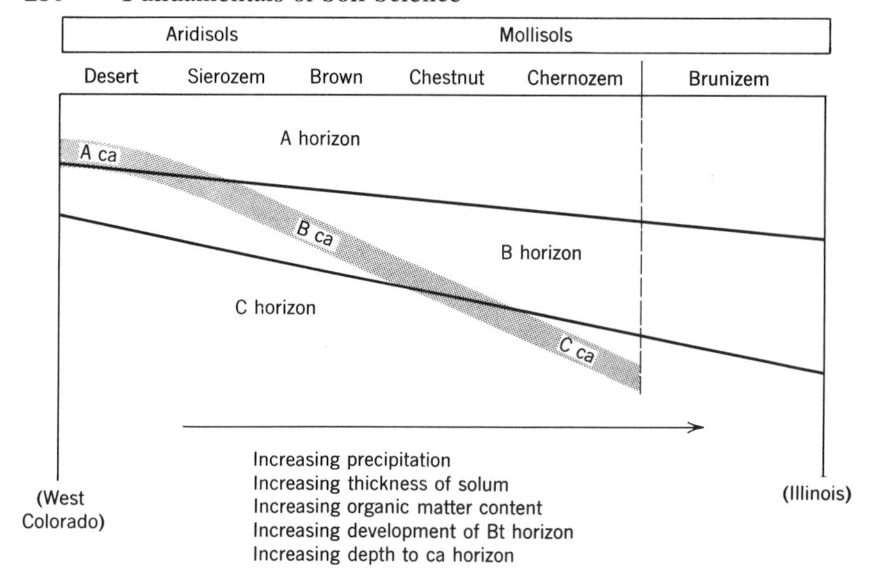


**Fig. 10–7** Generalized relationships between Mollisols and Aridisols in many parts of the world showing a gradual change in soil properties with increasing precipitation. Specifically, the diagram illustrates soil relationships on the great plains of the United States.

7. A soil profile near the Mollisol-Aridisol border in Colorado is shown in Fig. 10–8 and illustrates many of the dominant features of Mollisols.

Grazing is a major form of land use on the drier Mollisols. The proportion of land planted to crops varies from virtually none in some areas to most in others. Dry-land farming is practiced extensively for wheat production (Fig. 10–9). Irrigation is utilized, particularly along the river valleys, and a wide variety of irrigated crops are grown. Fruits, vegetables, and sugar beets occupy smaller irrigated acreages.

Mollisols developed under more rainfall than Brown soils include Chestnut, Chernozem, and Brunizem (Prairie), in order of increasing rainfall. Wheat becomes more important and grazing less important in the Chestnut and Chernozem zones as compared to the Brown-soil zone. Considerable corn is grown in the most humid part of the Chernozem areas (if not too far north) and corn becomes the dominant crop grown on Brunizems. Brunizem type Mollisols combine high natural fertility with adequate moisture for crop production. The two major areas of Brunizems occur in the Corn Belt of the United States and the pampa of Argentina.

Wheat is the most important food crop in the world in terms of acreage and production. Corn is perhaps third in importance after

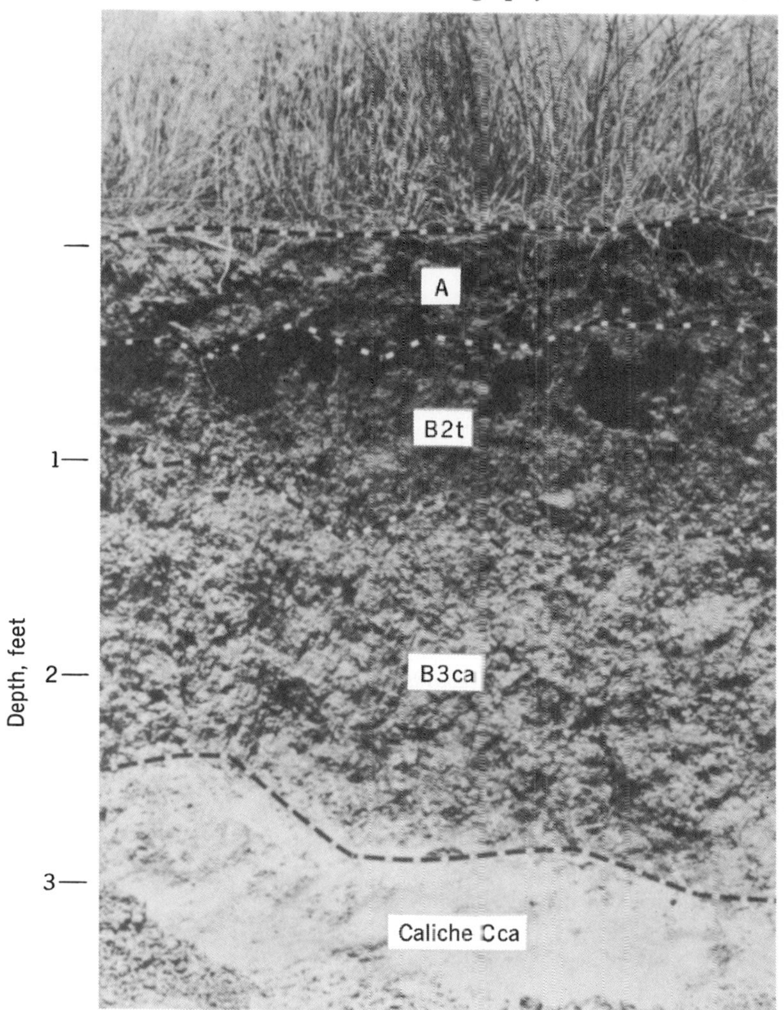


**Fig. 10–8** Photograph of a soil profile near the Mollisol-Aridisol border in eastern Colorado. Obvious lime accumulation zone can be seen. (Caliche is a more or less cemented deposit of calcium carbonate.)

rice. Compare the maps in Fig. 10–10, which show the production of wheat and corn in the world. The Soviet Union is the major wheat producer and the United States is the major corn producer. This correlates with the abundance of Brunizem-type Mollisols in the United States and their general absence in the Soviet Union. In the United States corn production is centered in the Corn Belt on "moist" Mollisols and the wheat production is concentrated more nearly on the

**Fig. 10–9** Wheat is a major crop near the Mollisol-Aridisol border and the limited rainfall necessitates the use of fallowing for water conservation.

"drier" Mollisols. A similar comparison can be made for the pampa of Argentina. Over half of the world's production of corn and about 15 percent of the world's wheat is produced in the United States, which is a partial indication that a considerable amount of the Mollisols in the world are located in the United States. Mollisols occupy only 9 percent of the world's surface (Table 10–4), but produce a much larger percentage of the world's food.

A very significant amount of the soil in the "moist" Mollisol (Brunizem) region in the United States developed under the influence of poor drainage and were called Humic Gley (1949 system). Humic Gley soils develop a Bg or Cg horizon that is recognized by its gray color and is indicative of saturation of the soil part of the year. Mollic horizons of these soils tend to be very thick, very dark-colored, and contain a high content of organic matter. Leaching has been minimal because of the presence of a high water table much of the time. They require drainage for crop production, and when properly drained, become some of the most productive soils for agriculture. In Illinois the most abundant single soil, the Drummer silty clay loam, is a Mollisol of the Humic Gley type and much of the Corn Belt's reputation of corn production has resulted from the large acreages of Mollisols developed on flat lands under poor drainage.

Some Mollisols have developed under trees where small animals were important in carrying organic matter into the soil where the organic matter decomposed. See Fig. 10–11 for such a landscape in eastern Michigan where the soils are intensely farmed for production of beans, sugar beets, and many other crops.

**Fig. 10–11** Mollisol landscape with Humic Gley soils that developed under hardwood forest on a lake plain in eastern Michigan. Note the light-colored streaks where drainage tile were recently installed. The soils are very productive for cash crops, especially beans and sugar beets.

## Spodosols

Most of the soils formerly called Podzols belong in the Spodosol order. All Spodosols have a spodic horizon. Spodic horizons are illuvial subsurface horizons where amorphorus materials composed of organic matter, aluminum, and iron have accumulated and are about comparable to Bhir horizons. All Spodosols form in a humid climate and mostly from sandy (siliceous) parent material. They are found from the tropics to the boreal regions, but the major areas in the world are just south of the Tundra in North America and Europe-Asia. Here, glaciation left large areas of sandy parent materials and the climate is humid. No major area of Spodosols is shown in the southern hemisphere on the map of Fig. 10–2. Trees are the common vegetation, although some of the most intensely developed Spodosols develop under heath vegetation. Ashy gray A2 horizons (albic) are a major feature of most Spodosols, but not a requirement (see Fig. 10–12).

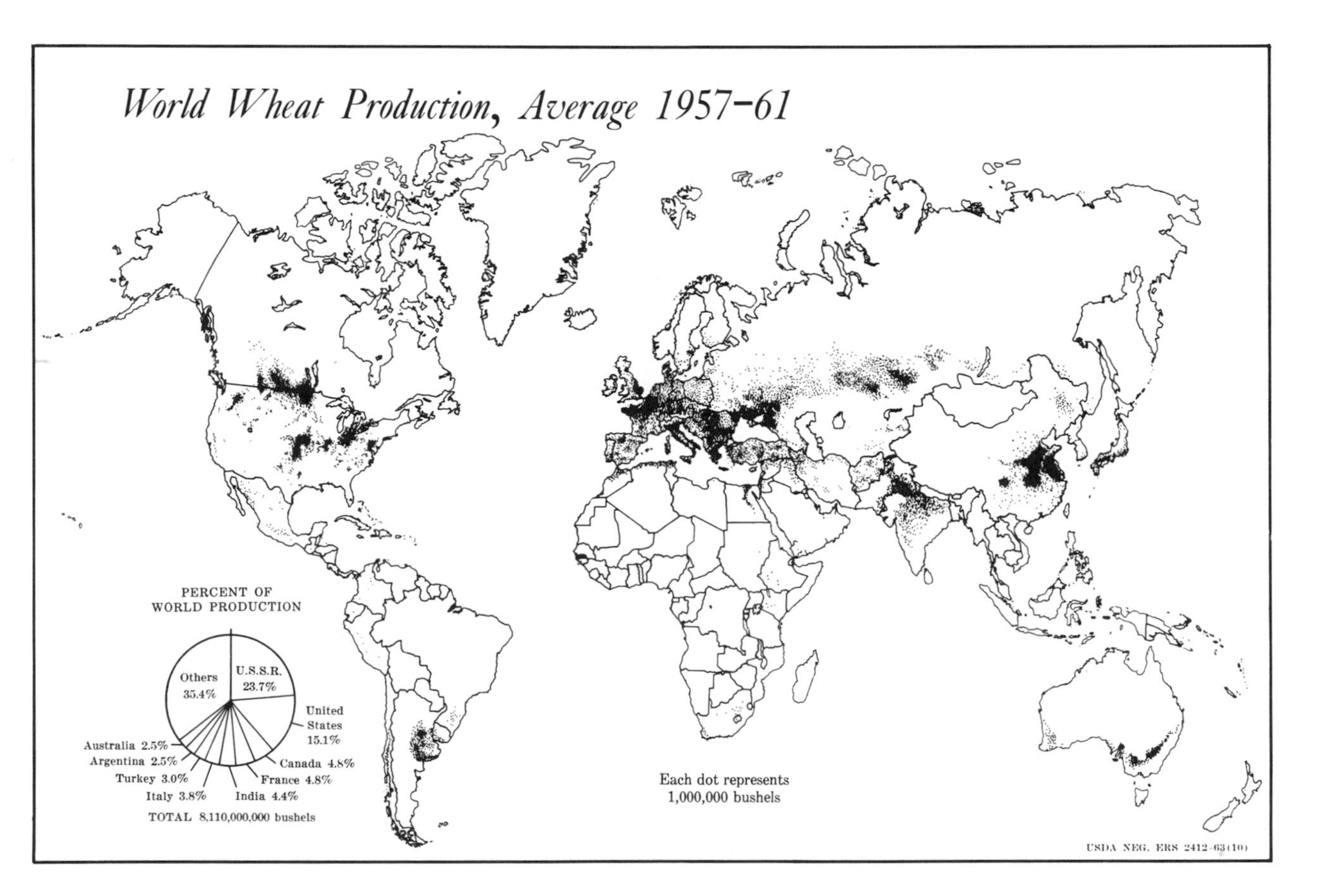


**Fig. 10–10** World production of wheat and corn. The Soviet Union is the major producer of wheat and the United States is the major producer of corn. This reflects the large acreages of "drier" Mollisols in the Soviet Union and large acreages of more humid region Mollisols in the United States.

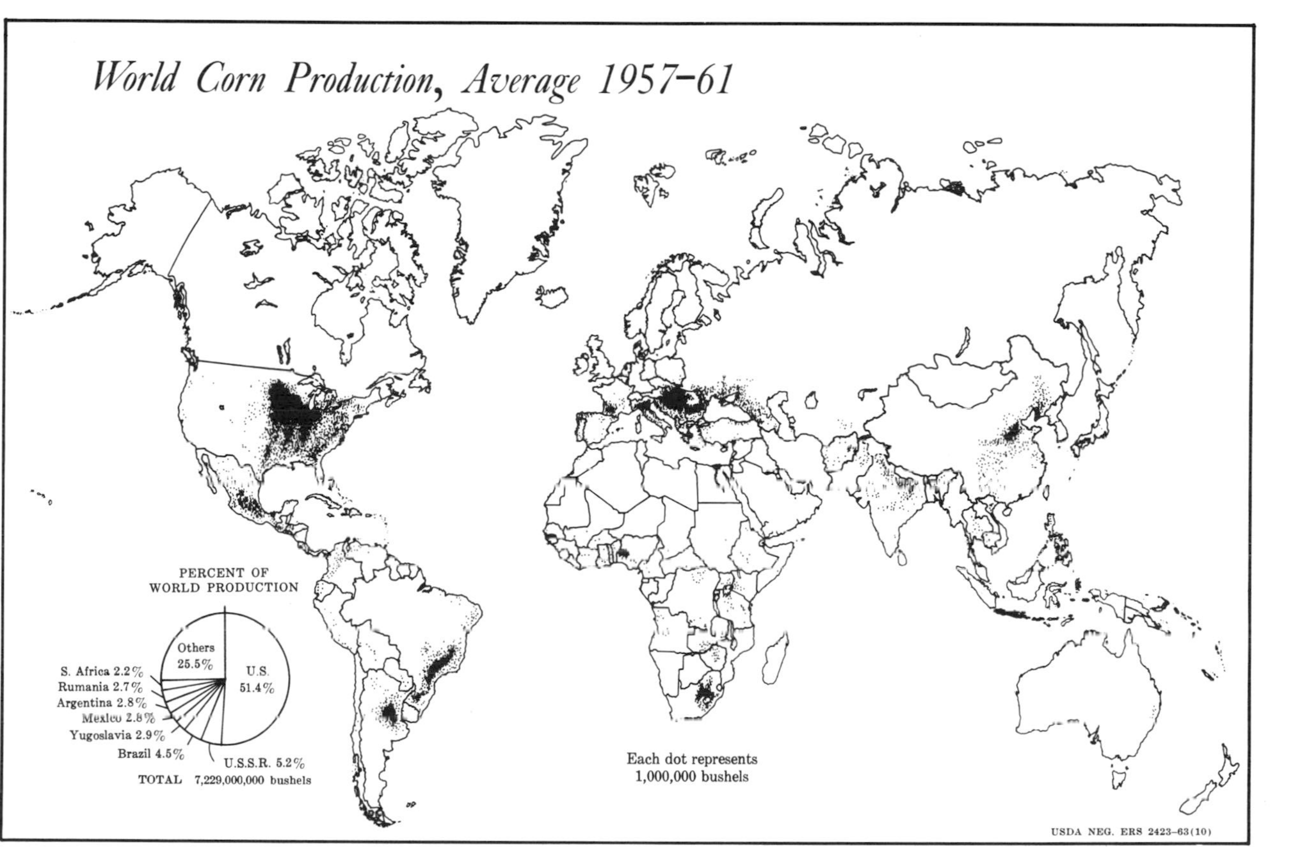
World Corn Production, Average 1957–61
PERCENT OF
WORLD PRODUCTION
Others
25.5%
U.S.
51.4%
S. Africa 2.2%
Rumania 2.7%
Argentina 2.8%
Mexico 2.8%
Yugoslavia 2.9%
Brazil 4.5%
U.S.S.R. 5.2%
TOTAL 7,229,000,000 bushels
Each dot represents
1,000,000 bushels
USDA NEG. ERS 2423-63(10)

Fig. 10–10

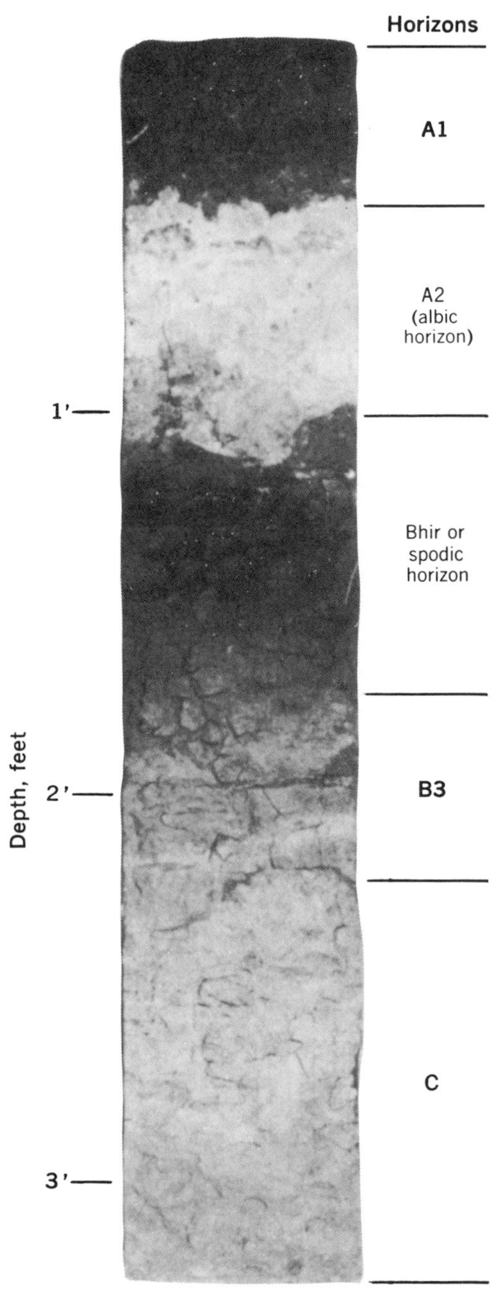


**Fig. 10–12** A Spodosol (Podzol) profile showing a nearly "white" A2 horizon and an ortstein-a cemented Bhir or spodic horizon. All mineral horizons have a sand texture, which is a common characteristic of Spodosols.

Spodosols have solums that are very acid throughout, and they have low cation-exchange capacity (except where humus has accumulated) and low base saturation percentages. The base saturation of some horizons is frequently less than 10 percent. Spodosols have limited capacity to store water and are naturally very infertile for most crops. Properties of a typical Spodosol are presented in Table 10–5.

**Table 10–5** Some Properties of the Horizons of a Spodosol (Podzol)

| Depth, Inches | Horizon | Cation-Exchange Capacity (me/100 gms) | Percentage Base Saturation | pH | Percentage Organic Matter | Percentage Clay |
|---|---|---|---|---|---|---|
| +2–0 | O2 | . . . | . . . | 3.6 | 45.6 | . . . |
| 0–4 | A2 | 7.1 | 10 | 3.8 | 0.8 | 2 |
| 4–9 | B2hir | 14.3 | 4 | 4.4 | 7.2 | 4 |
| 9–15 | B2ir | 4.1 | 22 | 4.8 | 2.0 | 3 |
| 15–28 | B3 | 6.5 | 9 | 5.2 | 0.8 | 2 |
| 28–50 | C | 4.6 | 4 | 5.0 | 0.1 | 7 |

Adapted from *Soil Survey Laboratory Memorandum* 1, Beltsville, Maryland, 1952. Soil is profile No. 39, Worthington loam from Coos County, New Hampshire.

The unsuitability of Spodosols for agriculture in colonial New England was described by C. L. W. Swanson, the former chief soil scientist of Connecticut.

"The virgin soil under a long-established forest is not always good. When the settlers cleared New England forests 300 years ago, the topsoil they found was only 2 to 3 inches thick. Below this was sterile subsoil and when the plow mixed the two together, the blend was low in nearly everything a good soil should have. It was not the lavish virgin soil of popular fancy. Such soil could not support extractive agriculture which takes nutrients out of the soil and does not replace them. Many New England lands that were treated in this way soon went back to forest."[3] Similar conditions existed in other places such as the Lake States. Large acreages of Spodosols were settled by farmers after logging removed the timber, but low soil fertility and droughtiness caused many farmers to abandon the land as shown in Fig. 10–13. The first bulletin published by the Michigan Agricultural Experiment Station was devoted to solving the problems of farmers on the "sand plains" of northcentral Michigan. Today the northern

[3] "The Road to Fertility," C. L. W. Swanson, *Time Magazine*, January 18, 1954.

Spodosol regions are characterized by sparse farming generally, but in localized areas intensive production of fruits and vegetables occurs. The cool summers of the region attract many summer tourists and many cities owe their existence to mining and lumbering.

**Fig. 10–13** These disintegrating buildings tell the story of forest removal, several generations of farming, and then abandonment of land in the Lake States with Spodosols (Podzols) of very low inherent fertility. It was not that the land was misused, necessarily, but rather that high production costs have made it noncompetitive for general farming.

Many Spodosols have developed from sandy parent material of marine origin along the southeastern coast of the United States and on the Florida peninsula. Use of modern technology has resulted in successful use of these soils for vegetable and cattle production.

## Alfisols

The moister Mollisols (Brunizems) occur in humid regions where trees are the natural vegetation. Many theories have been advanced to account for the extensive grasslands that exist in Iowa and Illinois. Along this wetter Mollisol boundary are large areas of soils developed under trees with ochric epipedons, argillic subsurface horizons (illuvial horizons of silicate clay accumulation), and similar or only slightly lower base saturation than nearby Mollisols. These soils are Alfisols and formerly were called Gray-Brown Podzolic and Gray Wooded. Some Planosols and Noncalcic Brown soils are also Alfisols. See Fig. 3–5 for physical property data of an Alfisol.

Most Alfisols have argillic horizons and occur in humid regions where the soil is moist at least part of the year. The requirement for over 35 percent base saturation in the argillic horizon of Alfisols means that bases are being released in the soil by weathering about as fast as bases are being leached out. Thus, Alfisols rank only slightly lower than Mollisols for agriculture.

Three large areas of Alfisols in the world are shown in Fig. 10–2. These areas are northeastern United States and southeastern Canada, northwestern Europe, and northeastern China plus most of Japan. All three areas are intensely cultivated, and the European and Asian areas support some of the densest human populations. Favorable climate and soils with fairly good fertility and physical properties make Alfisols one of the most productive of the soil orders for agriculture (Fig. 10–14). In all areas farm animals have played a very important role for power, food, and the production of manure that was carefully conserved and applied to the land to maintain soil fertility. Today, the availability of fertilizers has greatly reduced the farmers dependence on farm animals for manure. Wheat is a major crop on Alfisols in all three areas, but only in the United States is the climate warm enough for good corn production (see Fig. 10–10). Potatoes are a major crop in Europe and rice and wheat are in Asia.

**Fig. 10–14** A landscape in the Alfisol (Gray-Brown Podzolic) region of eastern United States where much of the land is used for general farming and livestock production.

## Ultisols

All of the soil orders discussed so far have *not* shown evidence of extreme weathering or age. Weathering and soil evolution is limited by precipitation, hardness of rock, high water table, or just the short period of time that has been available since the parent material was formed. Recent glaciation has played an important role in the existence of vast areas of slightly or only moderately weathered soils in the northern part of the northern hemisphere. As one approaches the humid tropics, ancient landscapes with very long periods of weathering coupled with abundant rainfall and high temperature have created two unique soil orders found in the humid tropics. These orders are Ultisol and Oxisol. The Oxisol order will be considered after Ultisol.

The word Ultisol comes from the Latin *ultimus*, meaning last, or in the case of Ultisol, soils that are the most weathered and that show the ultimate effects of leaching. Ultisols have argillic horizons with low base saturation, being less than 35 percent. High amounts of exchangeable aluminum are usually present. They occur in the warmer parts of the world where the mean annual soil temperature is 47° F or more and have a period each year when rainfall is considerably in excess of evapotranspiration. Few weatherable minerals usually exist in the soil to release bases and trees play a major role in transporting nutrients from the lower part of the soil to the upper part of the solum. Agriculture can be maintained only by shifting cultivation or by the use of fertilizers.

Most soils now classified as Ultisols were formerly called Red-Yellow Podzolic (1949 system). The gross morphology and horizon sequence is similar to the Gray-Brown Podzolics now classified as Alfisols. In fact, evidence points to the fact that Ultisols can be Alfisols before they become sufficiently weathered to be Ultisols. Some properties of Ultisols located in the southeastern United States are presented in Table 10–6. Several things can be noted from the data: (1) the clay content shows the development of an argillic horizon, (2) the organic-matter content of all horizons, except the very thin A1, is quite low, (3) cation-exchange capacity is relatively low, expressing the low organic-matter content and presence of low cation exchange capacity clays such as kaolinite, and (4) the amount of exchangeable bases and the percentage base saturation are very low, except for the very thin A1 horizon. The addition and decomposition of residues from the vegetation play an important role in maintaining the higher base saturation in the upper part of the solum (A1 and A2 horizons). Generally, Ultisols have a very low fertility level for food crops, but respond well to fertilization owing to their desirable physical properties.

**Table 10–6** Selected Characteristics of an Ultisol

| Horizon | Depth, Inches | Clay, Per-centage | Organic Matter, Per-centage | Cation-Exchange Capacity | Exchange-able Bases, me per 100 gm | Base Satura-tion, Per-centage |
|---|---|---|---|---|---|---|
| A1 | 0–1 | 9 | 5.2 | 14 | 4.1 | 29 |
| A2 | 1–5 | 8 | 1.8 | 6 | 0.9 | 15 |
| A3 | 5–7½ | 8 | 1.2 | 4 | 0.4 | 9 |
| B1t | 7½–12 | 17 | 0.4 | 5 | 0.4 | 8 |
| B2, 3t | 12–32 | 35 | 0.3 | 10 | 0.8 | 8 |
| IIC | 32–47 | 48 | 0.1 | 13 | 0.8 | 6 |

Data of Profile 92 from "Soil Classification, A Comprehensive System," USDA, 1960. Soil from Jenkins County, Georgia that developed from coastal plain alluvium under mixed forest of pines, oak, sweetgum, and blackgum.

When settlers moved across the United States in the eighteenth and nineteenth centuries, the farmers who settled on the Alfisols and Mollisols of the then "northwest" found good soil, and a surplus of agricultural products quickly followed. In 1827 Timothy Flint wrote:

"Everyone who was willing to work had an abundance of the articles which the soil produced, far beyond the needs of the country, and it was a prevalent complaint in the Ohio Valley that this abundance greatly exceeded the chances for a profitable sale."[4]

The farmers became agitators for the development of waterways that would permit shipment of excess food to markets. By the time the population of United States reached 50 million (about 1880), enough food was being shipped to Europe to feed 25 million people. Interestingly, the Ohio River is an approximate boundary between the Alfisolic and Mollisolic soils of the North and the Ultisolic soils of the South. An entirely different fortune befell the early settlers who moved south and east where Ultisols dominated the landscape. The plight of early Virginians and one farmer in particular, Edmund Ruffin, has been aptly described as follows.

"Virginians by the thousands, seeing only a dismal future at home, emigrated to newer states of the South and Middle West. The rate of population growth dropped from 38 per cent in 1820 to less than 14 percent in 1830, and then to a mere two percent in 1840. As Ruffin himself later described it, 'All wished to sell, none to buy.' So poor and exhausted were his lands and those of his neighbors that they

[4] *Virgin Land,* H. N. Smith, Harvard University Press, 1957. Permission granted by Harvard University Press.

averaged only ten bushels of corn per acre and even the better lands a mere six bushels of wheat."[5]

Ruffin was determined to make good on the plantation he inherited from his father in Virginia. By chance he read Sir Humphrey Davy's book *Elements of Agricultural Chemistry* published in London in 1813. He was particularly intrigued by Davy's statement to the effect that a soil which contains salts of iron or any other acid matter could be ameliorated by the application of quick lime. Ruffin experimented with marl and found it beneficial. Not only were the Ultisols very low in fertility, the low pH made manure ineffective, though its influence on nutrient availability and aluminum toxicity was a problem in the argillic horizon. Ruffin's discovery was a turning point in the use of very weathered Ultisols and led to the move away from soil exhaustion followed by abandonment to permanent agriculture. Today we can easily add fertilizers and alter soil pH. With the long year-round growing season and rainfall, this region that proved so troublesome at first has the potential to become one of the most productive agricultural and forestry regions in the United States. One big problem remaining to be solved is the development of low-cost methods of incorporating lime into subsoils to raise the pH and to reduce the Al toxicity.

One other major area of Ultisols is shown in Fig. 10–2 in southeastern Asia. Many smaller areas of Ultisols occur in regions where Oxisols are common. On many of these Ultisols shifting cultivation is still practiced.

## Oxisols

All soils with oxic horizons belong to the Oxisol order. Oxic horizons are subsurface horizons consisting of a mixture of hydrated oxides of iron and/or aluminum and variable amounts of 1:1 lattice clays. Few other minerals exist in oxic horizons except some that are highly insoluble. More specifically, the oxic horizon has (1) a thickness of 12 inches or more, (2) a cation-exchange capacity of 16 milliequivalents or less for each 100 grams of clay, (3) none or only a trace of minerals that can weather to release bases, (4) little if any water-dispersible clay, and (5) diffuse boundaries with adjacent horizons. An Oxisol profile is shown in Fig. 10–15 that has very diffuse horizon boundaries and has laterite or plinthite, a common feature of Oxisols (see also Fig. 1–7). In the 1949 system these soils were called Latosols.

Oxisols exist only on ancient land surfaces in the humid tropics and

[5] From the preface of *An Essay on Calcareous Manures,* Edmund Ruffin, The Belknap Press of Harvard University, 1961. Original book was published in 1832.

**Fig. 10–15** Profile of an Oxisol (Latosol) which contains chunks of laterite (or iron stone). Very little change is visible in the upper 4 to 6 feet. Scale at the right is divided into 6-inch lengths.

contain no reserve of bases beyond those on the exchange sites. Agriculture on Oxisol soils utilizes shifting cultivation similar to that on Ultisol soils. Ultisols and Oxisols commonly exist in the same landscapes and probably owe at least part of their differences to the tendency for Oxisols to develop from more basic rocks in which the minerals are more weatherable and with less tendency for silicate clays to form. As a result Oxisols are richer in iron and have fewer weatherable minerals still remaining in the soil. The soil aggregates are very stable and the soils are very erosion resistant.

Two very large areas of Oxisols exist: one in South America (including the Amazon Valley) and the other in Africa (including the Congo). Other large acreages exist in eastern India, Burma, and surrounding regions (Fig. 10–2). The total acreage is comparable to that of Mollisols, but few people, by comparison, live on the Oxisols. The great Amazon Basin, the Sahara Desert, and the Tundra region all have very low population densities of less than 2 persons per square mile.

## Vertisols

Vertisols are mineral soils that (1) are over 20 inches (50 cm) thick, (2) have 30 or more percent clay in all horizons, and (3) have cracks at least 1-cm wide to a depth of 20 inches (unless irrigated) at some time in most years. Conditions that give rise to the development of Vertisols are parent materials high in, or that weather to form, large amounts of montmorillonitic (expanding) clay and a climate with a wet and dry season. The typical vegetation in natural areas is grass or herbaceous annuals, although some Vertisols support drought-tolerant woody plants.

**Fig. 10–16** Large cracks develop in Vertisols during the dry season. Photograph of Guanica clay in the Lajas Valley of southwestern Puerto Rico.

The central concept of Vertisols is one of soils that crack widely in dry seasons as shown in Fig. 10–16. After the cracks develop in the dry season, surface soil material sloughs off into cracks. The soil rewets in the wet season from water that quickly runs into the cracks and is held in the soil by impermeable underlying layers. Repeated drying or rewetting periods cause a "humping up" of areas between the cracks to produce a microrelief called *gilgai*. Repeated cycles of expansion and contraction cause a gradual inverting of the soil, thus, they are called Vertisols. Expansion and contraction in the subsoil with wetting and drying produces shiny ped surfaces, *slickensides* (see Fig. 10–17). The expansion and contraction causes a misalignment of fence and telephone posts. Pipelines may be broken and road and building foundations destroyed. G. W. Olson, soil scientist at Cornell University, found that the Mayans in Guatemala avoided Vertisol areas for the construction of temples.

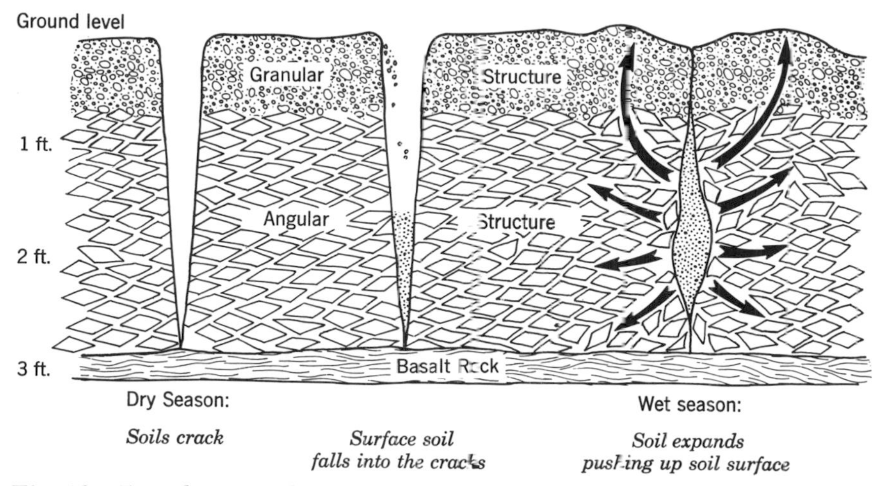


**Fig. 10–17** Schematic drawing showing the formation of Vertisols (Grumusols). From left to right (1) cracks develop in dry season, (2) loose material falls into cracks, and (3) wetting of the soil in the wet season causes expansion and movement of soil in the lower part of the soil to produce angular or wedge-shaped peds with shiny surfaces (slickensides) and a microrelief called gilgai. (Adapted from "Soils of Arizona," *S. W. Boul, Agr. Exp. Sta. Tech. Bul.*, **171**, 1966.)

Some properties of the Houston clay from the Blackland Prairie region of Texas are presented in Table 10–7 to illustrate some of the important properties of Vertisols relative to their nature and use.

There is no evidence of clay migration and the content of clay is very high in all horizons. All horizons have a clay texture. The high content

**Table 10–7** Some Properties of the Houston Clay—a Vertisol

| Horizon | Depth, Inches | Clay, Percentage | Organic Matter, Percentage | $CaCO_3$, Percentage | CEC, me per 100 grams |
|---|---|---|---|---|---|
| A11 | 0–18 | 58 | 4.1 | 17 | 64 |
| A12 | 18–40 | 58 | 2.1 | 20 | 58 |
| AC | 40–60 | 58 | 1.0 | 26 | 53 |
| C | 60–78 | 59 | 0.4 | 32 | 47 |

Data from "Houston Black Clay, the Type Grumusol: II Mineralogical and Chemical Characterization," G. W. Kunze and E. H. Templin, *Proc. Soil Sci. Soc. Am.* **20**:91–96, 1956. Average of 5 profiles.

of expanding clay makes the soil very sticky when wet and very hard when dry. Permeability to water is very low when the soil is wet. The organic matter decreases gradually with increasing soil depth. Lime was present in the parent material, and the high amount of lime still remaining in the upper soil horizons is evidence of the "closed-system" nature of the soil and the limited opportunity for any soluble material to be leached out of the bottom of the profile. The high cation-exchange capacity reflects the high content of montmorillonitic clay.

Vertisols are widely distributed in the world between 45° north and south latitude (see Fig. 10–2). The three largest areas of Vertisols in the world are in Australia (70 million acres), India (60 million acres), and the Sudan (40 million acres). Agriculturally the soils have great potential where power tools, fertilizers, and irrigation are available. The natural fertility level can be considered quite high, although the use of nitrogen and phosphorus is beneficial. Tillage of the soil is difficult with primitive tillage tools. The "Blacklands" of Texas and Alabama are some of the best agricultural lands in the United States. Worldwide Vertisols are used mainly for cotton, wheat, corn, sorghum, rice, sugar cane, and pasture.

Historically speaking, the classical concept of Rendzina as shallow soil overlying chalk or limestone was developed in Europe. The term Rendzina was later applied to deep, clayey soils like those in the Blacklands of Texas and Alabama in America. Oakes and Thorp proposed that these deep clayey soils be called *Grumusols* in 1950. In various parts of the world these soils have been called Black

Cotton, Tirs, Black Earths and Regur. Now, most of these soils would be classified as Vertisols.

### Histosols

Organic soils are classified as Histosols. Most Histosols are recognized by a histic epipedon that is over 12 inches thick, saturated with water at least 30 consecutive days a year, and contains at least 20 percent organic matter. Florida and Minnesota are two states in the United States that have large areas of these soils, however, Histosols are found throughout the world. Their total extent is less than 1 percent of the land surface of the world (Table 10–4)

Histosols develop where the soil is saturated from at least one month each year to continuous saturation. The characteristics of Histosols depends primarily on the nature of the vegetation that was deposited in the water and the degree of decomposition. In relatively deep water remains of algae and other aquatic plants give rise to highly colloidal material which shrinks greatly on drying. As the lake gradually fills, rushes, wild rice, water lilies, and similar plants flourish. The partially decayed remains of these plants are less slimy and colloidal. Gradually sedges, reeds, and eventually grasses are able to grow. Peat from such plants is much more fibrous than that produced from plants growing in deeper water. Shrubs and trees follow in time and produce a woody type of peat. Changes in water depth may cause a recurrence of deeper-water plants, and hence layers of more pulpy material may occur over fibrous peat, etc. The following plant succession in the filling of a Minnesota lake has been suggested by Soper.[6]

1. Stonewort: waterweed stage.
2. Pondweed: water lily stage.
3. Rush: wild rice stage.
4. Bog: meadow stage.
5. Bog: heath stage.
6. Tamarack: spruce stage.
7. Pine association.

When organic soils are drained in such a way as to remove excess water rapidly yet maintain the water level at a relatively shallow depth, they may be used for very intensive types of crop production. In the northern states these soils are used for the production of onions, celery, mint, potatoes, cabbage, cranberries, carrots, and other root crops. Corn is produced to some extent, and considerable areas are

[6] "The Peat Deposits of Minnesota," by E. K. Soper, *Minnesota Geological Survey Bull.* 16, 1919.

used as pasture. Late-spring or summer and early-fall frosts are the greatest hazard to crop growth in the temperate region. Other hazards include fires and wind erosion (Fig. 10–18). A great variety of special crops is grown on the organic soils of the South and East. Special methods of tillage, coupled with careful application of fertilizer, are required to bring these soils to their highest state of productivity.

**Fig. 10–18** Histosol or organic soil landscape. Crop on the left is grass sod to be used for landscaping. The lightness of the soil makes it ideal for sod production but this also contributes to its susceptibility to wind erosion. Note windbreak of trees at far end of field.

## SOIL CLASSIFICATION CATEGORIES BELOW THE ORDER

The discussion of the soil orders permitted an introduction to soil geography on a worldwide basis. Within each order there is much diversity and a need for much more precise definition and classification of use to permit wise utilization. These additional categories will be considered next.

### Suborders

The orders are divided into suborders primarily on the basis of chemical and physical properties that reflect either the presence or absence of waterlogging or genetic differences due to climate and its partially associated variable, vegetation. The distribution of suborders in the United States is given in Fig. 10–19 along with a legend defining the suborders. For example, the Aqualfs are "wet" (aqu for aqua) Alfisols saturated with water sometime during the year. Borolls are Mollisols of the cool regions. The suborder names all have two syllables with the last syllable indicating the order, such as *alf* for Alfisol and *oll* for Mollisol. Formative elements in names of suborders are given in Table 10–8.

**Table 10–8** Formative Elements in the Names of Suborders

| Formative Elements | Derivation of Formative Element | Connotation of Formative Element |
|---|---|---|
| alb | L. *albus*, white | Presence of albic horizon (a bleached eluvial horizon) |
| and | Modified from Ando | Andolike |
| aqu | L. *aqua*, water | Characteristics associated with wetness |
| ar | L. *arare*, to plow | Mixed horizons |
| arg | Modified from argillic horizon; L. *argilla*, white clay | Presence of argillic horizon (a horizon with illuvial clay) |
| bor | Gk. *boreas*, northern | Cool |
| ferr | L. *ferrum*, iron | Presence of iron |
| fibr | L. *fibra*, fiber | Least decomposed stage |
| fluv | L. *fluvius*, river | Flood plains |
| hem | Gk. *hemi*, half | Intermediate stage of decomposition |
| hum | L. *humus*, earth | Presence of organic matter |
| lept. | Gk. *leptos*, thin | Thin horizon |
| ochr | Gk. base of *ochros*, pale | Presence of ochric epipedon (a light surface) |
| orth | Gk. *orthos*, true | The common ones |
| plag | Modified from Ger. *plaggen*, sod | Presence of plaggen epipedon |
| psamm | Gk. *psammos*, sand | Sand textures |
| rend | Modified from Rendzina | Rendzinalike |
| sapr | Gk. *sapros*, rotten | Most decomposed stage |
| torr | L. *torridus*, hot and dry | Usually dry |
| trop | Modified from Gk. *tropikos*, of the solstice | Continually warm |
| ud | L. *udus*, humid | Of humid climates |
| umbr | L. *umbra*, shade | Presence of umbric epipedon (a dark surface) |
| ust | L. *ustus*, burnt | Of dry climates, usually hot in summer |
| xer | Gk. *xeros*, dry | Annual dry season |

## Great Groups and Subgroups

Suborders are divided into great groups (do not confuse with the great soil groups of 1938 system). Great group names are coined by prefixing one or more additional formative elements to the appropriate suborder name. The prefixes are used to indicate the presence or absence of certain diagnostic horizons. The formative elements with their meaning and connotation are shown in Table 10–9. As an example, a Fragiaqualf is a "wet" Alfisol with a fragipan. Subgroup names indicate to what extent the central concept of the great group is expressed. A typic Fragiaqualf is a soil that is typical for the Fragiaqualf great group.

**Table 10–9** Formative Elements for Names of Great Groups

| Formative Element | Connotation | Formative Element | Connotation |
|---|---|---|---|
| acr | Extreme weathering | moll | Mollic epipedon |
| agr | Agric horizon | nadur | See *Natr* and *Dur* |
| alb | Albic horizon | natr | Natric horizon |
| and | Ando-like | ochr | Ochric epipedon |
| anthr | Anthropic epipedon | pale | Old development |
| aqu | Wetness | pell | Low chroma |
| arg | Argillic horizon | plac | Thin pan |
| calc | Calcic horizon | plag | Plaggen horizon |
| camb | Cambic horizon | plinth | Plinthite |
| chrom | High chroma | quartz | High quartz |
| cry | Cold | rend | Rendzina-like |
| dur | Duripan | rhod | Dark-red colors |
| dystr, dys | Low base saturation | sal | Salic horizon |
| eutr, eu | High base saturation | sider | Free iron oxides |
| ferr | Iron | sombr | A dark horizon |
| frag | Fragipan | sphangno | Sphagnum-moss |
| fragloss | See *frag* and *gloss* | torr | Usually dry |
| gibbs | Gibbsite | trop | Continually warm |
| gloss | Tongued | ud | Humid climates |
| hal | Salty | umbe | Umbric epipedon |
| hapl | Minimum horizon | ust | Dry climate, usually hot in summer |
| hum | Humus | | |
| hydr | Water | verm | Wormy, or mixed by animals |
| hyp | Hypnum moss | vitr | Glass |
| luo, lu | Illuvial | xer | Annual dry season |

### Family and Series

The families indicate features that are important to plant growth such as texture, mineralogical composition, or temperature. The series gets down to the individual soil and the name is that of a natural feature or place near where the soil was first recognized. Familiar series names include Amarillo, Carlsbad, and Fresno and obviously refer to soils located in Texas, New Mexico, and California, respectively.

### The Pedon

A pedon is the smallest volume that we should describe and sample to represent the nature and arrangement of horizons of a soil and variability in its other properties that are preserved in samples. It is comparable in some ways to the unit cell of a crystal. A pedon has three dimensions. Its lower limit is the somewhat vague limit between the soil and the "not-soil" below. The lateral dimensions are large enough to represent the nature of any horizons and variability that may be present. A horizon may be variable in thickness or in composition, or it may be discontinuous. The area of a pedon ranges from 1 to 10 square meters, depending on the variability in the soil.

# 11

# Nutrient Requirement and Mineral Nutrition of Plants

The growth and development of plants are determined by numerous factors of soil and climate and by factors inherent in the plants themselves. Some of these factors are under the control of man, but many are not. Man has little control over air, light, and temperature, for example, but can influence the supply of plant nutrients in the soil. He may increase the supply of available nutrients by modifying soil conditions through good management or by making additions in the form of fertilizers. Anyone dealing directly with the growth of plants is particularly concerned with their nutrient requirements and the role of the nutrients in plant growth.

## ELEMENTS USED BY PLANTS

If a soil is to produce crops successfully, it must have, among other things, an adequate supply of all the necessary nutrients which plants take from the soil. Not only must required nutrient elements be present in forms that plants can use, but also there should be a rough balance between them in accordance with the amounts needed by plants. If any of these elements is lacking or if it is present in improper proportions, normal plant growth will not occur. Elements required by plants are *essential.*

## Characteristics of an Essential Element

For many centuries man knew that substances such as manure, ashes, and blood had a stimulating effect on plant growth. The effect was found to result basically from the essential elements contained in the materials. As recently as 1800, however, man did not know which elements removed from the soil were indispensable. Discovery of the chemical elements and techniques for their determination were prerequisites for determining which nutrients were essential for plant growth. The search is still continuing. Two of the criteria commonly used in establishing the essentiality of a plant nutrient are (1) its necessity for the plant to complete its life cycle and (2) its direct involvement in the nutrition of the plant apart from possible effects in correcting some unfavorable condition in the soil or culture medium.

Sixteen elements are commonly listed as essential for plant growth. It is expected that this list will grow as more refined techniques and materials become available for experimentation. As was pointed out in Chapter 2, well over 40 different elements in addition to the essential ones have been detected in plants. Some are known to stimulate growth without meeting the requirements of an essential element. In some instances, part of the plants requirement may be satisfied by another element as is the case where sodium substitutes for some of the potassium in sugar beets. Nutrients shown to be able to substitute for essential elements or that are sometimes needed by some plants, or that benefit plants include cobalt, vanadium, silicon, and sodium.

## The Essential Elements

The sixteen essential elements are divided into two groups on the basis of the *amount* plants require. The macronutrients are required in relatively large amounts of 10 to 100 pounds or more per acre. The micronutrients are required in small amounts. Difficulty is sometimes encountered in growing plants so that a micronutrient deficiency can be produced in the laboratory. Both groups are listed as follows:

| Macronutrients | Micronutrients |
|---|---|
| Carbon | Manganese |
| Hydrogen | Copper |
| Oxygen | Zinc |
| Nitrogen | Molybdenum |
| Phosphorus | Boron |
| Potassium | Chlorine |
| Calcium | Iron |
| Magnesium | |
| Sulfur | |

## Sources of Plant Nutrients

Plants get their nutrient elements from three sources: air, water, and soil. Carbon and some of the oxygen are obtained from the air, whereas hydrogen, some oxygen, and possibly some carbon are taken from the soil solution. Legumes inoculated with effective nodule bacteria obtain some of their nitrogen from the air. Some sulfur may diffuse into the leaves as sulfur dioxide and be utilized. Other nutrient elements must be taken from the soil under natural conditions.

## Quantities of Nutrients in Crops

The nutrient content of several crops is given in Table 11–1. In considering these quantities of nutrients, it is important to keep in mind that they are taken from the more readily available supply in the soil. Furthermore, the quantities removed by a single crop may seem rather small in some instances; but, when the quantities contained in all the crops of a rotation are totaled or when the amounts removed by crops for several years are considered, the necessity of supplying plant nutrients in the form of fertilizer and manure to maintain soil fertility is apparent.

The quantities of elements given in Table 11–1 do not represent the total quantities that crops require during growth, but rather the quantities contained in the harvested material. Roots and other portions of the plant which may not be harvested require considerable quantities of nutrients.

# NUTRIENT UPTAKE

The surfaces of leaves and stems, as well as roots, can absorb nutrients. Any exposed plant surface appears to be able to function in this regard. Carbon dioxide absorption by the leaves is the major avenue for obtaining carbon. The processes involved in the transport of a nutrient ion from the soil environment into the root and its translocation and distribution within the plant are complex and interrelated. Discussion in this chapter will concentrate on the absorption of mineral nutrients from soils by roots.

## The Process of Nutrient Uptake

Two well-established phenomena are known as regards nutrient absorption by plants. First metabolic energy is required. If root respiration is curtailed, the net uptake of nutrients is minor even from a concentrated solution. Second, the process is selective. Plants have a capacity to selectively absorb certain ions over a wide range of

**Table 11–1** Approximate Pounds per Acre of Nutrients Contained in Crops

| Crop | Acre Yield | Nitrogen | Phosphorus as P | Potassium as K | Calcium | Magnesium | Sulfur | Copper | Manganese | Zinc |
|---|---|---|---|---|---|---|---|---|---|---|
| GRAINS | | | | | | | | | | |
| Barley (Grain) | 40 bu. | 35 | 7 | 8 | 1 | 2 | 3 | 0.03 | 0.03 | 0.06 |
| Barley (Straw) | 1 ton | 15 | 3 | 25 | 8 | 2 | 4 | 0.01 | 0.32 | 0.05 |
| Corn (Grain) | 150 bu. | 135 | 23 | 33 | 16 | 20 | 14 | 0.06 | 0.09 | 0.15 |
| Corn (Stover) | 4.5 tons | 100 | 16 | 120 | 28 | 17 | 10 | 0.05 | 1.50 | 0.30 |
| Oats (Grain) | 80 bu. | 50 | 9 | 13 | 2 | 3 | 5 | 0.03 | 0.12 | 0.05 |
| Oats (Straw) | 2 tons | 25 | 7 | 66 | 8 | 8 | 9 | 0.03 | ... | 0.29 |
| Rice (Rough) | 80 bu. | 50 | 9 | 8 | 3 | 4 | 3 | 0.01 | 0.08 | 0.07 |
| Rice (Straw) | 2.5 tons | 30 | 5 | 58 | 9 | 5 | ... | ... | 1.58 | ... |
| Rye (Grain) | 30 bu. | 35 | 5 | 8 | 2 | 3 | 7 | 0.02 | 0.22 | 0.03 |
| Rye (Straw) | 1.5 tons | 15 | 4 | 21 | 8 | 2 | 3 | 0.01 | 0.14 | 0.07 |
| Sorghum (Grain) | 60 bu. | 50 | 11 | 13 | 4 | 5 | 5 | 0.01 | 0.04 | 0.04 |
| Sorghum (Stover) | 3 tons | 65 | 9 | 79 | 29 | 18 | ... | ... | ... | ... |
| Wheat (Grain) | 40 bu. | 50 | 11 | 13 | 1 | 6 | 3 | 0.03 | 0.09 | 0.14 |
| Wheat (Straw) | 1.5 tons | 20 | 3 | 29 | 6 | 3 | 5 | 0.01 | 0.16 | 0.05 |
| HAY | | | | | | | | | | |
| Alfalfa | 4 tons | 180 | 18 | 150 | 112 | 21 | 19 | 0.06 | 0.44 | 0.42 |
| Bluegrass | 2 tons | 60 | 9 | 50 | 16 | 7 | 5 | 0.02 | 0.30 | 0.08 |
| Coastal Bermuda | 8 tons | 185 | 31 | 224 | 59 | 24 | ... | 0.21 | ... | ... |
| Cowpea | 2 tons | 120 | 11 | 66 | 55 | 15 | 13 | ... | 0.65 | ... |
| Peanut | 2.25 tons | 105 | 11 | 79 | 45 | 17 | 16 | ... | 0.23 | ... |
| Red Clover | 2.5 tons | 100 | 11 | 83 | 69 | 17 | 7 | 0.04 | 0.54 | 0.36 |
| Soybean | 2 tons | 90 | 9 | 42 | 40 | 18 | 10 | 0.04 | 0.46 | 0.15 |
| Timothy | 2.5 tons | 60 | 11 | 79 | 18 | 6 | 5 | 0.03 | 0.31 | 0.20 |
| FRUITS AND VEGETABLES | | | | | | | | | | |
| Apples | 500 bu. | 30 | 5 | 37 | 8 | 5 | 10 | 0.03 | 0.03 | 0.03 |
| Beans, Dry | 30 bu. | 75 | 11 | 21 | 2 | 2 | 5 | 0.02 | 0.03 | 0.06 |
| Cabbage | 20 tons | 130 | 16 | 108 | 20 | 8 | 44 | 0.04 | 0.10 | 0.08 |
| Onions | 7.5 tons | 45 | 9 | 33 | 11 | 2 | 18 | 0.03 | 0.08 | 0.31 |
| Oranges (70 Pound Boxes) | 800 boxes | 85 | 13 | 116 | 33 | 12 | 9 | 0.20 | 0.06 | 0.24 |
| Peaches | 600 bu. | 35 | 9 | 54 | 4 | 8 | 2 | ... | ... | 0.01 |
| Potatoes (Tubers) | 400 bu. | 80 | 13 | 125 | 3 | 6 | 6 | 0.04 | 0.09 | 0.05 |
| Spinach | 5 tons | 50 | 7 | 25 | 12 | 5 | 4 | 0.02 | 0.10 | 0.10 |
| Sweet Potatoes (Roots) | 300 bu. | 45 | 7 | 62 | 4 | 9 | 6 | 0.03 | 0.06 | 0.03 |
| Tomatoes (Fruit) | 20 tons | 120 | 18 | 133 | 7 | 11 | 14 | 0.07 | 0.13 | 0.16 |
| Turnips (Roots) | 10 tons | 45 | 9 | 75 | 12 | 6 | ... | ... | ... | ... |
| OTHER CROPS | | | | | | | | | | |
| Cotton (Seed and Lint) | 1500 lbs. | 40 | 9 | 13 | 2 | 4 | 2 | 0.06 | 0.11 | 0.32 |
| Cotton (Stalks, Leaves & Burs) | 2000 lbs. | 35 | 5 | 29 | 28 | 8 | ... | ... | ... | ... |
| Peanuts (Nuts) | 1.25 tons | 90 | 5 | 13 | 1 | 3 | 6 | 0.02 | 0.01 | ... |
| Soybeans (Grain) | 40 bu. | 150 | 16 | 46 | 7 | 7 | 4 | 0.04 | 0.05 | 0.04 |
| Sugar Beets (Roots) | 15 tons | 60 | 9 | 42 | 33 | 24 | 10 | 0.03 | 0.75 | ... |
| Sugarcane | 30 tons | 96 | 24 | 224 | 28 | 24 | 24 | ... | ... | ... |
| Tobacco (Leaves) | 2000 lbs. | 75 | 7 | 100 | 75 | 18 | 14 | 0.03 | 0.55 | 0.07 |
| Tobacco (Stalks) | – | 35 | 7 | 42 | ... | ... | ... | ... | ... | ... |

Adapted from "Our Land and Its Care," The Fertilizer Institute, Washington, D.C. 1962.

conditions while effectively excluding others. Any theory of ion absorption must take into account these two phenomena.

The roots of plants are more or less surrounded by the soil solution and are in intimate contact with soil particles at many sites. Root cells have an "outer" space into which ions from the soil solution and from the exchange sites of the soil can diffuse. Diffusion of ions into these spaces is reversible and occurs without regard to the plants metabolic activities. It is a passive activity so far as the plant is concerned. This is a prelude to the irreversible transport of the ions across a seemingly impermeable membrane that requires an expenditure of energy.

The interior surface of the "outer" space has binding sites where carrier molecules are believed to be located. The carriers combine with the ions and together they migrate across a membrane that is impermeable to the ion alone. Once across the membrane the carrier molecule and ion separate as the ion is deposited in the "inner" space of the cell, commonly called the vacuole. The carrier transport requires energy that is derived from respiration and the process enables the cell to achieve an ionic concentration in the cell which may be many times that of the external soil solution (see Fig. 11–1).

This theory explains the selective absorption of ions since the carriers are specific for a given ion or group of ions. Nitrate and phosphate are both anions but require separate carriers. In fact, different carriers are required for $H_2PO_4^-$ and $HPO_4^{2-}$. Calcium and magnesium are transported by different carriers. The carrier for potassium can also transport cesium while still another can transport sodium and lithium. Anions and cations are absorbed by the same mechanism but using different carriers.

The most rapid uptake of nutrients occurs near the tip of newly formed roots or root hairs. This is where the rate of respiration is greatest. As roots mature, morphologically speaking, they lose the capacity to take up nutrients and the large, suberized roots so readily seen in the soil serve largely as a plumbing system for nutrient (and water) transport, rather than for uptake. It is common for new roots to originate along such mature roots so a revival in nutrient uptake can occur. In the spring, small white newly formed roots appear along the large tap roots of established alfalfa plants.

### Factors Affecting Nutrient Uptake

Those factors that affect metabolism, and thereby the availability of respiratory energy, will directly affect nutrient uptake. These include the supply of respiratory substrate, temperature, and oxygen supply.

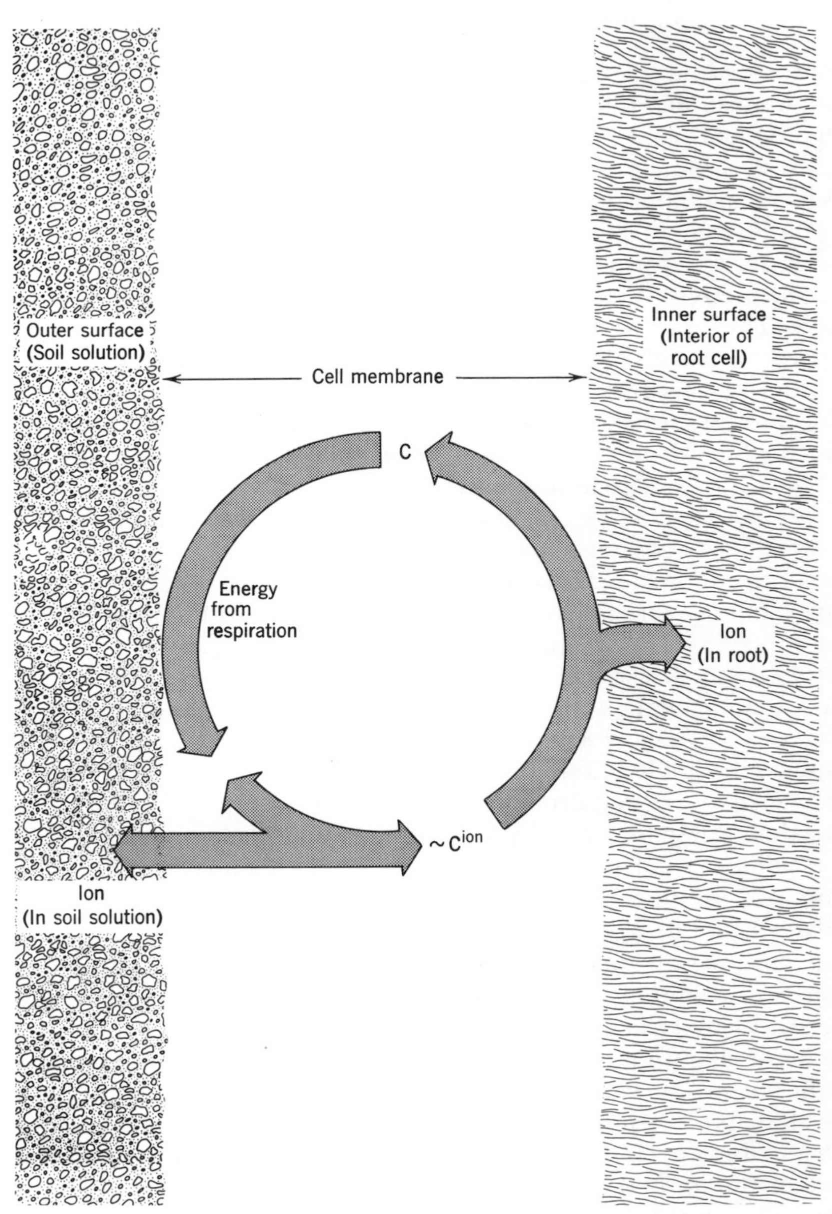


**Fig. 11–1** A diagrammatic representation of the carrier theory of nutrient uptake. An organic carrier, *C*, links up with an ion from the soil solution at the exterior surface of the root cell. The ion is carried across the cell's membrane and deposited in the interior of the cell in the roots. (From "Roots, Selectors of Plant Nutrients," J. B. Hanson, *Plant Food Review*, Spring 1967, p. 8.)

The oxygen supply can be significantly altered by management practices. Soil compaction can reduce nutrient uptake through its affect on the availability of oxygen for root respiration. It is interesting that not all nutrients are reduced to the same degree. Lawton established that soil compaction reduced the uptake of potassium more than it did phosphorus or nitrogen, and that calcium was least affected in the case of corn.[1] Plants grown in water culture usually obtain maximum growth when air is bubbled through the water.

An increase in the concentration of the nutrient in the environment external to the root will favor an increase in nutrient uptake when the initial concentration is low.

The moisture content of the soil is important because it influences the rate of movement and diffusion of ions into the "outer" spaces of the root cells. For example, it has been observed that drying the soil reduces phosphorus uptake. This is expected because phosphorus is so slightly soluble and has such low mobility in the soil.

The density and distribution of roots in the soil is also important. Roots that enter a zone with a high nitrogen and phosphorus level tend to proliferate. It seems that when phosphate fertilizer is placed with the nitrogen in the soil, greater uptake of phosphorus occurs. The density of roots and extent of root surfaces becomes more important as the mobility of the nutrient in the soil decreases. Thus, root proliferation is expected to be less important for nitrate uptake than for phosphate uptake. Crops with deeply penetrating root systems, in general, require less fertilization than those with shallow root systems.

## FACTORS AFFECTING PLANT COMPOSITION

The mineral composition of the plant is the net result of many complex and interrelated processes which operate simultaneously. Striking plant nutrient deficiency symptoms and a below normal concentration in the plant tissue can occur for a single nutrient when it is the only one in short supply. Animals fed entirely on plants grown in such an instance may develop a disease resulting from the lack of a nutrient in their diet as shown in Fig. 11–2. Cobalt is not essential for plants but it is for ruminants and a shortage in the diet can produce the condition seen in Fig. 11–3. Where animals consume feed from a wide variety of sources, the relationship between animal nutrition and soil fertility is not so direct.

[1] "The Influence of Soil Aeration on the Growth and Absorption of Nutrients by Corn Plants," K. Lawton, *Soil Sci. Soc. Am. Proc.*, **10**:253–268, 1945.

**Fig. 11–2** In different soil areas animals may suffer from a deficiency of different nutrients in the locally grown feed. "A calf showing an advanced stage of salt sick or nutritional anemia. Note emaciation, lack of condition as indicated by the hair, appearance of the eye and that the animal shows evidence of diarrhoea." The animal "recovered when given access to the iron-copper supplement." (Courtesy of Florida Experiment Station.)

Some interesting studies have been conducted to compare the feeding value of crops produced on a soil of low fertility with that grown on a high fertility soil.[2] A first impression might be that high fertility soils produce more nutritious crops. Facts do not support such a view. Crops grown on low fertility soils will probably be less well-developed, but if the nutrients in both soils are in balance, little difference in plant composition occurs. The striking deficiency symptoms of plants and very low percentages in plant tissue are not so much the result of soil fertility level as they are the balance of nutrients with regard to each other. It is sometimes easy to produce a deficiency of nitrogen by the addition of large amounts of phosphorus and potassium to a soil that normally would produce a good yield.

[2] "Nutritive Values of Crops and Cows Milk as Affected by Soil Fertility," *Mich. Agr. Exp. Sta. Quart. Bull.* **32,** 352–359, 1950.

**Fig. 11–3** Two views of the same cow. Above, when fed a ration deficient in cobalt. Below, after a cobalt supplement was included in the ration. (Courtesy of Dairy Department, Michigan State University.)

## Plant Species and Plant Composition

Perhaps, one of the most notable differences in plant composition exists between the grasses and legumes (Table 11–2). Legumes contain a higher percentage of nitrogen and calcium and a lower percentage of potassium than grasses grown under comparable conditions.

Tobacco is known for its high potassium content which commonly exceeds 4 percent in the leaves. Certain varieties of cereals which have high resistance to disease and insects have a silica encrusting layer in the epidermis and may contain up to 15 percent silica.[3]

The higher base content of the foliage of hardwoods, as compared to spruce, was used to explain the higher pH of soils developed under the hardwood forest in Chapter 9. These examples support the fact that the composition of plants is influenced by genetic factors.

**Table 11–2** Average Composition of Legumes and Grasses

| | Percentage Composition | | | |
|---|---|---|---|---|
| Plant | Nitrogen | Potassium | Calcium | Magnesium |
| Grasses | 0.99 | 1.54 | 0.33 | 0.21 |
| Legumes | 2.38 | 1.13 | 1.47 | 0.38 |

Data of Snider and cited by H. P. Cooper et al., *Soil Sci. Soc. Am. Proc.*, **12**:359–363, 1947.

## Level of an Individual Nutrient

An increase in the amount of a nutrient in the soil that is available or readily soluble may or may not cause an increase in the percentage of the nutrient in the plant. It depends on the extent to which the total growth of the plant is increased. Several possibilities exist. If the nutrient is in short supply and the growth of the plant is limited by it, addition of the nutrient would probably result in a great increase in the amount absorbed. A correspondingly large increase in plant growth could occur so that the percentage composition of the plant remained the same (range b, Fig. 11–4). Where growth is increased more than uptake, a decrease in percentage composition can occur (range a in Fig. 11–4). An increase in growth and percentage composition of the tissue may occur simultaneously as shown by the part of the curve labelled *c* in Fig. 11–4.

[3] "Fertilizer Use, Nutrition and Manuring of Tropical Crops," A. Jacob and H. V. Uexküll, p. 58, Verlagsgesellschaft für Ackerbau mbH, Hannover, 1960.

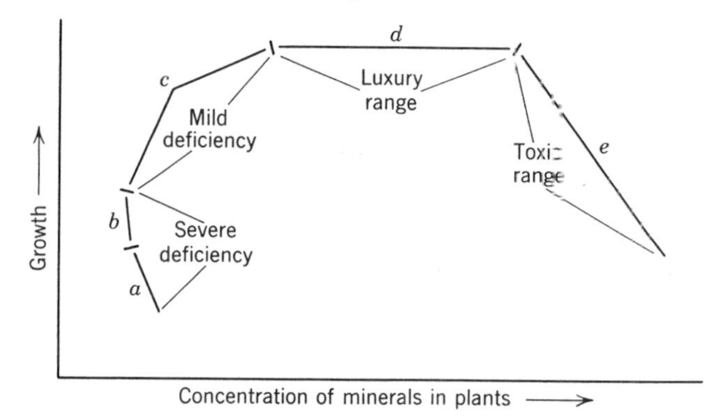


**Fig. 11–4** Diagram showing the relationship between the mineral concentration of plant tissue and growth. (From "Mineral Analysis of Plant Tissue," P. F. Smith, *An. Rev. Pl. Phy.*, Vol. **13**, p. 102, 1962. Used by permission.)

A point is finally reached where increasing increments of a nutrient, added as a fertilizer, are ineffective in increasing yield. Increasing amounts of the nutrient in available form encourage increased uptake and cause a rapid increase in percentage composition. This is the luxury consumption range *d*. The toxic range is where growth declines and percentage composition continues to increase.

This principle has been used to effectively increase the protein content of crops. Nitrogen fertilization was used to increase the quality of wheat grain for milling purposes by increasing its protein content in Kansas. Fifty pounds of nitrogen per acre profitably increased yields. An additional fifty pounds of nitrogen per acre did not increase the grain yield but it increased the protein content and quality enough to make it a profitable practice.[4]

## Cation Content of Plants a Constant

Potassium, calcium, and magnesium all play some of the same roles in the plant, for instance, their role in the buffer system of plant cells. In this regard, they can substitute for each other. This view is consistent with the common observation that the milliequivalents of these three cations in plant tissue tends to be a constant. Further support of the cation constancy in plants is that a large application of potassium sometimes produces magnesium deficiency. Note in Fig. 11–5 that increasing the amount of potassium in the nutrient solution caused large and comparable decreases in the amount of calcium and mag-

[4] "Nitrogen Utilization by Wheat After Fallow," F. W. Smith, *Agron. Abstracts*, p. 42, 1962.

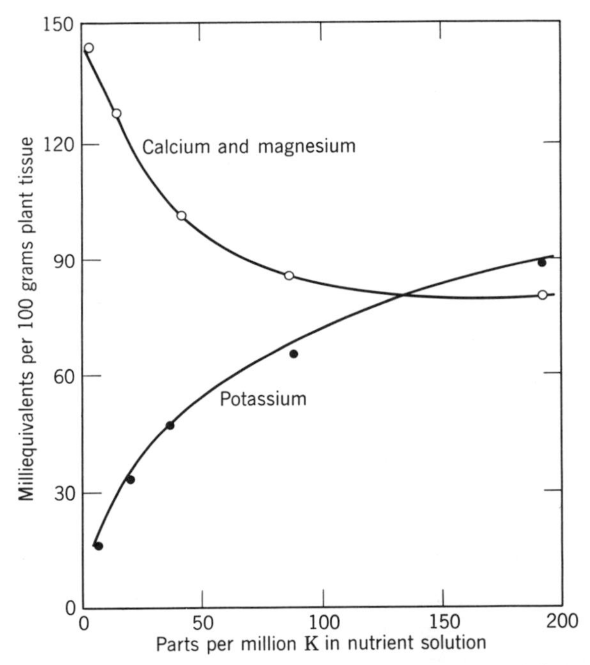


**Fig. 11–5** The calcium and magnesium and the potassium content of alfalfa plants as a function of the potassium concentration in the nutrient solution. (Adapted from data in "Further Evidence Supporting Cation-Equivalent Constancy in Alfalfa," A. Wallace et al., *Jour. Am. Soc. of Agron.*, Vol. **40**, p. 82, 1948.)

nesium in alfalfa plants. Note that the total amounts of potassium, calcium, and magnesium in the tissue remained the same.

Corn grown on the highly calcareous Harpster soils of northcentral Iowa commonly has a severe potassium deficiency symptom so that during the summer these areas of soil can readily be located by observing the growth of the corn. The potassium deficient plants were found to contain as much as ten times more calcium and magnesium than potassium. Contrast this with the fact that grasses tend to contain as much or more potassium as the sum of calcium and magnesium (Table 11–2).

## Plant Composition Changes with Age

Nutrient accumulation occurs at a faster rate than plant weight when the plant is young whereas the reverse is true when the plant approaches maturity. This causes a declining concentration of nutrients in plants with increasing age. It is a common practice for dairy

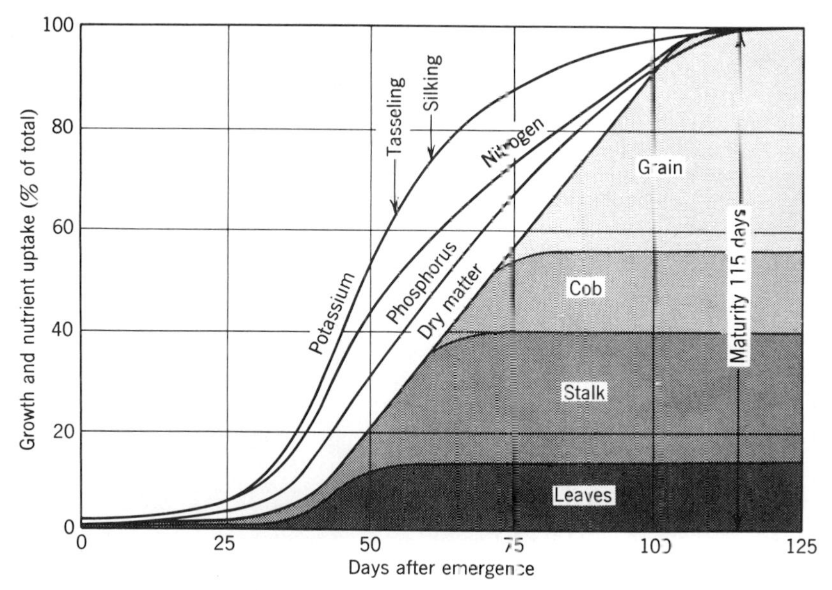


**Fig. 11-6** Uptake of nutrients in relation to the accumulation of dry matter by corn plants. (From "Growth and Nutrients Uptake by Corn," J. J. Hanway, Iowa State Univ. Coop. Ext. Pamphlet 277, 1960.)

farmers to cut forages when they are relatively immature in order to provide more nutritious feed.

The accumulation of dry matter in the various parts of the corn plant and the accumulation of nitrogen, phosphorus, and potassium through the season are given in Fig. 11-6. In this instance, corn plants had absorbed 40 percent of their nitrogen, 30 percent of their phosphorus, and over 50 percent of their potassium by the time they had made 20 percent of their growth.

## Composition Varies with Plant Part

The pattern of organ development for corn is also shown in Fig. 11-6. The grain accounted for nearly half of the weight of mature plants and was largely produced during the last one-third of the growing season. During this time the plant accumulated over 40 percent of its dry matter and only 25 percent of its nitrogen. Thus, the production of the grain occurred at the expense or loss of some nitrogen from the earlier formed organs—stems, leaves, and roots (Fig. 11-7). The translocation of many nutrients and food is a continuing process in plant growth and different organs have different priority for the materials.

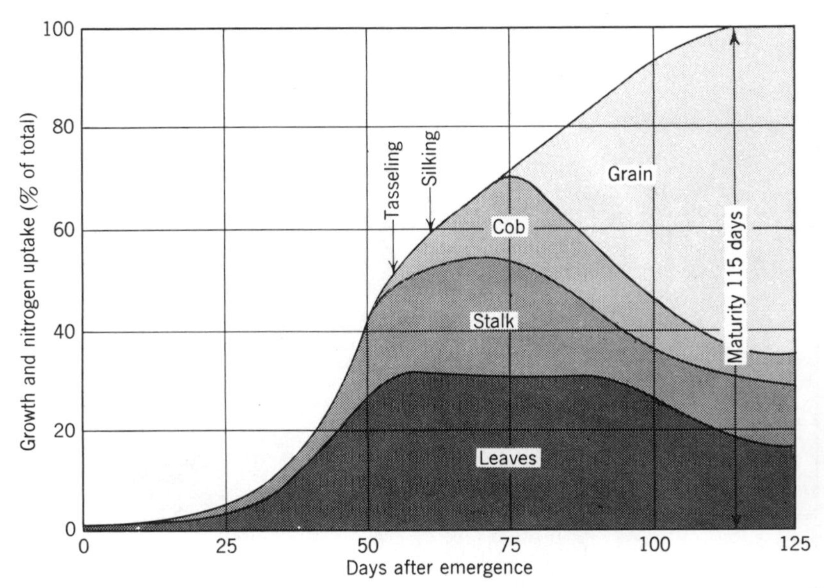


**Fig. 11–7** Part of the nitrogen, as well as phosphorus, potassium, and some other nutrients, is translocated from vegetative tissue to the grain during the later part of the growing season. An amount of nitrogen equal to about half of the nitrogen in the corn grain was absorbed during the time the grain was produced. A similar amount of nitrogen was translocated from the other organs to the grain during the same period. (From Iowa State Univ. Coop. Ext. Pamphlet 277, 1960.)

Fruit or seed production has the highest priority and it is natural for nutrients accumulated in the vegetative parts to be translocated and used later for seed production. Nutrients in excess of fruit growth remain in other organs of the plant. This causes the seeds of plants to have a similar composition when grown under widely different conditions, whereas the composition of the vegetative parts may vary greatly.

The difference in the mobility of the nutrients in plants will be discussed in more detail in the next section. It is important to recognize that some nutrients, iron for example, are relatively immobile. When stored in the lower levels of the plant, they may not be available for the production of other organs later in the season.

## FUNCTIONS AND EFFECTS OF THE ESSENTIAL ELEMENTS

The function of all elements is not well understood, but it is believed that they affect plant growth in one or perhaps more of the following ways: (1) are constituents of plant tissue; (2) act as catalysts or stimulants; (3) effect oxidation-reduction processes in the plant; (4) may aid in regulating the acid content of the plant; (5) may affect the plant osmotically; (6) may affect the entrance into the plant of other elements; or (7) may aid plant growth by providing a more favorable environment for the plant roots.

Each nutrient performs definite duties within the plant, and no one nutrient can be completely substituted for another. Although each element performs certain specific functions, all must work together to produce the best results. It must be remembered, therefore, that the effect of any particular nutrient on plant growth is governed by the supply of the other essential elements, and hence the effects of any one cannot be interpreted on the basis of the activity of that element alone. A balanced supply of nutrients tends to produce plants more resistant to diseases and attacks by insects.

## Roles of Essential Elements in Plant Nutrition

**Carbon, Hydrogen, and Oxygen.** These three elements serve largely in the synthesis of carbohydrates, proteins, fats, and related compounds. They are the major constituents in most plant compounds.

**Nitrogen.** Perhaps no element has received so much attention as has nitrogen in studies relative to plant nutrition. It is found in greater quantities in young, growing parts of plants than in the older tissues and is especially abundant in the leaves and seeds. Nitrogen is a constituent of every living cell and hence its contribution to plant and animal life is evident. It is a part of many proteins which serve as enzymes and also is a part of the chlorophyll molecule.

**Phosphorus.** The importance of phosphorus in plant and animal nutrition is perhaps well recognized by every one. It is present in seeds in larger amounts than in any other part of plants, although it is found extensively in the young growing parts. Like nitrogen, it is a constituent of every living cell. It is a constituent of phospholipids, nucleoproteins, and phytin, the latter being a storage form of phosphorus in seeds.

This element plays an important role in energy transformation in the cells of both plants and animals. As such it is necessary for normal transformations of carbohydrates in plants—the changing of starches to sugar, for example. Phosphorus is also necessary for the assimilation of fats, and apparently it increases the efficiency of the chloroplastic mechanisms.

**Potassium.** This element plays an important part in many of the vital physiological processes in the plant, although the exact nature of the mechanism by which potassium functions is not definitely known. It is essential in all cell metabolic processes and apparently has a specific role in influencing the uptake of certain other mineral elements, in regulating the rate of respiration, in affecting the rate of transpiration, and perhaps also in influencing the action of enzymes, and aiding the synthesis and translocation of carbohydrates. It is unique in that it does not become a component of plant compounds but remains in ionic form in the plant. For this reason it is readily lost from plant foliage by leaching.

**Calcium.** Calcium in the form of calcium pectate is a part of cell walls and is necessary for the growth of meristems. It also exists in plants in the oxalate form.

**Magnesium.** This element is active in enzymes systems, is a part of the chlorophyll molecule, and aids in the translocation of phosphorus in plants.

**Sulfur.** Sulfur appears to be important in plants mainly as a constituent of compounds including amino acids (cystine and methionine), and in the oil of plants of the crucifer and mustard families. Cabbage and turnips have a high sulfur requirement.

**Iron and Manganese.** Both iron and manganese play important roles in plant enzyme systems. They are required for chlorophyll synthesis. Their utilization in the plant is interrelated: an excess of manganese causes an inactivation of the iron.

**Boron.** The role of boron in plants is less well known than the roles of most of the other elements. Death of the meristems and actively growing portions of the plant occur as a result of a deficiency. This causes reactions which may or may not be related to the role of boron. It appears, however, to play a role in calcium utilization.

**Copper and Zinc.** Copper and zinc are components of enzymes. Both apparently are necessary for the formation of growth promoting substances.

**Molybdenum.** The reduction of nitrates in plants is dependent on molybdenum. A deficiency results in a piling up of nitrates and an interference in protein synthesis. Nitrogen fixation in legumes is also dependent on the presence of molybdenum.

**Chlorine.** This element was the last to be added to the essential list as it is rarely, if ever, deficient for plant growth. Functions attributed to chlorine include those of a regulator of osmotic pressure and cation balance. Its role is not well established.

## Effects of Nutrients on Plant Growth

An excess or deficiency of almost any nutrient will delay maturity. In general, whenever a nutrient deficiency or excess is corrected by fertilization, there will be a tendency to a more rapid and efficient development of the plant and a hastening of maturity. Thus, it can be said that in this way the nutrients affect the plant in a general way. In this section consideration will be given to some of the specific effects of nitrogen, potassium, and phosphorus which are important in crop production.

**Nitrogen.** A deficiency of nitrogen is evidenced by a gradual loss of chlorophyll, which results in a light green to yellow color, and by a slow and stunted growth. An abundance of nitrogen promotes rapid growth with a greater development of dark green leaves and stems. Although one of the most striking functions of nitrogen is the encouragement of aboveground vegetative growth, this growth cannot take place except in the presence of adequate quantities of available phosphorus, potassium, and other essential elements.

An ample supply of available nitrogen during the early life of the plant may stimulate growth and result in earlier maturity. However, the presence of an excess of nitrogen throughout the growing season of the plant frequently prolongs the growth period. This effect is especially significant for certain crops in regions having a short growing season or in areas where an early-fall freeze may do great damage to fruit trees whose season's growth period has been prolonged.

A large supply of available nitrogen encourages the production of soft, succulent tissue which is susceptible to mechanical injury and the attack of disease. Either effect may decrease the quality of the crop. However, the development of softness in the tissues may be desirable or undesirable, depending on the kind of crop. For vegetables used for their leaves, pronounced succulence, tenderness, and crispness are desired. Other vegetables and some fruits may have their keeping and shipping qualities impaired when they are grown with an excess of available nitrogen. An excess of nitrogen may encourage lodging in grains, which frequently decreases the quality, but a normal amount of nitrogen usually increases plumpness in grains.

**Potassium.** Potassium has a counterbalancing effect on the results of a nitrogen excess. It enhances the synthesis and translocation of carbohydrates, thereby encouraging cell wall thickness and stalk strength. A deficiency is sometimes expressed by stalk breakage or lodging. It also increases the sugar content of sugar beets and sugar canes. Highest dry matter yields of these two crops can be obtained with very high rates of nitrogen fertilization, but the greatest production of sugar results from moderate nitrogen applications and a sufficient level of available potassium. Root crops like potatoes also have a high potassium requirement. Less succulent foliage is promoted by good supplies of potassium and reduces disease. There is some evidence which indicates that alfalfa is less susceptible to frost injury when well fertilized with potassium.

**Phosphorus.** The effects of too little or too much phosphorus on plant growth are less striking than those of nitrogen or potassium. It appears to hasten maturity more than most nutrients, as an excess stimulates early maturation. Phosphorus deficiency is characterized by stunted plants having about equally affected root and top growth. Many soils produce forage deficient in phosphorus in terms of the nutritional requirement of animals, and fertilization with sufficient phosphorus to increase the phosphorus content of the forage improves forage quality in these cases.

As more knowledge about the effect of the nutrients on plant growth is obtained, more emphasis in the future will be placed on their role in the improvement of color, flavor, and storage quality. (See Fig. 11–16).

## NUTRIENT DEFICIENCY SYMPTOMS

When plants are starving for any particular nutrient, characteristic symptoms usually appear on these plants. If crops are not vigorous and healthy, it is important to know and understand the cause. If the unhealthy appearance is due to disease, it may be possible to save the crop by spraying, or, if it is a nutrient deficiency, fertilizers may be applied as a top-dressing on the soil or on the foliage as a spray in time to save the crop. These deficiency symptoms appear only when the supply of a particular element is so low that the plant can no longer function normally, and then it will usually be profitable to apply fertilizer long before the symptoms indicating acute starvation actually appear.

### Nutrient Mobility in Plants and Deficiency Symptoms

Translocation of nutrients within the plant is an ever continuing process. In this regard there is considerable difference in the mobility of the various nutrients. When a shortage of a mobile nutrient occurs, it is removed from the older first formed tissues and translocated to the growing points. This causes the symptoms to appear on the lower leaves. Nitrogen is very mobile in plants and deficient plants have yellow-colored lower leaves and green upper leaves (Fig. 11–14). Other nutrients which are mobile in the plant include phosphorus, potassium, magnesium, and zinc. Those with limited mobility which produce symptoms on the new leaves or growing points include calcium, boron, iron, copper and manganese (Fig. 11–8).

**Fig. 11–8** Manganese-deficient bean plants. Note the healthy lower leaves and the light-colored intervein areas of the upper leaves. A progression of more severe symptoms occurs from the bottom to the top of these plants. That is related to the immobility of manganese in plants.

## Deficiency Symptoms

It should be pointed out that these deficiency symptoms are not always easily diagnosed. Some of them might be mistaken for discoloration or abnormal characteristics produced by diseases, or they may be due to a deficiency of some other element or factor of plant growth. But information concerning these symptoms has been accumulating rapidly, and they have become a valuable aid in determining the need for certain nutrients, especially when used in conjunction with tissue and soil tests.

**Nitrogen.** The need for more nitrogen is indicated by a light green to yellow appearance of the leaves. As a rule, the older bottom leaves start to turn light green, then turn yellow at the tip. The entire leaf may turn yellow, even though the tissues are alive and turgid. In the corn plant the yellowing extends up the midrib of the leaf, with the outer edges remaining green the longest. A nitrogen-starved cucumber may have a small or pointed blossom end; a deficiency of nitrogen may cause the kernels of cereals to become shriveled and light in weight. In fruit trees the early shedding of leaves, death of lateral buds, poor set of fruit, and development of unusually colored fruit are indications of a lack of nitrogen. Extreme nitrogen starvation is most likely to occur in sandy soils or waterlogged soils, although fine-textured soils with low humus content usually need additional nitrogen.

**Phosphorus.** If phosphorus is deficient, cell division in plants is retarded and growth is stunted. A dark green color associated with a purplish coloration in the seedling stage of growth is a symptom of phosphorus deficiency. Later, plants become yellow. The yellowing is associated with early maturity but is definitely a symptom of phosphorus starvation. Occasionally a pale or yellowish green color develops when the lack of phosphorus inhibits the utilization of nitrogen by the plant. Bronze or purple leaves sometimes are observed at the top of new shoots of phosphorus-starved apple trees. In the absence of sufficient phosphorus, general maturity of the crop and seed formation are usually delayed. With corn, poor pollination frequently is associated with phosphorus starvation. Perhaps the most characteristic symptom of phosphorus deficiency, among plants in general, is the stunted growth.

**Potassium.** A deficiency of potassium usually shows up as a "leaf scorch" in most plants. Corn indicates a need for potassium by a yellowing of the tips and margins of the lower leaves (Fig. 11–11). This coloration does not move up the midrib as with a nitrogen deficiency but gradually spreads upward and inward from the leaf tip and edges. This leaf scorch is frequently mistaken for "burning" or "firing" and is ascribed to a deficiency of moisture during dry weather. When insufficiently supplied with potassium, alfalfa frequently develops a series of white spots near the margin of the older leaves. Sometimes this spotting effect is accompanied by a yellowing of the leaf edges, and at times the leaf margins turn yellow without the formation of white spots. The edges of the leaves finally dry up and curl under. Potato plants indicate a potassium deficiency by a marginal scorch of the lower leaves, and frequently the areas between the veins of potato leaves bulge out, giving a wrinkled appearance. A cucumber starved for potassium grows with a small stem end.

**Boron.** Many physiological diseases of plants, such as the internal cork of apples, yellows of alfalfa, top rot of tobacco, cracked stem of celery (Fig. 11–18), and heart rot and internal black spot of beets, are associated with a deficiency of boron. In sugar beets, boron deficiency appears as a stunting and curling or twisting of the petioles, associated with a crinkling of the heart leaves. They have unusually dark green and thicker leaves which wilt more rapidly under drought conditions. The older leaves frequently become chlorotic, and rotting of the beets, starting in the crown, may occur. Girdle or canker of table beets occurs as a cracking of the outer skin of the beets near the soil surface, followed by a breakdown of the root tissue.

**Calcium.** A deficiency of calcium is characterized by malformation and disintegration of the terminal portion of the plant. Calcium is not readily removed from the older tissues to be used for new growth when a deficiency occurs. The deficiency symptoms have been established for many plants, by the use of greenhouse methods, but they are seldom seen in the field.

**Copper.** There is little evidence to indicate that copper is lacking in soils except those of a high organic content. Growth abnormalities of many plants that are produced on peat and muck soils have been corrected by the application of copper compounds, and in general the copper additions produce better color of such crops as onions, spinach, lettuce, and carrots, increase sugar content of beets and carrots, and improve the flavor of most crops.

The upper leaves of tobacco plants grown without copper are unable to maintain their vigor, and they wilt badly. The leaves are permanently wilted and do not gain turgidity even in the presence of sufficient moisture. It has also been observed that the amount of seed set is reduced and the seed stalk is unable to stand erect if the supply of copper is inadequate.

**Iron.** Deficiency symptoms for iron are striking and are commonly seen on plants growing on calcareous or alkaline soils. Iron-deficient plants have a light-yellow leaf color, which is most evident on the younger leaves. The intervein areas are most affected and the veins retain a darker color. *Iron-chlorosis* is the name given to this condition.

**Magnesium.** Magnesium is a constituent of chlorophyll. As with several other nutrient elements, a deficiency of magnesium results in a characteristic discoloration of the leaves. Sometimes a premature defoliation of the plant results from magnesium deficiency. The chlorosis of tobacco, known as "sand drown," is due to a magnesium deficiency. Cotton plants suffering from a lack of this element produce purplish red leaves with green veins (Fig. 11–12). Leaves of sorghum and corn become striped; the veins remain green, but the areas between the veins become purple in sorghum and yellow in corn. The lower leaves of the plant are affected first. In legumes the deficiency is shown by chlorotic leaves.

**Manganese.** In the absence of sufficient manganese, tomato, bean, oat, tobacco, and various other plants are dwarfed. Associated with this dwarfing is a chlorosis of the upper leaves of the plant. The veins, however, remain green (Fig. 11–9). The "gray speck" of oats has been attributed to a shortage of this element in some soils.

It has frequently been observed that the leaves of onions growing in alkaline muck soils become dwarfed and curled during growth and that the bulbs remain immature at harvest time. On similar soils celery becomes yellow; spinach, lettuce, and potatoes are chlorotic and frequently unmarketable.

**Zinc.** Zinc deficiencies frequently have been encountered in Florida. Pecan rosette, the yellows of walnut trees, the mottle leaf of citrus, the little leaf of the stone fruits and grapes, white bud of corn, and the bronzing of the leaves of tung trees are all ascribed to zinc deficiencies. In tobacco plants a zinc deficiency is characterized by a spotting of the lower leaves, and in extreme cases almost total collapse of the leaf tissue may occur. Zinc deficiency frequently delays the maturity of white beans when grown on alkaline soil in Michigan.

**Sulfur.** A deficiency of sulfur is similar to one of nitrogen. Plants are stunted and light green to yellow in color.

**Molybdenum.** This element is needed for nitrogen fixation in legumes and when it is deficient legumes show symptoms of nitrogen deficiency. When molybdenum causes other metabolic disturbances in the plant, other symptoms are seen. A deficiency of molybdenum in cauliflower causes a cupping of the leaves. It is owing to the reduced rate of expansion near the leaf margin compared to that in the center of the leaf. The leaves also tend to be long and slender, giving rise to the symptom called "whiptail." Interveinal chlorosis, stunting of plant, and general paleness are also exhibited, depending on the kind of plant.

## SITUATIONS WHERE SOIL-NUTRIENT DEFICIENCIES ARE LIKELY TO EXIST

The nutrients that plants obtain from the soil have their origin mainly in mineral and organic compounds. Their sufficiency or deficiency for crops will be related to those processes that cause organic matter decomposition and the weathering of minerals. Some nutrients will be deficient because soil reaction favors their continued existence in compounds of low solubility.

Fig. 11-9 Bean leaves yellowed from deficiency of manganese and potassium compared with a normal leaf on the left. The center leaf yellowed first at the tips and along the edges. The crinkled appearance was caused by the continued growth of the interior portions after growth had ceased along the leaflet edges. The right leaf is uniformly yellow and has green veins due to insufficient manganese. (Michigan Spec. Bull. 353.)

Fig. 11-10 Tobacco leaves indicating deficiency of iron and manganese. The center leaf is normal. The left leaf has lost its color except along the principal veins because of insufficient iron. The checkered appearance and associated dead spots in the right leaf are due to manganese deficiency. (*From* Hunger Signs in Crops, *Kodachrome by J. E. McMurtrey.*)

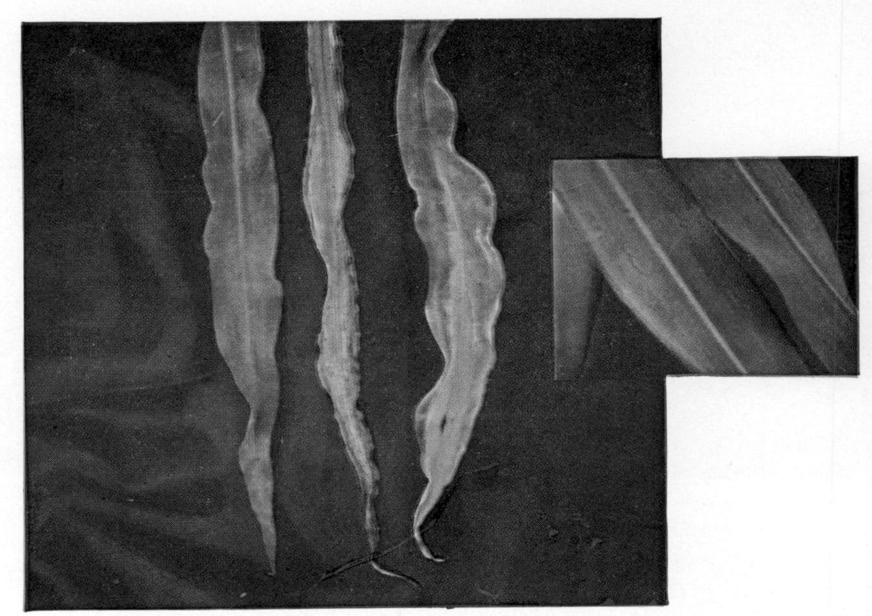

Fig. 11-11 Corn leaves deficient in nitrogen, potassium, and phosphorus. The left leaf is normal. The yellowed tip and margins of the center leaf arise from lack of potassium. In the next leaf insufficient nitrogen caused yellowing to affect the tip and then proceed up the midrib. The insert on the right shows the reddish purple edges of a phosphorus-deficient leaf next to a normal leaf. (Michigan Spec. Bull. 353.)

Fig. 11-12 The cotton leaf on the left is normal. The other two show a purplish red color with green veins because of insufficient magnesium. (*From* Hunger Signs in Crops, *Kodachrome by J. E. McMurtrey.*)

Fig. 11-13 (Left) Alfalfa and clover leaves showing potassium deficiency. The silt loam soil was fertilized with phosphate. The white dots scattered through the yellow edges of the leaflets are unmistakable evidence of potassium shortage.

Fig. 11-14 (Right) Wheat grown on clay loam soil. The yellow, dried-up lower leaves on the left bunch are characteristic of nitrogen deficiency. Tissue tests showed a high nitrate content in the bunch of plants to the right and no nitrate in those of the left bunch. (Michigan Spec. Bull. 353.)

Fig. 11-15 Sugar beet leaves deficient in (1) manganese, (2) potassium, and (3) nitrogen. Leaf number 4 is normal. Note that when nitrogen is deficient the veins do not remain green as they do when manganese is deficient. The marginal yellowing of the potassium-deficient leaf is characteristic. (Michigan Spec. Bull. 353.)

Fig. 11-16 Cineraria plants, potassium deficient on the left, normal in the center, and nitrogen deficient on the right. A normal color in the foliage of flowering plants is important from a sales standpoint. (Michigan Spec. Bull. 353.)

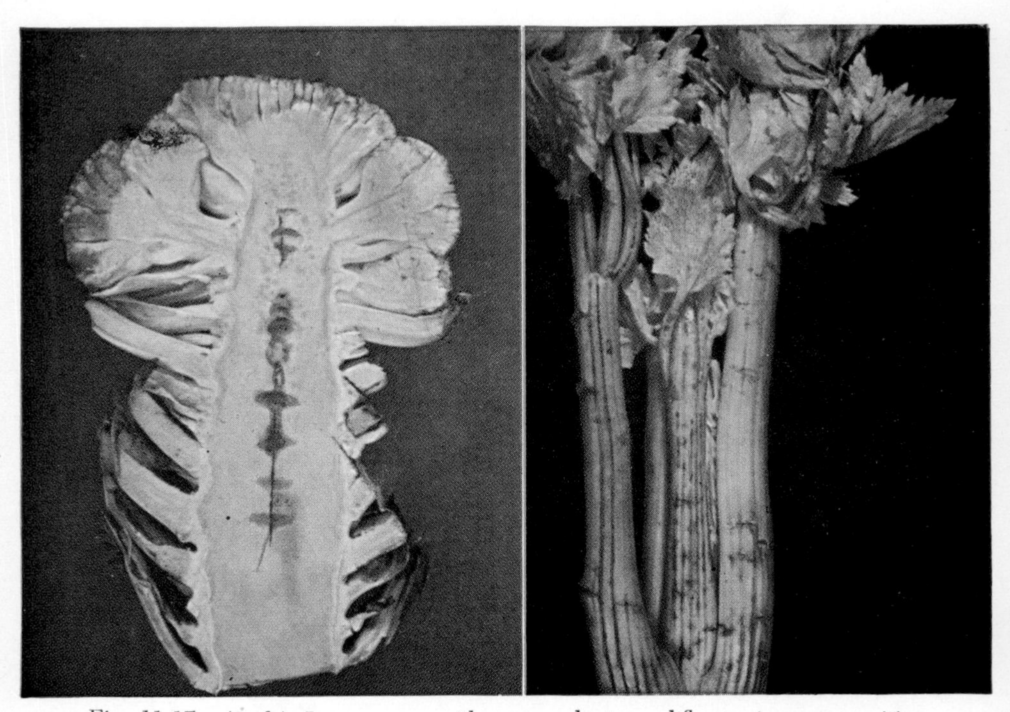

Fig. 11-17 (Left) In common with many plants cauliflower is very sensitive to boron deficiency. The broken-down tissue in the stalk and brown color of the edible portion of the head are noteworthy. (Michigan Spec. Bull. 353, *Kodachrome by E. K. Walroth, Eastern States Farmers Exchange.*)

Fig. 11-18 (Right) A deficiency of boron causes "crack stem" in celery. Stalks of this kind have no market value. (Michigan Spec. Bull. 353, *Kodachrome by E. K. Walroth, Eastern States Farmers Exchange.*)

## Nitrogen Supplying Power of Soils

The organic matter is the major nitrogen reservoir. Generally speaking, the soil organic matter contains about 5 percent nitrogen. A soil that contains 3 percent organic matter would contain about 3000 pounds of nitrogen per acre furrow slice. From less than 1 to more than 4 percent of the organic matter is mineralized each year. Thus, if a soil contained 3 percent organic matter and the mineralization rate was 2 percent, 60 pounds of nitrogen would be released per year as ammonium. This is not sufficient for high yields of many non-leguminous crops (Table 12–1). Full yields of a good grass sward, yielding 4 to 5 tons per season, are achieved in Britain only when 300 to 350 pounds of nitrogen are applied per acre. With no nitrogen fertilizer production drops to about one ton per acre.[5]

Quantities of organic matter in various soils and factors affecting the rate of decomposition were discussed in Chapter 6. Crops grown on organic soils provide an illustration of a situation where nitrogen fertilizer is frequently unnecessary.

## Fixation of Phosphorus as a Major Factor in Its Availability

Plants utilize a large amount of phosphorus in comparison to the total amount present in the soil. The maintenance of a good supply of available phosphorus for plant growth is a problem in both acid and alkaline soils. The problem arises from the tendency of phosphorus to combine with many different constituents in the soil to form insoluble compounds by processes collectively called phosphorus fixation.

Other factors contribute to the problem of maintaining an ample amount of available phosphorus. Less total phosphorus exists in the soil solution as the soil dries out because the volume of water in which phosphorus compounds can dissolve becomes smaller. As much as one-third of the phosphorus in the soil may be in organic combination, and the factors that affect nitrogen mineralization also affect the release of organic phosphorus.

## Potassium Supply as a Function of Mineral Weathering

The failure of potassium to become an integral part of plant compounds accounts for the failure of potassium to accumulate in soil

[5] Cited in "Fruits of Research in Fertilizer Use," G. W. Cooke, Span, 5:89, No. 2, 1962.

organic matter. The potassium relationships in the soil are closely linked to the mineral fraction which follows an equilibrium as follows:

$$\text{Nonexchangeable K} \rightleftharpoons \text{Exchangeable K} \rightleftharpoons \text{Soil Solution K}$$

Removal of potassium from the soil solution causes a shift to the right in the equilibrium. Exchangeable potassium moves to the soil solution, which, in turn, brings about an increase in the rate of conversion of nonexchangeable to the exchangeable form. If these rates of conversion from one form to another can meet plant needs, potassium fertilization is unnecessary. Such a situation exists in many loess-derived soils in the Chernozem and more arid regions. The high content of potassium minerals and limited leaching are the major factors contributing to the situation. Most soils of the humid regions require potassium fertilization for continued high yields.

When soluble potassium fertilizers are applied to soils, the soil solution potassium will increase, causing an increase in exchangeable potassium and ultimately an increase in the nonexchangeable. The conversion of exchangeable potassium to the nonexchangeable form is called potassium fixation. In effect, this is desirable as the utilization of the more soluble forms will encourage the transformation of "fixed" potassium back to the exchangeable form. This equilibrium mechanism provides for the conservation of potassium supplies and minimizes the possibilities for luxury consumption. The potassium in the illite clay of Fig. 7–13 is "fixed."

Potassium deficiency is common for crop production on organic soils since they have little or no mineral material to serve as a source of potassium.

### Calcium and Magnesium Supply as Related to Soil Reaction

The relationships of these nutrients and their supply in soils have been discussed in Chapters 7 and 8. It suffices at this point to state that their need for crop production will be closely related to the degree of leaching, soil pH, and percentage base saturation.

### Sulfur as a Constituent of Rainfall

Sulfur is used in large amounts by most plants but is usually found in considerable quantities in the soil. In areas close to industrial centers sufficient sulfur to supply crops is brought down by rain and snow from the atmosphere. The ordinary grades of superphosphate contain 40 to 50 percent of calcium sulfate, and several other fertilizers and fertilizer ingredients also contain sulfur. They are all important sources of sulfur in the fertilizer-consuming areas of the United States.

### Micronutrient Availability Closely Related to Soil Reaction

Molybdenum is deficient in some acid soils and in some cases the deficiency has been remedied by the application of lime which causes an increase in pH. Possibly the increased hydroxyl concentration makes more molybdenum available through anion exchange. Boron is also likely to be deficient in acid soils owing mostly to excessive leaching.

Nutrients whose availability is reduced in alkaline soils include iron, manganese, copper, zinc, and boron. Low solubility of compounds containing these nutrients is the major reason for low availability.

Great variation exists between crops in their ability to overcome conditions of low nutrient availability. Plants differ greatly in the amounts they require and in their capacity to obtain them from a given soil. Their use in fertilizers depends as much on the type of crop grown as on the soil conditions.

### Chlorine as a Unique Fertilizer Element

The chlorine requirement of crops is very small, even though chlorine may be one of the most abundant anions in plants. It is unique in that there is little probability that it will ever be needed as a fertilizer because of its addition to the atmosphere from ocean spray and, consequently, widespread addition of it to soils in precipitation. The presence of chlorine in most of the potassium fertilizer used today is another important manner in which chlorine is added to soils.

# 12

# Composition, Manufacture, and Use of Fertilizers

The history of man, in one sense, has been a record of efforts to increase the food supply by increasing the available nutrient supply. For thousands of years man has used lime, marl, ashes, bones, manures, mud, and legumes to add nutrients to soils. The story of the Indian, Squanto, showing the Pilgrims how to grow corn by placing a dead fish near each hill is well known. In Chapter 6 the use of trees for nutrient accumulation under shifting cultivation was discussed. Rapid progress in the development of chemical fertilizers occurred after the discovery of the major essential plant nutrients a little over a century ago. Now, it has been estimated that at least a fourth of man's total food supply can be attributed to the use of chemical fertilizer.[1] The present importance of fertilizers has been well summarized by Lamer who wrote, "Under present economic and political conditions, in all countries of the world, fertilizers are one of the most important strategic weapons of modern agriculture. Agricultural history has passed through various stages in its development; at present, it is in the fertilizer epoch."[2] This chapter presents information for gaining an understanding of the nature, manufacture, and efficient use of

[1] "Human Food Production as a Process in the Biosphere," Lester R Brown, *Scientific American*, Volume 223, No. 3, p. 168, 1970.

[2] From "The World Fertilizer Economy," Mirko Lamer, California, Stanford Press, 1957.

fertilizers—one of the most important means of increasing the food supply in a world where the number of underfed people increases daily.

## FERTILIZER MATERIALS

Fertilizers, in a broad sense, include all materials that are added to soils to supply certain elements essential to the growth of plants. However, the term *fertilizer* usually refers to chemical fertilizers. Chemical fertilizers do not contain plant nutrients in elemental form as nitrogen, phosphorus, or potassium, but the nutrients exist in compounds that provide the ionic forms of nutrients that plants can absorb. The major kinds of nitrogen fertilizers will be discussed first.

### Nitrogen Fertilizer Materials

The atmosphere is about 79 percent nitrogen, and as noted in Chapter 5, there are about 34,500 tons of nitrogen over every acre. Yet it is ironic that man's food supply is more limited by a lack of nitrogen than any other plant nutrient. The major source of essentially all industrial nitrogen (including fertilizer nitrogen) results from the fixation of atmospheric nitrogen according to the following generalized reaction:

$$N_2 + 3H_2 \xrightarrow[\text{pressure and catalysts}]{\text{proper temperature}} 2NH_3 \text{ (anhydrous ammonia)} \quad (1)$$

The source of hydrogen is usually natural gas ($CH_4$). A process flow chart of ammonia synthesis is presented in Fig. 12–1.

Anhydrous ammonia is the most concentrated nitrogen fertilizer, being 82 percent nitrogen. At normal temperature and pressure anhydrous ammonia is a gas and is stored and transported as a liquid under pressure. After direct application of $NH_3$ to the soil, the $NH_3$ absorbs a hydrogen ion and is converted to $NH_4^+$, a stable ammonium ion.

About 98 percent of the nitrogen fertilizer produced in the world is ammonia or one of its derivatives. The manufacture of ammonia and five of its derivatives is shown diagrammatically in Fig. 12–2 and a brief summary of some properties of nitrogen fertilizers is given in Table 12–1.

Frequent attempts have been made to determine the relative efficiency of nitrogen fertilizers by applying equal quantities of nitrogen per acre, in the various materials, for a given crop. Since so many factors, such as temperature and moisture conditions, soil reaction, leaching, kind of crop, and time and method of application, affect the

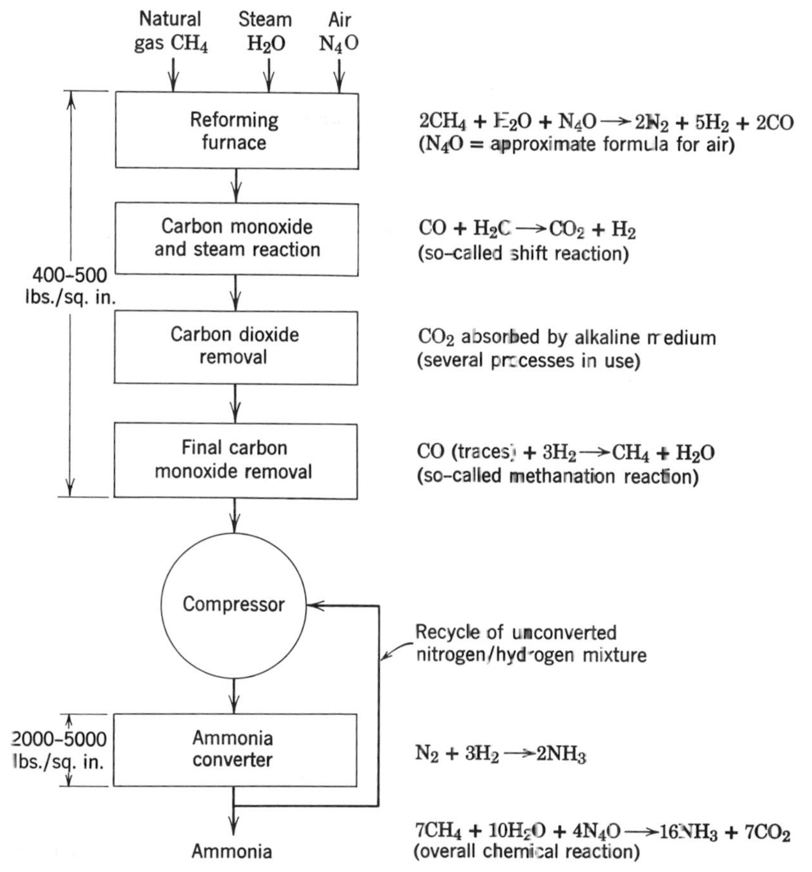


**Fig. 12–1** Process flow chart of ammonia synthesis. From "The World Food Problem," Vol. 2, A Report of the President's Science Advisory Committee, The White House, May 1967, p. 385.)

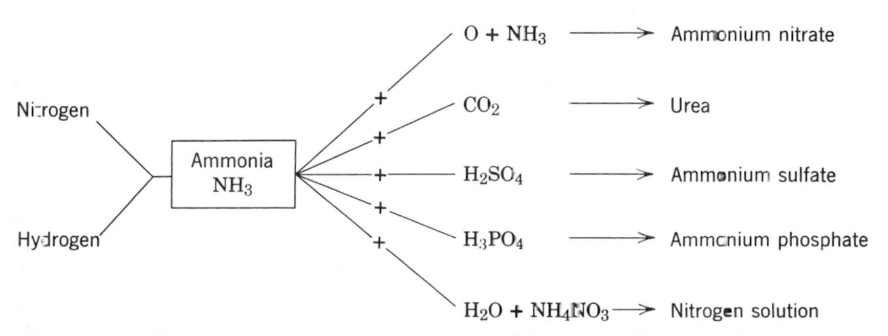


**Fig. 12–2** Diagrammatic representation of the manufacture of ammonia and some of its derivatives.

action of any nitrogen fertilizer, relative fertilizing values so obtained may be misleading.

**Table 12–1** The Principal Fertilizers Supplying Nitrogen

| Nitrogen Carrier | Nitrogen, Percentage | Remarks |
|---|---|---|
| Anhydrous ammonia | 82 | Used directly, or for ammoniation, nitrogen solutions, etc. |
| Ammonium nitrate | 33 | Conditioned to resist absorption of water |
| Ammonia liquor (aqua ammonia) | 24 | Formed when ammonia is absorbed in water |
| Nitrogen solutions | variable | Many kinds formed from solutions of aqua ammonium, ammonium nitrate, urea, etc. |
| Urea | 45 | $CO(NH_2)_2$, is hydrolyzed to ammonium in soils |
| Urea-formaldehyde | 35–40 | Contains insoluble, slowly available nitrogen |
| Ammonium sulfate | 20 | Also produced as a by-product of the coking of coal. |
| Sodium nitrate | 16 | Also a naturally occurring salt in Chile |

Nitrogen, in the nitrate form, is readily soluble in water and is rather quickly used by most crops. Nitrates are easily leached from the soil by rains because of their high solubility and because they are not fixed or held in the soil to any appreciable extent. Although the ammonium form of nitrogen is soluble in water, it is not leached out as readily as nitrate because rather large quantities of ammonium can be absorbed and held by the soil. Plants can use ammoniacal nitrogen, but most of it is converted to nitrates, as a result of the process of nitrification, before plants take it up. As a consequence, nitrate and ammonium forms of nitrogen are of about equal value for most crops and most situations.

Urea hydrolyzes rapidly in warm, moist soils to form ammonium carbonate. The ammonium may be used directly by plants or may be converted to nitrate and then used as nitrate. Urea-formaldehyde is one of the more recently developed nitrogen fertilizers and is not water soluble. The nitrogen in urea-formaldehyde is released in available form slowly to provide a continual supply of nitrogen through the growing season. Its high cost greatly limits its use.

### Phosphorus Fertilizer Materials

Historically, bones served as a source of phosphorus, yet as recently as 1840 the value of bones as a fertilizer was found to result largely from their phosphorus content. At about this same time, Liebig suggested that bones be treated with sulfuric acid to increase the solubility or availability of the phosphorus. This marked the beginning of the modern day fertilizer industry because it led, in 1842, to the patenting of a process for the manufacture of superphosphate by the treatment of mineral rock phosphate with sulfuric acid.

The only important source of mineral phosphate used to manufacture fertilizers today is rock phosphate (see Fig. 12–3). The production of ordinary superphosphate by the acidulation of rock phosphate with sulfuric acid is shown in Equation 2.

$$\underset{\text{(rock phosphate)}}{[Ca_3(PO_4)_2]\cdot 3\,CaF_2} + \underset{\text{sulfuric acid}}{7H_2SO_4} \rightarrow \underset{\text{monocalcium phosphate}}{3Ca(H_2PO_4)_2} + \underset{\text{gypsum}}{7CaSO_4} + \underset{\text{hydrogen fluoride}}{2HF} \qquad (2)$$

The ordinary super phosphate produced by reaction 2 consists of about half monocalcium phosphate and about half gypsum. The hydrogen fluoride can be recovered and in some cases is used to fluorinate water. Ordinary superphosphate has a phosphorus content of about 9 percent P or 20 percent $P_2O_5$.

Under proper conditions the reaction of rock phosphate with sulfuric acid will yield phosphoric acid. By treating rock phosphate with phosphoric acid, a more concentrated superphosphate can be produced as follows:

$$\text{rock phosphate} + 14H_3PO_4 \rightarrow 10Ca(H_2PO_4)_2 + 2HF \qquad (3)$$

The same phosphorus compound is produced with sulfuric acid and phosphoric acid, but without the production of any gypsum when phosphoric acid is used. This more concentrated superphosphate contains about 20 percent phosphorus or the equivalent of 45 percent $P_2O_5$. Concentrated superphosphate is commonly called triple superphosphate. Both types of superphosphates are of about equal quality as fertilizers when the same amount of phosphorus is applied.

Ammonium phosphates are produced by neutralizing phosphoric acid with ammonia. The two popular kinds produced are monoammonium phosphate and diammonium phosphate. Ammonium phosphates are good sources of both phosphorus and nitrogen, the phosphorus being water soluble. Some ammonium phosphate is produced as a by-product of the coking industry by using the ammonia produced in the coking of coal to neutralize sulfuric or phosphoric acid.

**Fig. 12–3** Rock phosphate is mined from surface deposits by hydraulic pressure. It is then washed and crushed in preparation for acidulation. (Courtesy of Agricultural Bureau of American Agricultural Chemical Co.)

Basic slag, sometimes referred to as Thomas phosphate, is produced as a by-product of the iron industry. Phosphorus is contained in certain iron ores, and steel made from them is brittle if most of the phosphorus is not removed. The slag is produced by oxidizing the phosphorus in molten iron by a blast of air blown through it. The molten iron is contained in a converter lined with lime, and the oxidized phosphorus combines with this lime. The slag so produced rises to the surface and is drawn off, cooled, and ground so finely that most of it will pass a 100-mesh screen. The phosphorus in slag is soluble in citric acid and is considered available to crops. A summary of these and other phosphate materials is summarized in Table 12–2.

## Potassium Fertilizer Materials

Settlement of the eastern seaboard of the United States was stimulated in the early 1600s by a need in England for several important products including woodashes and potash. To obtain potash, the ashes of trees were leached with water to remove the potassium compounds ($K_2CO_3$), dried, and then calcined to produce potassium oxide. Later, huge mineral deposits of potassium salts were discovered in Germany and Germany supplied the world market until after World War I. During the 1920s the Carlsbad, New Mexico potash mines were opened and, more recently, in 1959 the mining of potash was begun in Saskatchewan, Canada (see Fig. 12–4).

**Table 12–2** The Principal Phosphatic Materials

| Material | Percentage Available | | Remarks |
|---|---|---|---|
| | $P_2O_5$ | P | |
| Rock phosphate | 25–35[a] | 11.0–15.4[a] | Effectiveness depends on degree of fineness, soil conditions, and crop grown |
| Superphosphate (ordinary) | 20 | 8.7 | Made by treating ground phosphate rock with $H_2SO_4$ |
| Triple superphosphate | 46 | 20 | Made by treating ground phosphate rock with liquid $H_3PO_4$ |
| Monoammonium phosphate | 48 | 21 | Made by neutralizing $H_3PO_4$ with ammonia |
| Diammonium phosphate | 53 | 23 | Made by neutralizing $H_3PO_4$ with ammonia |
| Basic slag | 5–20 | 2.2–8.8 | By-product obtained in the manufacture of steel |
| Bone meals | 17–30[a] | 7.5–13.2[a] | Includes raw as well as steamed bone meals |
| Colloidal phosphate | 18–23[a] | 7.9–10.1[a] | A finely divided, relatively low-grade rock phosphate or phosphatic clay |

[a] Total phosphoric acid instead of amount that is available.

Common minerals in the potash deposits include sylvite (KCl), sylvinite (mixture of KCl and NaCl), kainite ($MgSO_4KCl \cdot 3\ H_2O$) and langbeinite ($K_2SO_4 \cdot 2\ MgSO_4$). Processing of the ore consists of separating KCl from the other products in the ore. The flotation method is commonly used to separate KCl from the ore mixture. The ore is ground, suspended in water, and treated with a flotation agent that adheres to the KCl crystals. As air is passed through the suspension, the KCl crystals float to the top and are skimmed off (Fig. 12–5). After some further purification, the nearly pure KCl is dried and screened for particle size. The fertilizer material is called muriate of potash (KCl) and is about 60 percent $K_2O$.

Although the great bulk of the potassium comes from mines and about 95 percent of the potassium fertilizer is KCl, several other sources and materials are worthy of mention. Some potassium is obtained from brine lakes at Searles Lake in California and Salduro Marsh in Utah. If needed, the sea represents an "endless" supply of potassium, since each cubic mile of sea water contains about 1.6

**Fig. 12–4** A powerful continuous mining machine at work more than half a mile underground at Esterhazy, Sask., Canada. (Courtesy of American Potash Institute and International Minerals and Chemical Company.)

**Fig. 12–5** A flotation plant for the recovery of potassium minerals. (Courtesy of American Potash Institute and the International Minerals and Chemical Company.)

million tons of potassium. The second most widely used potassium fertilizer, although used to only a minor extent, is $K_2SO_4$, which is about 40 percent K or 48 percent $K_2O$. A very small amount of $KNO_3$ is also produced.

All the potash fertilizer salts are soluble in water and are considered readily available. In general, it can be said that there is very little difference in their effects on crop production except in rather special cases. Discrimination is sometimes made against carriers having a high percentage of chlorine for special crops like potatoes and tobacco. The sulfate, or carbonate if available, is usually preferred for tobacco, especially where large amounts are to be added because a crop of superior burning quality is produced. The muriate is just as efficient as the sulfate for potato production, according to most of the experimental evidence.

### Materials Containing Micronutrients

Although the bulk of fertilizer consists of carriers containing nitrogen, phosphorus, and/or potassium, micronutrients are sometimes added to fertilizers to supply one or more of the micronutrients. Some common micronutrient carriers are listed in Table 12–3.

**Table 12–3** Some Common Micronutrient Carriers

| Carrier | Nutrient Composition |
|---|---|
| Borax ($Na_2B_4O_7 \cdot 10H_2O$) | 11% B |
| Copper sulfate ($CuSO_4 \cdot 5H_2O$) | 25% Cu |
| Ferrous-sulfate ($FeSO_4 \cdot 7H_2O$) | 20% Fe |
| Manganese oxide (MnO) | 48% Mn |
| Manganese sulfate (variable hydration) | 23–25% Mn |
| Zinc sulfate ($ZnSO_4 \cdot 7H_2O$) | 35% Zn |

## MIXED FERTILIZERS

Soils vary greatly in their ability to supply crops with available nutrients, and the mineral requirements of different crops are also quite variable. In order to supply nutrient deficiencies in soils and to meet the various requirements of different crops, fertilizers containing two or more essential elements are prepared in many different grades. They are known as *mixed fertilizers* and are made by mixing two or more of the separate fertilizer carriers.

## Preparation of Mixed Fertilizers

The preparation of mixed fertilizers can be a relatively simple operation, especially if the mixture is to be of low grade, that is, a fertilizer containing a comparatively low percentage of nutrients. It consists essentially in mixing suitable materials in the correct proportion to give the desired grade or analysis. There are many bulk mixing plants in the United States which follow this simple procedure. On the other hand, many of the larger fertilizer factories follow much more involved processes which require careful chemical and temperature control, particularly in the preparation of some of the carriers of materials used in making the final mix.

After the acidulation of rock phosphate in the manufacture of superphosphate, the product is allowed to stand in a large pile for a considerable time to "cure" before being ground. If these phosphates are not properly "cured," the chemical reactions involved are not completed and the fertilizer made from them will harden in the bags. Mixtures of superphosphate and potash carriers are often made and stored for a considerable time before being used in the final fertilizer mixture.

Fertilizers in storage tend to assume an unfavorable mechanical condition, chiefly the result of "setting up," which is essentially a cementation as in plaster of paris. It may also be due to the surface tension effects of moisture which forms films around the particles of fertilizer.

The drilling qualities of certain mixed fertilizers made from fertilizer salts which take up moisture readily can be appreciably increased by adding an organic filler such as muck or diatomaceous earth. Fertilizers are commonly formulated today so that the only material in them classified as filler is what is needed to insure a good physical condition.

## The Fertilizer Grade or Guaranteed Analysis

The fertilizer grade or guaranteed analysis expresses the nutrient content. In all states the grade is expressed in the order N—$P_2O_5$—$K_2O$. The total nitrogen is expressed as elemental nitrogen (N); the phosphorus is expressed as available phosphoric acid ($P_2O_5$); and the potassium is given as water-soluble potash ($K_2O$).

Considerable interest has developed in recent years in changing the basis for expressing the fertilizer grade. In 1955 the Soil Science Society of America passed a resolution favoring a change from the N—$P_2O_5$—$K_2O$ basis for expression to the N—P—K or elemental basis. They outlined the following advantages for such a change. First, greater uniformity would exist in the expression of the nutrient content of fertilizers, of soils, and of plants. Plant composition in present-day literature is usually expressed on the elemental basis. The same is true in the field of animal nutrition and biochemistry, where feeding standards are usually expressed in terms of elemental amounts of phosphorus and calcium. Second, it would result in greater simplicity. The elemental basis requires fewer symbols for expression. It is a common practice to refer to N—P—K fertilizer but to refer to the calculations of the amounts of plant nutrients on the N—$P_2O_5$—$K_2O$ basis. This leads to confusion and creates an obstacle to learning in education programs for farmers. Third, accuracy of expression of the true ratio of the major nutrient elements in a fertilizer requires that the elemental basis be used. Reference to a 10—20—10 fertilizer tends to lead one to believe that the ratio of N—P—K is 1—2—1. The ratio of N—$P_2O_5$—$K_2O$ is 1—2—1, but the actual N—P—K ratio is 1.0—0.88—0.83. A scale for converting $P_2O_5$ to P and $K_2O$ to K is given in Fig. 12–6.

### Calculation of Fertilizer Formulas

In calculating formulas for mixtures it is necessary to decide first what percentages of nitrogen, available phosphoric acid, and water-soluble potash are desired in the fertilizer mixture and then what materials are to be used in making the mixture.

For example, to make one ton of a 6—12—12 fertilizer using the following ingredients:

| | |
|---|---|
| Ammonium nitrate | 33 percent nitrogen |
| Superphosphate | 20 percent available phosphoric acid |
| Muriate of potash | 60 percent water-soluble potash |

The problem is to find out how much of each of these materials is needed. This may be done by use of the equation

$$X = \frac{A \cdot B}{C}$$

in which $X$ equals pounds of carrier required, $A$ equals pounds of mixed fertilizer required, and, with nitrogen, $B$ equals the percentage of nitrogen desired in the mixture, and $C$ equals the percentage of nitrogen in the carrier (ammonium nitrate). By substituting these values in the above equation, the result is easily determined.

$$X = \frac{2000 \times 6}{33}$$

$X = 364$ pounds, the amount of ammonium nitrate required

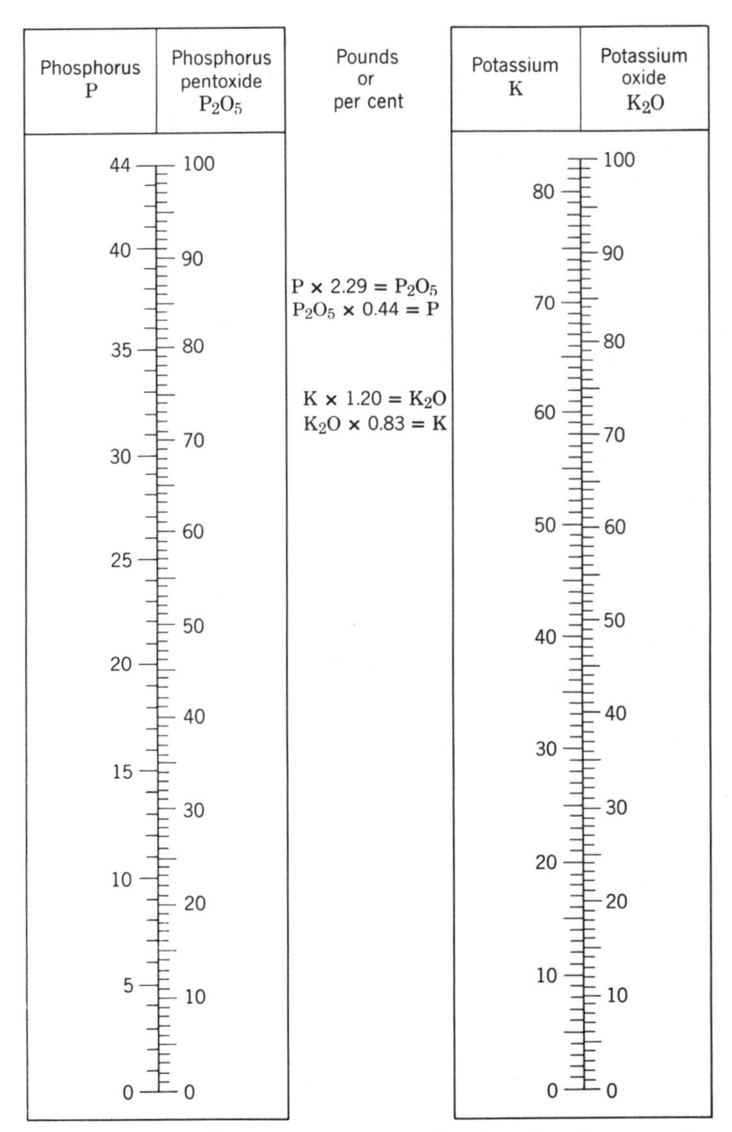


**Fig. 12–6** Fertilizer conversion scales for phosphorus and potassium.

In a like manner the required amounts of superphosphate and muriate of potash may be determined.

$$X = \frac{2000 \times 12}{20}$$

$X = 1200$ pounds, the weight of superphosphate required

$$X = \frac{2000 \times 12}{60}$$

$X = 400$ pounds, the weight of muriate of potash required

The total amount of materials used in this fertilizer mixture (364 + 1200 + 400) equals 1964 pounds. It is necessary to add 36 pounds of filler or physical conditioner to make a ton of the required mixture.

## Liquid Fertilizer

A liquid fertilizer is simply a solution containing one or more water-soluble forms of nutrients. Materials similar to those used in the manufacture of liquid fertilizers have been added to soils for many years by dissolving them in irrigation water and as components of conventional dry fertilizers. Advantages of liquid fertilizers over the dry fertilizers include (1) the saving of labor in handling where pumps and pipes can be used, (2) their convenience as foliar sprays, and (3) the convenience of adding pesticides. Disadvantages include (1) the increased fixation of the phosphorus, especially in mixed rather than banded application, (2) corrosion of metal containers and equipment, and (3) the need for special equipment for storage and application. A liquid-fertilizer attachment on a four-row planter is shown in Fig. 12–7.

A popular method used to manufacture liquid fertilizer consists of the neutralization of phosphoric acid with ammonia. Fertilizers with various ratios of nitrogen and phosphorus can be manufactured by varying the degree of neutralization. Potassium can be added to the fertilizer by the addition of KCl. The amount of potassium that can be added, without causing "salting out," is rather limited and makes the development of high-potash liquid fertilizers impractical.

The time of application and the placement of liquid fertilizer follow closely the principles that apply to the dry forms.

## Efficiency of Different Grades of Mixed Fertilizers

The efficiency of a fertilizer is determined by the uniformity with which it can be distributed and by its quality or its chemical composi-

**Fig. 12–7** Four-row planter with liquid-fertilizer attachment. (Courtesy of John Deere, Moline, Illinois.)

tion. The drillability of a fertilizer is determined by its physical properties, the two most important of which are the relative sizes of particles and their moisture content or hygroscopic properties. A fertilizer is said to be of good quality if it does not cause injury to plants when applied in the usual manner, does not leave any harmful residual effect in the soil, and contains the nutrient elements in proper balance. High-grade mixed fertilizers are now on the market which meet these requirements, and they are being made in granular form to insure good drillability. Thus it can be said that a decided increase can be made in the plant-nutrient content of many of the popular low-analysis mixtures without bringing about any marked changes in their chemical or physical properties. In general, a pound of plant nutrients in high-analysis fertilizer mixtures is just as efficient as a pound of plant nutrients in low-analysis fertilizers.

## General Nature of Fertilizer Laws

In general, the nature of the laws controlling fertilizer sales in the various states is similar. They all require periodical registration of brands or analyses offered for sale and accurate labeling of the bags or packages. Most of the states require that there be printed on each fertilizer bag, or on an attached tag, the following information:

1. Name, brand, or trademark.
2. Analysis (guarantee) or chemical composition.
3. Net weight of fertilizer.
4. Name and address of manufacturer.

Fertilizers offered for sale may be sampled by inspectors any time during the year and at any point in the particular state. The samples are sent to the control laboratory and are analyzed to determine whether the goods are up to the guarantee. The results are checked against the guaranteed analysis, and by this means the purchaser is protected from loss through the activities of unreliable companies. The inspection and analysis may be in the hands of the state department of agriculture, of the state agricultural experiment station, or of a state chemist. Once a year, in most states, the findings of the chemists are published in bulletin form, and copies of this bulletin may be obtained on request by any interested person.

## EVALUATION OF FERTILIZER NEEDS

The natural forests and grasslands that existed before man modified the landscape by agricultural pursuits is evidence that essentially all of the soils of the world are capable of supporting plants, provided the climate is favorable. Several important factors help to account for this situation. There is great diversity in terms of plant needs and tolerances to toxic elements. Nutrient recycling under natural conditions results in repeated use of the nutrients for plant growth. In the natural ecosystem there may also be abundant vegetation, which is the product of slow growth over many years. When man establishes agriculture, the need for additional nutrients is inevitable if yields are to be maintained above a meager annual yield of 8 to 10 bushels of grain per acre. Cultivated crops also have different nutrient requirements than natural plants, and many of the nutrients are removed from the land by the harvesting of crops. As a consequence, a need is created for evaluating the kind and amount of fertilizer to use, since unnecessary fertilizer is costly and the wrong fertilizer may be harmful.

### Deficiency Symptoms and Tissue Tests for Evaluating Fertilizer Needs

Plants, like animals, exhibit peculiar symptoms that are associated with particular nutrient deficiencies. Color photographs of some of these are found in Chapter 11. When the sap of a plant showing a deficiency symptom is tested, a low or deficient amount of the deficient nutrient is usually found. In this way, deficiency symptoms and rapid tissue tests can be used to diagnose plant growth problems. By the time deficiency symptoms appear on a crop, however, the yield potential may be greatly reduced. In addition, tissue tests do not provide information on the quantity of fertilizer to apply. For these reasons, soil tests provide the soundest basis for evaluating the fertilizer needs of most field and vegetable crops.

### Soil Tests for Fertilizer Recommendations

Soil testing is based on the concept that a crop's response (growth) to fertilizer will be related to the amount of *available* nutrients in the soil. For example, the exchangeable potassium is generally a good indication of the available potassium. The more exchangeable potassium there is in a soil, the less plants will respond to potassium fertilizer as shown in Fig. 12–8.

Considerable experimental work is needed to be able to make fertilizer recommendations for specific situations. Fertilizer recommendations must take into account the different nutrient needs of crops, types of soil, and climatic conditions. In the United States the state agricultural experiment stations have taken the responsibility to work on the crops, soil conditions, and climatic variations that are locally important. The experimental work to develop fertilizer recommendations never "ends" because new crop varieties are developed, soil-testing techniques are continually improved, and crop-yield expectations are generally increasing. Periodically, the latest research results are integrated with previous knowledge and are used to construct reference tables for making fertilizer recommendations.

The fertilizer recommendations in Table 12–4 take into account five important factors. These are soil test levels (in this case, pounds of exchangeable potassium per acre), type of crop, yield expectation, type of soil texture, and the climate (since the table is applicable for Michigan). For example, the potassium recommendation for sugar beets expected to yield 24–28 tons per acre growing on soils with loam or finer texture that have 120 pounds of exchangeable potassium per acre furrow slice is 100 pounds of K fertilizer expressed as $K_2O$ or 83 pounds expressed as K.

One of the weakest steps in getting good fertilizer recommendations is obtaining a representative sample. A single soil sample of one pint in size, when used to represent 5 to 10 acres, means that a pint of soil is used to characterize 5000 to 10,000 tons of soil. Persons in-

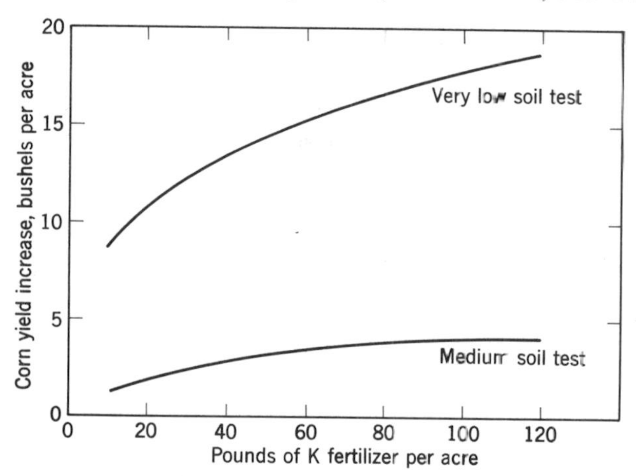


**Fig. 12–8** The effect of rate of application of K fertilizer on the yield of corn on soils testing very low and medium in available K. (Adapted from "Understanding Soil Testing," A. Bauer, L. Hanson, and J. Grava, *Better Crops with Plant Food*, Reprint D-10-60, Am. Potash Institute.)

terested in testing soil should consult a testing lab for proper sampling procedures and care of samples after they have been collected.

## Foliar Analysis for Evaluating Fertilizer Needs of Tree Crops

There are several conditions that limit the effectiveness of soil tests for making fertilizer recommendations for tree crops. The roots of trees are distributed through such a large volume of soil that sampling the upper soil layer or plow layer is of limited usefulness. Furthermore, trees conserve nutrients by the movement of nutrients from the leaves into the wood late in the year before the leaves fall. Deep roots of trees in unfrozen soil in cold winters may accumulate nutrients all winter long. Since the trees produce over many years, it is possible to evaluate the nutritional status by foliar analysis and find success by adding fertilizers. As with soil tests, foliar analysis must be correlated with yield response before good recommendations can be made. In many cases this research has been carried out and foliar analysis has become an important tool for horticulturists and foresters. Researchers have also determined the proper part of the plant to select for testing. Persons interested in use of foliar analysis should obtain instructions for sampling from a testing laboratory.

## Fertilizer Nutrient Interactions

Frequently, the plant response to a fertilizer nutrient is affected by the use of some other fertilizer nutrient or the level of some other soil nutrient. If nitrogen fertilizer is applied and plant growth is stimulated, this increases the demand for all other plant nutrients. It is not uncommon to find that the response to one nutrient is related to the level of some other nutrient in the soil.

**Table 12–4** Potash-Potassium Recommendations for Field Crops Growing on Loams, Clay Loams, and Clays in Michigan

| Available Soil Potassium[a] – Pounds of K per Acre | | | | | Pounds per Acre Annually Recommended | |
|---|---|---|---|---|---|---|
| | | | | | $K_2O$ | K |
| | | | | Less than 60 | 250 | 208 |
| | | | Less than 60 | 60–109 | 200 | 166 |
| | | Less than 60 | 60–99 | 110–159 | 150 | 125 |
| | Less than 60 | 60–99 | 100–139 | 160–209 | 100 | 83 |
| Less than 60 | 60–99 | 100–139 | 140–179 | 210–249 | 75 | 62 |
| 60–99 | 100–149 | 140–179 | 180–219 | 250–299 | 50 | 42 |
| 100–139 | 150–199 | 180–219 | 220–239 | – | 25 | 21 |
| 140–179 | – | – | – | – | 15 | 12.5 |
| 180+ | 200+ | 220+ | 240+ | 300+ | 0 | 0 |
| Barley 40–69 Bu.<br>Buckwheat<br>Corn 60–79 Bu.<br>Cover crops<br>Field beans 15–29 Bu.<br>Oats 50–79 Bu.<br>Pasture<br>Rye<br>Soybeans 30–50 Bu. | Alfalfa-3T<br>Barley 70–100 Bu.<br>Clover<br>Corn 80–119 Bu.<br>Field beans 30–50 Bu.<br>Oats 80–120 Bu.<br>Sudangrass<br>Wheat 40–49 Bu. | Alfalfa-4T<br>Corn 120–140 Bu.<br>Sugar beets 18–23 T.<br>Wheat 50–70 Bu. | Alfalfa-6T<br>Sugar beets 24–28 T. | Potatoes 250–350 cwt. | | |

From "Fertilizer Recommendations for Michigan Vegetables and Field Crops," Ext. Bull. E-550, Michigan State University, Nov. 1966.

[a] Available soil potassium determined by the 1*N* ammonium acetate method or 0.13*N* hydrochloric acid method.

An interesting nitrogen and sulfur interaction exists for some crops in the rural areas of southeastern United States. Several factors are involved that affect the supply of sulfur in the soil and the ability of plants to increase their growth when nitrogen fertilizer is used. In the rural areas only small amounts of sulfur are added to soils in precipitation (near cities rain washes out sulfur from industrial gases). In recent years ordinary superphosphate, which contains about half calcium sulfate, has been increasingly replaced by treble superphosphate, which does not contain sulfur. As a consequence, the effectiveness of nitrogen fertilizer is limited in many cases by an insufficient supply of soil sulfur.

An interesting phosphorus-zinc interaction is shown in Fig. 12–9. Previous use of phosphorus fertilizer was associated with zinc deficiencies because the phosphorus reduced the plants' ability to utilize soil zinc. Interactions represent a complicating factor in making fertilizer recommendations and must be considered.

## USE AND APPLICATION OF FERTILIZERS

For the farmer, fertilizer use is a matter of profit. As a result, little or no fertilizer is used on many of the world's most infertile soils, although considerable amounts of fertilizer are used on many of the world's best agricultural soils. Regardless of the situation, once a fertilizer recommendation has been made and fertilizer is to be used, the problem of proper use arises. This section considers several items that are important in efficient use of fertilizers.

### Efficiency in the Use of Fertilizers

Most soil types, particularly in the humid regions, respond favorably to the use of well-chosen fertilizers. Fertilizers should not be expected to make up for every shortcoming of the soil and crop, such as poor seed, unadapted varieties of crops, unfavorable weather conditions, poor tillage practices, weeds, poor drainage, bad physical condition of the soil, low organic content, or insufficient lime. All these factors are important, as all affect the efficiency of any fertilizer for any crop grown on any given soil. In other words, the proper use of fertilizer is only one, though a very important one, of the many phases of the scientific management of soil.

The most profitable return from fertilizer is nearly always obtained from those soils that are in the best physical condition for plant growth. The most profitable results from fertilizers cannot be expected when they are used on soils that are too compact or too loose, too dry or too wet.

### Application and Placement of Fertilizers

Another important factor in the efficiency of fertilizers is the method of applying them. The most efficient placement is direct application to the foliage. This avoids the problems of fixation, leaching, and denitrification which are encountered in soil application. One pound

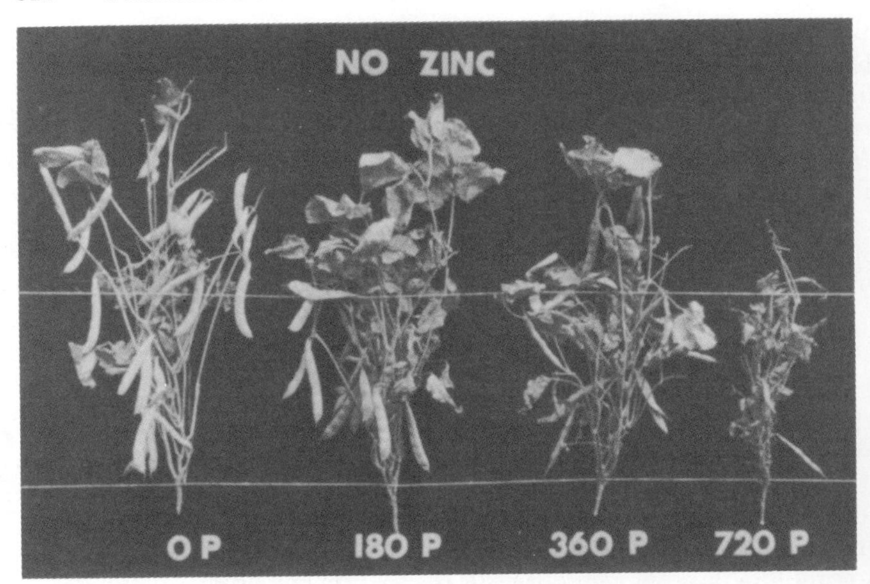


**Fig. 12–9** Samples of white beans grown on soil which had received from 0 to 720 pounds of phosphorus per acre in fertilizer during the previous three years. The poorer growth and delayed maturity of plants on which phosphorus had been used was owing to a zinc deficiency caused by the use of phosphorus fertilizer.

of manganese applied as a spray was as effective as ten pounds applied to the soil for vegetable production on an organic soil at the Michigan Agricultural Experiment Station. Foliar application is well suited to micronutrient fertilization where rapid fixation in the soil occurs and, particularly, where it can be combined with another operation (as spraying to control disease).

Limitations of foliar application are related to the amount that can be applied in a single application or the possible burning effect on the leaves, and their application must wait until crops are established. Perhaps the most intensive use of foliar application has been developed for the long-growing period crop of pineapple in Hawaii. Here, about three-fourths of the nitrogen and one-half of the phosphorus and potassium are sometimes applied this way.[3]

Most of the fertilizer used in the world is applied to the soil directly by machinery or airplane or as a component of irrigation water. The efficiency of applied fertilizer is dependent, among other factors, on

[3] Cited in "Foliar Absorption of Mineral Nutrients," S. H. Wittwer and F. G. Teubner, *An. Rev. Pl. Phy.*, **10**:23, 1959.

its placement in relation to the seeds or roots and on its need. The proper placement and time of application minimizes losses by leaching or dentitrification and conversion to unavailable forms by fixation reactions. The entire root zone need not be fertilized as a "few" roots are capable of absorbing the nutrients the plant needs (see Fig. 12–10). It is commonly recommended, for a row crop like corn, that a complete starter fertilizer be placed in a band about two inches to the side and two inches below the seed (Fig. 12–11) at planting time. It will be in the pathway of young roots and quickly utilized, yet, it will not cause injury or delay in germination. Band placement will minimize soil fertilizer contact and keep phosphorus and potassium fixation to a minimum. Further, root proliferation in the fertilizer band increases the amount of root surface for absorption (Fig. 12–11). Another application, usually only nitrogen, is applied after emergence as a part of a normal cultivation operation. Nitrogen applied at this time reduces overstimulation of vegetative growth and reduces the possibility of loss before it is absorbed by the plant. Obviously, with hay, pasture crops, and orchards this method of fertilizer application cannot be followed.

Other factors are to be considered in many instances. As the overall fertility level of the soil is increased, the benefit of localized placement is reduced. Fertilizer purchased in bulk form is less expensive and lower labor costs result in its use. This makes it possible for many farmers to apply more fertilizer in a less efficient manner and still obtain good results. When only nitrate nitrogen, which is mobile in the soil, is applied and is not fixed, almost any placement is satisfactory so long as the nitrogen permeates the soil and is not leached out of the rooting zone.

## Salt Effect of Fertilizers on Germination

When fertilizers are properly placed an inch or so from the seed, germination is not affected, but early growth of the plant is stimulated. However, if seed and fertilizer are placed together, germination may be delayed or prevented. This is because fertilizers are basically soluble salts and when dissolved in the soil solution, they increase the osmotic pressure of the soil solution. The effect on the osmotic pressure is related to the solubility of the fertilizer material. Potassium chloride and ammonium nitrate are very soluble and are very likely to prevent or delay germination of seed when placed in contact with seeds. In fact, dropping small amounts of these fertilizers in small areas on the lawn by accident or faulty spreading will kill the grass. Superphosphates have much less solubility, by comparison, and when

**Fig. 12–10** This soybean plant was grown by the split-root technique. The seed was germinated in quartz sand and the tiny seedling was transferred to this container filled with quartz sand. A root was pulled through the hole in the side of the can and immersed in a complete nutrient solution in the beaker at the time when transplanted. The growth of the plant shows that virtually all of the nutrients a plant needs can be absorbed by a very small portion of the total root system. The quartz sand in the can was kept moist and abundant root growth occurred in it.

placed with seed have little or no effect on seed germination if used at normal rates. The effect of placing some complete fertilizer with wheat seed on the germination of the wheat is shown in Fig. 12–12. Since the increased osmotic pressure of the soil solution produced by the fertilizer inhibits water absorption by seeds (and roots), one would expect the effects to be less damaging when the soil is kept moist, as is also shown in Fig. 12–12.

## Time of Application

The nutrients in fertilizers are available or water soluble and, therefore, are likely to react with the soil and become unavailable or

**Fig. 12–11** Fertilizer placed two inches to the side and two inches deeper than the seed will be in the pathway of early developing corn roots. Diffusion of nutrients into the soil surrounding the fertilizer granules has caused root proliferation in this location.

leach out of the root zone. Nitrate nitrogen is very soluble and mobile in soils and is subject to leaching. In fact, excessive use of nitrogen fertilizer can pollute the ground water. Phosphorus, by contrast, is very immobile in soils. Phosphate ions react with other ions in the soil solution to form insoluble, unavailable compounds. Potassium is intermediate in that it is adsorbed to the "exchange," which limits its mobility, but the potassium remains available to plants. For these reasons fertilizers are most effective if applied near the time when plant needs are greatest. This is not always practical since soil conditions, labor supply, and other factors affect the timing of fertilizer applications.

**Fig. 12–12** Placing a complete fertilizer with wheat seed was more damaging on germination in dry soil than in moist soil.

## Effect of Fertilizers on Soil pH

Normally, fertilizes do not produce significant changes in soil pH. However, there are some cases where high rates of nitrogen fertilizer may adversely increase soil acidity and, in some cases, where proper selection of fertilizers and their use can help to bring about desirable changes in soil pH. A summary of the effects of fertilizers on soil pH follows:

1. The common potassium fertilizers, such as the muriate and sulfate of potash, have no permanent effect on soil acidity.
2. Superphosphates in general will have no permanent effect on soil reaction. Basic slag, bone meal, and rock phosphate have a tendency to neutralize soil acidity.
3. Fertilizers containing nitrogen in the form of ammonia or in other forms subject to nitrification (being changed to nitrate) will produce acidity unless sufficient liming material is present in the fertilizer to neutralize the acid formed. Some experimental fields that have received applications of sulfate of ammonia fairly regularly for many years without being treated with lime have become too acid to grow clover (see Fig. 12–13). This effect is more pronounced on soils such as sands and sandy loams. Fertilizers containing sulfate of ammonia should not be discriminated against, however, since the increased quantity of lime needed to keep the

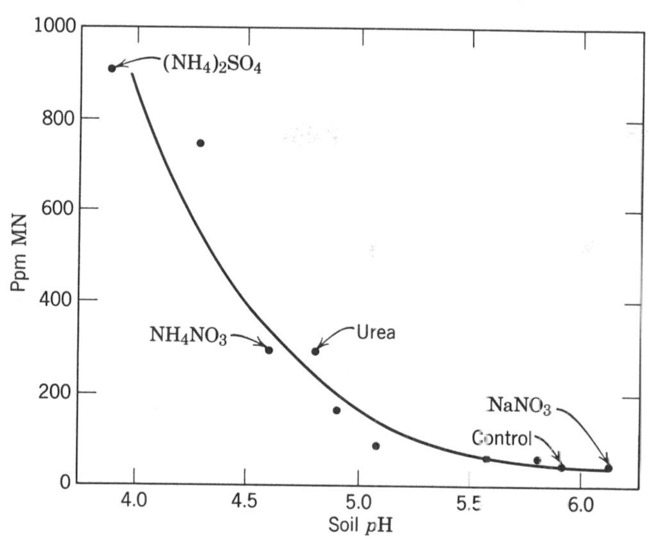


**Fig. 12–13** Manganese concentration in the midribs of corn leaves at tasseling time and soil pH. Differences in soil pH, and consequently manganese in plants, was caused by applying a total of 1500 pound of nitrogen over a five-year period on the Spinks loamy sand. The use of ammonium sulfate resulted in toxic concentration (over 400 to 500 PPM) of manganese in plants. (Michigan Agricultural Experiment Station.)

soil in condition for growing legumes in a normal system of soil management is of little practical significance.

4. Nitrogenous fertilizers in which the nitrogen is in the nitrate form and is combined with bases such as sodium or calcium will result, upon being utilized by plants, in decreased soil acidity (Fig. 12–13). Some of the fertilizers in this group are nitrate of soda, calcium nitrate, Cal-Nitro, and Calurea. Calcium cyanamide should be placed in this group although the nitrogen is not in the nitrate form, but the fertilizer carries a rather high content of lime. The acidity or basicity of several nitrogen fertilizers is given in Table 12–5.

In general, the systematic use of medium to large amounts of fertilizer at suitable times in the rotation will not appreciably affect soil acidity. Where it is desirable to increase soil acidity (to lower pH), ammonium sulfate can be used as a source of nitrogen.

## Effect of Fertilizers on Plant Growth

One of the most important factors affecting the growth of plants is the weather (temperature and quantity and distribution of rainfall), and the response that any particular crop will make to the application

**Table 12–5** Equivalent Acidity or Basicity of Several Nitrogen Fertilizers

| | | Equivalent Acidity or Basicity, pounds $CaCO_3$ | |
|---|---|---|---|
| Fertilizer Material | Nitrogen, Percentage | Per Unit of Nitrogen | Per 100 Lb of Fertilizer |
| Ammonium nitrate | 33 | 37 | 59 |
| Sulfate of ammonia | 20 | 107 | 110 |
| Ammo Phos | 11 | 107 | 59 |
| Anhydrous ammonia | 82 | 36 | 148 |
| Cal-Nitro | 20 | 0 | 0 |
| Calcium nitrate | 15 | 27[a] | 20[a] |
| Crude nitrogen solution | 44 | 24 | 53 |
| Nitrate of soda | 16 | 36[a] | 29[a] |
| Urea | 45 | 36 | 84 |
| Calcium cyanamide | 22 | 57[a] | 63[a] |

[a] Basicity.

of fertilizer is largely governed by the weather, particularly the moisture supply. In seasons when it is necessary to delay planting certain crops because of unfavorable weather conditions, the application of fertilizers may speed up the growth processes of the plant and thus offset somewhat the unfavorable effects of the season. But, looking at the fertilizer-weather relationships from another viewpoint, it is frequently observed that fertilizers in general stimulate early crop growth and, if dry weather prevails about midseason, the fertilizer may result in decreased yields. This is brought about because the soil moisture is more rapidly exhausted by the fertilized crop through increased growth and greater leaf development. On the other hand, fertilized crops may be more drought resistant due to more deeply penetrating root systems.

The early growth of a crop should not be taken as a measure of the effect of a fertilizer on yield. At times fertilizers may stimulate early crop growth, but as the season advances this difference disappears, and at harvest no increase is found. Fertilizers may also have little effect on the rate of growth of certain crops, but at harvest a decided increase in yield is noted.

Some of the most striking effects of fertilizers occur on tree crops. About one-half pound of nitrogen applied around the base of Christmas trees in June may change the foliage color from yellow green to a dark green by cutting time. The value of the tree may be increased by an amount equal to ten or more times the cost of the fertilizer.

In the forest nursery great savings occur if trees can be grown to plantable size a year sooner than usual. The pine plants shown in Fig. 12–14 and labeled 2-1 were grown as seedlings two years and as transplants in their present bed for one year. Proper use of fertilizer sometimes enables trees like these to be planted after growing for three years. Where plants can be grown to plantable size in three rather than four years, a savings of about $540.00 per acre may result.[4]

### Effect of Fertilizers on Quality and Composition of Crops

Some of the more general effects of nitrogen, phosphorus, and potassium on the quality of crops have already been discussed (Chapter 11). (See also Fig. 12–15.) The influence of fertilizers on the composition of the mineral matter of plants is exceedingly complex. Their influence is affected by the variety of crop, climate, water supply, and other environmental conditions. The influence which the presence of one element exerts on the absorptive powers of the plant for other mineral elements of the soil or fertilizer also complicates this problem.

Although it is possible to increase to a limited extent the content of mineral constituents of some crops by fertilization, one should not be misled by exaggerated claims for mineralizing human and animal foods. For a more detailed discussion of this subject see pages 282 and 283.

### Continuous High Rates of Use Not Harmful

Data can be cited from an experiment in New Jersey where fertilizer was applied annually at the rate of one ton per acre for a ten-year period.[5] The pH was lowered from 5.75 to 4.80 and this was accompanied by an increase in exchangeable hydrogen and a decrease in exchangeable calcium and magnesium. The potassium in the fertilizer resulted in an increase in the amount of exchangeable potassium. The conclusion from the experiment was that the effects could be readily overcome by the use of lime-containing magnesium.

## FERTILIZER USE AND ENVIRONMENTAL QUALITY

Algae are the most abundant plants growing in surface waters and their growth is commonly limited by a lack of nitrogen and/or phosphorus. Consequently, any activity which enriches surface waters with nitrogen or phosphorus might contribute to excessive algal

[4] "Fertilizers in Forestry," J. H. Stoeckeler and H. F. Arneman, *Advan. Agron.*, **12**:132, 1960.

[5] "The Effect of Increasing Fertilizer Concentration on Exchangeable Cation Status of Soils," F. E. Bear, *Soil Sci. Soc. Am. Proc.*, **16**:327–330, 1952.

**Fig. 12–14** High investment cost per acre makes it desirable to use fertilizer to reduce the number of years required to grow trees to plantable size. These plants were grown as seedlings two years and have been at their present location one year (2-1).

growth or an algal bloom. Further, nitrogen as nitrate can be harmful in surface or ground waters used for human or animal consumption. This section will consider some important factors related to the pollution of waters resulting from use of nitrogen and phosphorus fertilizers.

## Natural Nitrate Content of Soil Percolate

Nitrogen fertilizers are mostly water soluble and the nitrate form of nitrogen is mobile in soils. During rains nitrogen fertilizer is readily carried into the soil. For this reason, nitrogen fertilizers are not likely to be carried into surface waters by runoff water. The nitrate from nitrogen fertilizer, however, might migrate downward through the soil with percolating water and be carried to the underground water reservoir or ground water. In fact it is natural for some nitrate ions to move to the ground water in natural ecosystems. This is owing to the fact that nitrate is the most common form of available nitrogen in soils for plants and the nitrate may be carried to the water table in wet seasons when plants are relatively inactive. A look at some experimental data should be helpful.

In Chapter 4 we discussed some lysimeter data collected at Coshocton, Ohio which showed nutrient losses in soil percolate. One lysimeter in 1948 grew a grass crop and no fertilizer or manure was applied. This situation can be considered comparable to a natural grassland. The loss of nitrate nitrogen per acre per year was 4.55 pounds and the amount of soil percolate was 6.2 inches. Since there

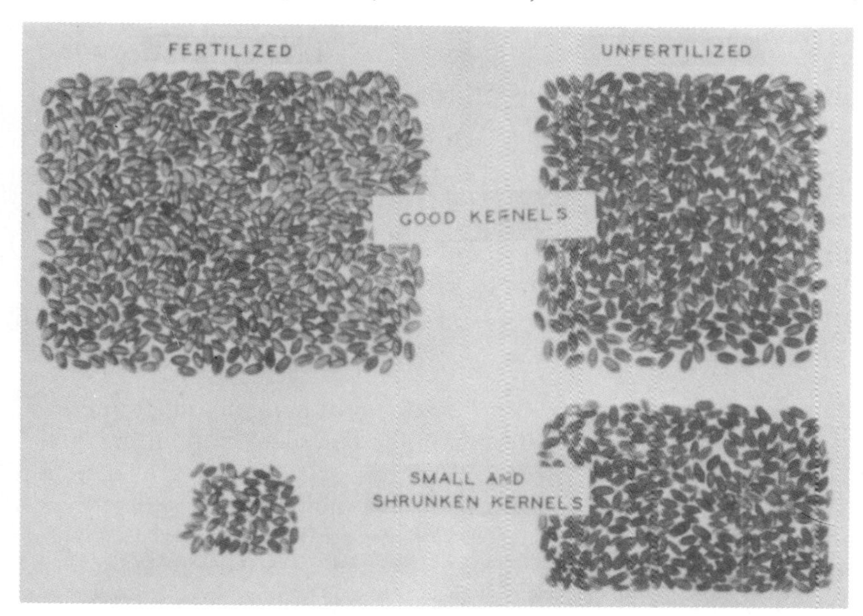


**Fig. 12–15** Properly chosen fertilizer improves the quality of many crops. Left: 20 grams (549 kernels) of wheat grown on fertilized Brookston silt loam soil contained 53 shrunken grains. Right: 20 grams (617 kernels) of wheat from unfertilized soil contained 268 shrunken grains.

are 4.4 acre inches of water in a million pounds, the parts per million (ppm) of nitrate nitrogen in the soil percolate can be calculated from the following equation:

$$\text{ppm nitrogen} = \frac{\text{pounds of N leached per acre} \times 4.4}{\text{inches of soil percolate}} \tag{4}$$

Substituting in the above equation and solving shows that the "natural" soil percolate had 3.3 ppm of nitrate nitrogen. This value is very reasonable since 2 to 3 ppm of nitrogen are considered adequate for algal growth and this is also the nitrogen content commonly found in springs and shallow wells. The U.S. Health Service considers 10 ppm of nitrate nitrogen the maximum allowable in water used to make baby formula.

## Factors Affecting Nitrate Pollution of Ground Water

The factors that affect the amount of nitrogen in drainage water or soil percolate can be represented in equation form as follows:

$$\underset{\text{percolate}}{\text{N in soil}} = \underset{\text{as nitrate}}{\text{amount of N available}} - \text{amount of N immobilized} \tag{5}$$

*Factors affecting available nitrate*
a. organic matter mineralization
b. fixation from atmosphere
c. N added in precipitation
d. N added as manure
e. N added in fertilizer

*Factors affecting N immobilized*
a. kind of crop grown
b. yield or growth of crop
c. amount and distribution of rainfall

A modest amount of nitrogen fertilizer could have no effect, increase or decrease the amount of nitrogen in the percolate, depending on the particular situation. If, however, an excessive amount of nitrogen fertilizer was applied and the soil's immobilization capacity was greatly exceeded, nitrate pollution of the ground water would occur. Thus, we see that farmers should avoid using too large amounts of nitrogen fertilizer, particularly when plant growth is inactive. The danger of nitrate pollution is also greatest on sandy soils with high percolation capacity. In these cases it is important to use modest amounts of nitrogen fertilizer and take care to apply fertilizer only when crops are actively growing.

## Phosphate Pollution of Surface Waters

Phosphate from fertilizer, in contrast to nitrate, reacts with soil constituents to form insoluble compounds that are immobile in soils. For this reason there is little possibility that ground waters would become polluted from the use of phosphorus fertilizers. Erosion of soil particles can, however, carry phosphate adsorbed on soil particles into surface waters. This type of contamination occurs naturally as well as a result from erosion of agricultural land. In fact, considerable concern exists where unprotected sloping land lies exposed for long periods of time where highways or subdivisions are constructed. The real problem here is not so much the use of phosphorus fertilizer but rather erosion.

Erosion is a selective process. The finer particles are removed and these finer particles are richer in plant nutrients. Research at the University of Wisconsin showed that the fine eroded soil material had 3.4 times more phosphorus than the soil itself. Where fertilizers promote a more vigorous plant cover and reduce soil erosion, fertilizers reduce the danger of enriching surface waters with phosphate.

# 13

# Animal Manures and Wastes

After centuries during which animal manures were highly valued for their fertilizing quality, manure on many American farms has become a liability. Today, the nutrients in the manures can be purchased in fertilizers cheaper than manure can be stored and handled. Furthermore, farms have become more specialized and bigger so that most of the manure is produced on fewer larger farms instead of a large number of small farms. A farm with 10,000 head of cattle produces about 260 tons of manure a day. The magnitude of the manure-disposal problem is dramatized by the fact that the domestic animals in the United States produce body waste equivalent to 1.9 billion people. Emphasis has changed from application of manure on soils to increase yields to the use of the soil for animal waste disposal. This chapter considers the production, composition, storage, and handling of animal manures.

## PRODUCTION AND COMPOSITION OF MANURE

A large share of the nitrogen and minerals in the crops fed and, to a lesser extent, of the organic matter, is excreted as manure by the animals. Since the animal must be heated, have energy for work, and have nutrients for growth and body maintenance, it is obvious that a portion of the nutrients and organic matter is taken from the feed consumed by the animal. To obtain a better understanding of the real nature of manure it is necessary to consider the chemical changes of feed in the digestive tract of animals, the components of manure, the quantity and composition of excrements from different kinds of livestock, and the proportion of fertilizing constituents recovered from the feed eaten by the animals.

### Chemical Changes in Feed in the Digestive Tract of Animals

Complex chemical changes occur in the food in the digestive tract of animals. These changes are brought about partly by digestive enzymes and partly by the numerous bacteria which live in the intestinal tract. Twenty to 30 percent of the dry weight of the solid excrement consists of living and dead cells of bacteria.

The various constituents of feeds undergo different rates and different degrees of decomposition. Sugars and starches are easily broken down; celluloses and hemicelluloses are less easily decomposed, and lignins are very resistant. Proteins vary considerably in their ease of breakdown. Proteins in "concentrates" are usually much more highly digestible than proteins in "roughages." Approximately one-half of the organic matter in the feed is decomposed during digestion.

Most of the potassium in feed is absorbed from the digestive tract and excreted in the urine. Only a small fraction of the phosphorus, except in hogs, is so absorbed. Consequently, most of the phosphorus in manure is carried in the solid fraction.

### Components of Manure

Farm manure consists of two components, the solid and the liquid. The solid excrement, on the average, contains one-half or more of the nitrogen, about one-third of the potassium, and nearly all the phosphorus that is excreted by the animal. The nitrogen in the feces exists largely in two forms: first, the residual proteins that have resisted decomposition in the digestive processes; and, second, the proteins that have been synthesized in the cells of bacteria. Over one-half of the nitrogen may be present as synthesized protein, and this form is readily broken down when added to soils, so that the nitrogen is available to plants. The solid excrement also contains large quantities of lignin. In other words, a large share of the organic matter in the feces is humified; a compound is formed similar to the humus that is formed in soils. As much as 50 percent of the organic matter in the solid excrement may be in the humified state, and the nitrogen contained therein is only slowly available to plants when added to soils.

The liquid fraction or urine contains those plant nutrients which have been digested and utilized in the animal body and are later excreted. All the plant nutrients in this fraction are soluble and either are directly available to plants or readily become so. The liquid portion of manure differs from the solid not only in regard to the availability of nutrients but also in its low content of phosphorus and in its high content of potassium and nitrogen. The distribution of plant nutrients between liquid and solid portions of manure is shown in Fig. 13–1.

In general, the more digestible the food consumed by an animal, the greater is the proportion of its plant nutrients that appears in the urine. Furthermore, as a rule, the higher the food is in nitrogen, the greater the digestibility of the nitrogen and the larger the amounts of nitrogen in the urine.

When voided, the nitrogen in urine exists largely as urea and hippuric and uric acids. These compounds are not volatile at ordinary temperatures, but manure contains organisms capable of rapidly breaking these compounds down with the formation of ammonia, which combines with water and carbon dioxide to make ammonium carbonate. This compound is unstable; even in solution it tends to decompose, losing ammonia, especially at higher temperatures [$(NH_4)_2CO_3 \rightarrow 2NH_3 + H_2O + CO_2$]. This compound may lose all its ammonia on drying. The unstable nature of the nitrogen in urine presents a major problem in the handling of manure.

## Quantity and Composition of Excrements

Many factors influence the quantity of manure produced and its composition, such as (1) the kind and age of the animal, (2) the kind and the amount of feed consumed, (3) the condition of the animal, and (4) the milk produced or the work performed by the animal. Wide variations are often found in the manure of animals even of a given

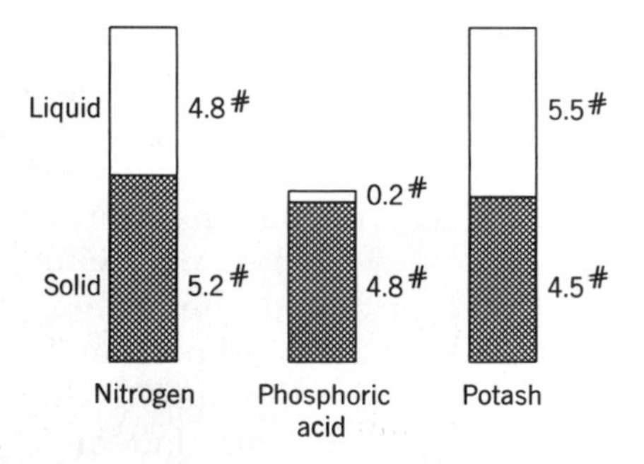


**Fig. 13–1** Distribution of plant nutrients between liquid and solid portions of a ton of average farm manure.

class. Animals of different ages and doing different kinds of work require different amounts and proportions of nutrients to maintain them. A young animal, for example, which is building muscle and bone needs considerable phosphorus, nitrogen, calcium, and other elements, and the manure produced by such animals will contain much less of these elements. Since the composition of manure is so variable, data such as those presented in Table 13–1 can only be approximate.

**Table 13–1** Quantity and Composition of Fresh Manure Excreted by Various Kinds of Farm Animals

| Animal | Excrement | Pounds per Ton | Water, Percentage | Nitrogen, Pounds | $P_2O_5$, Pounds | $K_2O$, Pounds | Tons Excreted[a] per Year |
|---|---|---|---|---|---|---|---|
| Horse | Liquid | 400 | ... | 5.4 | Trace | 6.0 | ... |
| | Solid | 1600 | ... | 8.8 | 4.8 | 6.4 | ... |
| | Total | 2000 | 78 | 14.2 | 4.8 | 12.4 | 9.0 |
| Cow | Liquid | 600 | ... | 4.8 | Trace | 8.1 | ... |
| | Solid | 1400 | ... | 4.9 | 2.8 | 1.4 | ... |
| | Total | 2000 | 86 | 9.7 | 2.8 | 9.5 | 13.5 |
| Swine | Liquid | 800 | ... | 4.0 | 0.8 | 3.6 | ... |
| | Solid | 1200 | ... | 3.6 | 6.0 | 4.8 | ... |
| | Total | 2000 | 87 | 7.6 | 6.8 | 8.4 | 15.3 |
| Sheep | Liquid | 660 | ... | 9.9 | 0.3 | 13.8 | ... |
| | Solid | 1340 | ... | 10.7 | 6.7 | 6.0 | ... |
| | Total | 2000 | 68 | 20.6 | 7.0 | 19.8 | 6.3 |
| Poultry | Total | 2000 | 55 | 20.0 | 16.0 | 8.0 | 4.3 |

Compiled from *Fertilizers and Crop Production*, Van Slyke, Orange Judd Publishing Co., pp. 218, 220, 1932.

[a] Clear manure without bedding; tons excreted by 1000 pounds of live weight of various animals.

The urine makes up only 20 percent of the total weight of the excrement of horses but 40 percent of that from hogs. These represent the two extremes. Since the urine makes up only 20 to 40 percent of the total weight of manure from any animal, and yet contains approximately two-thirds of the total potash and somewhat less than one-half of the nitrogen, it is evident that, pound for pound, the urine is more concentrated and hence more valuable than the solid portion.

On a percentage basis there is considerably more nitrogen in the urine from most animals than in the feces, although more than one-half of the total quantity excreted is in the solid manure. The phosphorus, although existing in relatively small quantities, is found largely in the solid. Potash, however, occurs in larger quantity and in higher concentration in the liquid than in the solid.

It is interesting to note that water is one of the most variable constituents in manure. Poultry and sheep manure, on a tonnage basis, contain much greater quantities of plant food that any of the other manures; yet these two types of manure contain much less water. The different manures, on a wet basis, listed in Table 13–1, contain from 22 pounds of plant food per ton in cow manure to 47.4 pounds in sheep manure. If the composition of the total manure excreted by each of the different animals per 1000 pounds live weight were calculated on a dry-weight basis, the manures would not vary so greatly in their plant-food content. Horse and sheep manures are considerably drier than cow or hog manures; this fact partially explains why horse and sheep manures heat more quickly than cow and hog manures in storage and why the former are preferred in the preparation of hotbeds.

## Fertilizing Constituents Recovered from Feed

The fertilizing constituents in feed consumed by animals are not lost from the animal body in any measurable quantity except through excretion in the urine and feces. Therefore, all the fertilizing constituents that are not built into animal tissue or secreted in milk are excreted in manure. The recovery of nutrients may be practically complete if the animal is not gaining or losing weight, producing milk, or supplying nutrients for developing young. Furthermore, animals that are not secreting milk and are losing weight may excrete more nitrogen than is contained in their feed as a result of animal-tissue breakdown. It has been indicated that young, growing animals and milking cows remove larger quantities of nutrients from feeds than do other classes of livestock. One thousand pounds of milk contain approximately 6.0 pounds of nitrogen, 1.7 pounds of phosphoric acid, and 1.7 pounds of potash; and about 2.5 pounds of nitrogen, 1.5 pounds of phosphoric acid, and 0.2 pounds of potash are contained in each 100-pound weight of a calf.

On an average, farm animals return in the manure from the feed consumed 75 to 80 percent of the nitrogen, 80 percent of the phosphoric acid, 85 to 90 percent of the potash, and 40 to 50 percent of the organic matter (Fig. 13–2). Variations in the recovery of nutrients from feeds are due to different conditions, such as the kind and concentration of feed, age of animal, kind of animal, milk production, and work performed by the animal. Consequently, values may vary considerably from those given above.

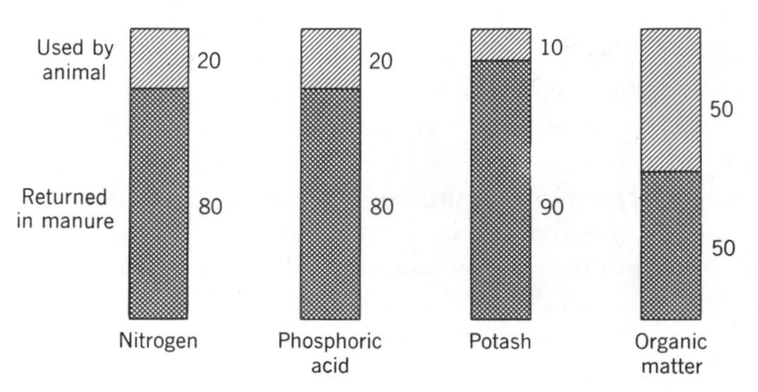


**Fig. 13–2** Average proportion of plant nutrients and organic matter in feed consumed by animals and excreted in manure.

## LOSSES OCCURRING IN MANURE HANDLING

Common methods of handling manure about the barn and delayed application of manure result in the dissipation of a large share of its value before it reaches the field. The degree to which manure decreases in value is determined mainly by the kind and by the methods used in collecting, storing, and applying it. Several processes contribute to losses of the fertilizing constituents and organic matter in manure and the pertinent facts about these losses will be discussed.

### Losses of Liquid Manure

The loss of urine from manure is serious from the standpoint of plant-nutrient content. This loss occurs mostly from failure to use ample bedding, leakage through stable floors, seepage into earth floors, or drainage from manure heaps.

On the basis of the total plant food in manure, about 50 percent of the value is in the urine. It can be readily understood that large quantities of plant nutrients are lost and that the portions so lost, as already pointed out, are the most readily available forms of plant nutrients in manure. It is probably safe to assume that more than one-half the liquid is lost on many farms. In terms of dollars and cents, this represents a very significant loss to the farmer.

### Leaching Losses

Frequently manure is thrown from stables into an adjoining lot, where it is unprotected from rain; sometimes it is thrown directly under the eaves of the barn. Leaching losses are greater when the manure is thrown in small, loose, and open piles.

In the course of six months, manure which is exposed to rain may lose more than half its fertilizing value, depending on the quantity of rain and the nature of the pile. The materials leached out are the most readily soluble and therefore the most quickly available plant nutrients. Leaching losses are confined not alone to the urine fraction but also to the nitrogen, phosphorus, and potassium compounds of the solid portion. Under alkaline conditions considerable quantities of soluble organic matter may leach out also; this is indicated by the dark color of leachings from manure piles.

## Losses by Volatilization

Losses incurred by volatilization fall principally on nitrogen and organic matter. Large quantities of ammonia are produced in manure from urea and other nitrogenous compounds, and in the earlier stages of manure decomposition the ammonia is combined largely with carbonic acid as ammonium carbonate and bicarbonate. These ammonium compounds are rather unstable, and gaseous ammonia may be readily liberated (Fig. 13–3). The tendency to lose ammonia nitrogen increases with the increase in concentration of ammonium carbonate and with the increase in temperature.

At ordinary temperatures little or no loss of ammonia from manure occurs at pH 7.0 or below. High temperatures, produced by aerobic decomposition in a loose manure pile, are conducive to very rapid loss of ammonia.

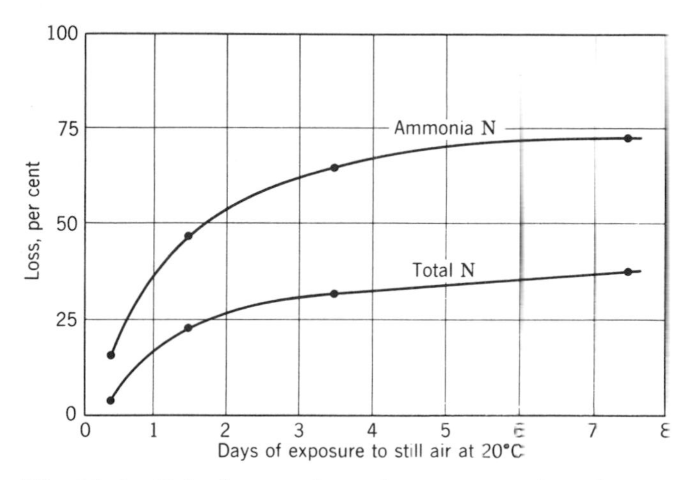


**Fig. 13–3** Volatilization loss of ammonia and total nitrogen from fermented cow manure exposed to drying. (Based on data from "Conservation and Availability of the Nitrogen in Farm Manure," A. Floyd Heck, *Soil Sci.*, Vol. **31**, p. 353, 1931.)

Freezing also tends to increase the loss of ammonia by increasing the concentration of the solution through the crystallization of the water. This loss may be considerable when manure is spread and becomes frozen.

Air movement greatly affects the loss of ammonia. Wind movements hasten evaporation of water, which decreases the capacity of the water to hold ammonia. Thus, manure which is permitted to dry out may lose appreciable quantities of ammonia. This fact emphasizes the importance of ammonia loss due to air circulation in loosely piled manure heaps and in overfermented manure that is forked frequently. Losses may also be considerable if manure is spread and permitted to dry before plowing under.

It has been pointed out that when manure decomposes it suffers important losses in organic matter. These losses occur mainly in the carbohydrate constituents. One of the important end products of carbohydrate decomposition is carbon dioxide, most of which is lost from manure by volatilization. Shrinkage that accompanies the partial decomposition of manure is evidence of organic-matter loss.

Little or no loss of phosphoric acid or potash from manure occurs through volatilization.

## MANAGEMENT OF MANURE FOR CROP PRODUCTION

When manure is used to maintain or to improve soil fertility, the primary objective in the care and handling of farm manure is to prevent loss of plant nutrients as much as possible. Even under the most favorable conditions, it is practically impossible to prevent some loss of nitrogen and also some loss of organic matter. There is no difficulty in conserving the phosphorus and potassium compounds, however, because they are not volatile.

### The Use of Litter

Bedding or litter is used primarily to furnish clean and comfortable places for animals to lie down. In relation to the value of manure, bedding is used principally for these purposes: (1) to prevent loss of urine by drainage, (2) to make manure easier to handle, (3) to adsorb and hold plant nutrients, and (4) to increase organic matter and plant-nutrient content.

The principal value of litter as a preserving agent lies in its ability to absorb the urine, which in general carries more than one-half the fertilizing value of manure. The efficiency of various litters can be stated in terms of the amount of liquid which a given weight of material will absorb. Ordinary straw can take up two or three times its weight of water; cut straw can take up about five times its weight of water, whereas good peat moss can absorb as much as ten times its weight of water.

## Hauling Manure Directly to the Field

Many farmers haul manure daily to the field, and this is usually a good practice. Manure spread on the land and worked into the soil is perhaps in its safest place. The soil has the capacity to rapidly fix large quantities of plant nutrients carried in the manure. As a rule very little loss occurs when manure is hauled directly to the field, except when it is spread on hillsides, where it may be washed down the slope, or when it is placed on frozen ground. If the liquid fraction has been thoroughly absorbed by bedding materials, there will be little loss by runoff. At any rate, it is better to have the liquid go into the soil for plant use than to let it soak in the soil of the barnyard or wash away in the drainage.

If manure is spread daily and is not worked into the soil, considerable loss of volatile ammonia may occur, owing to drying winds or freezing weather, or both, unless sufficient rain has fallen to wash the ammonia into the soil. To decrease losses due to drying or freezing, the land should be disked after the manure is spread if possible.

## Storage of Manure

In storing manure all practical precautions should be taken to keep losses at a minimum, and consideration must be given to costs involved. Handling manure not only increases the expense but also decreases the value of manure by exposing it to the air, thereby increasing decomposition losses.

Good storage of manure makes provision for keeping the manure heap (1) thoroughly compact, (2) sufficiently moist but not too wet, (3) under cover or shelter, and (4) undisturbed during storage.

Permitting the accumulation of manure under the feet of animals in stalls or in covered lots or pens is an efficient and practical means of storage. If plenty of bedding is used, the liquid manure is absorbed, the tramping of the animals compacts the manure, and fermentation losses are kept at a minimum. This type of storage also permits considerable flexibility in time of hauling. Some of the liquid manure may be lost when earth floors are used, and sometimes it may be economical to provide concrete floors to prevent this loss.

On many farms it is necessary to resort to outdoor open storage. The manure should be placed in compact piles with straight sides and flat tops and in such a position that it will receive neither roof nor surface water. Although this method usually results in considerable loss of plant food, loss is reduced if the manure is piled properly.

### Use of Superphosphate to Reduce Nitrogen Volatilization Losses

The ammonium carbonate produced during storage or handling of manure decomposes to release gaseous ammonia. Since ordinary superphosphate is about half calcium sulfate, the ammonium carbonate can be converted to the more stable ammonium sulfate by reacting ammonium carbonate with calcium sulfate. The reaction expected by the addition of ordinary superphosphate with the ammonium carbonate in manure is as follows:

$$Ca(H_2PO_4)_2 + 2CaSO_4 + 2(NH_4)_2CO_3 \rightarrow Ca_3(PO_4)_2 + 2(NH_4)_2SO_4 + 2H_2O + 2CO_2$$

In this reaction tricalcium phosphate is formed, which does not react with ammonium sulfate on drying, so that loss of ammonia is prevented. However, owing to the limited solubility of the gypsum and the limited contact it makes with the manure, the amount of nitrogen fixed is small.

## Application of Manure

Prompt spreading of manure is generally considered most effective, although when in good storage it is likely to lose less value than if spread on the field without being plowed under or worked into the soil immediately. Losses of applied manure may occur in three ways: (1) volatilization of ammonia nitrogen as a result of drying or freezing; (2) surface runoff water carrying soluble portions of all three nutrients; or (3) leaching of nutrients.

Much experimental work with commercial fertilizers indicates that their effectiveness is decreased if they are applied a considerable time before the seeding of the crop. This effect is attributed to leaching losses and to the fixation in less soluble forms of the plant nutrients by the soil. Fresh and properly stored manures contain large amounts of soluble nutrients, and the same principles apply. Manure applied to corn land in the spring is likely to give greater returns than a like amount of manure applied in the fall. This effect will probably apply only to the crop of the first year and not to the later crops. If manure has already lost its readily soluble nutrients, the time of application is less important.

Little loss of nutrients from manure may be expected when applied on level, medium- to fine-textured soils. These soils are safe for fall and winter spreading of manure, although they fix considerable quantities of the nutrients in less available forms. It is not advisable to place manure on sandy soils or on hilly fields much ahead of plowing time because of leaching and erosion losses. It must be remembered, however, that manure applied on sloping fields decreases erosion losses, thereby counterbalancing to some extent the loss of nutrients from the manure.

The rate at which manure should be spread will depend on the amount produced on the farm. As much of the cultivated land should be covered as possible. It is usually better to cover an entire field with a light application than to give only a part of the field a heavy coating. In other words, the crop response from 50 tons over 10 acres proves larger than from 50 tons on 5 acres. It is wise to gauge the rate of application so as to extend the acreage covered for each crop.

### Effect of Manure on Soil Erosion

Incorporating manure with soil may be effective in reducing soil erosion by increasing the permeability of the soil to water, thus decreasing runoff losses, and by increasing the density of the vegetative cover, which in turn decreases rate of surface runoff and increases water penetration. A large number of experiments has been reported, showing that manure is effective in reducing both water and soil losses. Care should be exercised on steeply sloping lands because manure may be washed into streams and may cause pollution.

### Residual Effects of Manure

An application of manure usually shows a favorable influence on crop yields for several years. These beneficial effects are distributed over a longer time than those of chemical fertilizers. Striking results in showing the long-continued effects have been obtained (Fig. 13–4) by making heavy applications of manure for several years in succession and then discontinuing the application.

The lasting effect is due in part to the slow availability of certain plant food constituents in manure, in part to the fact that a portion of the organic matter may remain for several years and aid in increasing the soluble plant nutrients in the soil, and in part to greater quantities of roots and stubble of the larger crops grown.

### Value of Fresh Versus Rotted Manure

In speaking of rotted manure, it is usually inferred that the original structure of the materials has disappeared more or less completely. In this connection, if we assume that the fresh manure is a normal

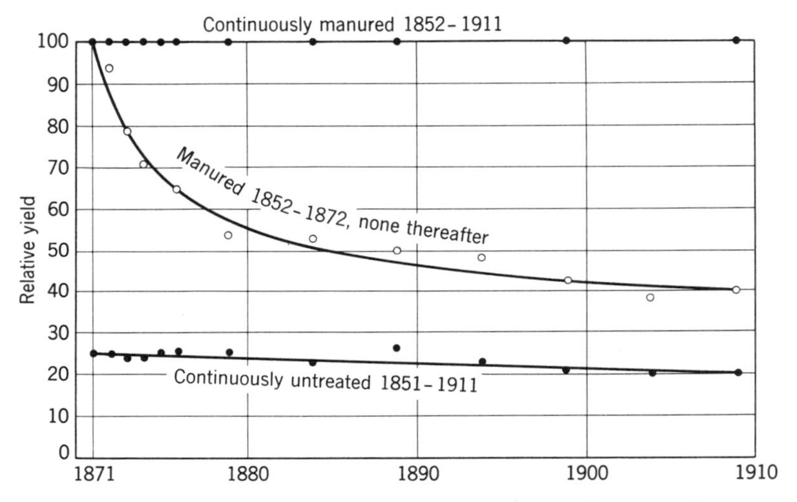


**Fig. 13-4** Residual effects of heavy applications of manure (14 tons annually) on the yield of barley grown continuously on the soil of the Rothamsted Experiment Station in England. (From *Ohio Exp. Sta. Bull.*, **605**.)

mixture of urine and feces and that conditions have been controlled, fresh and rotted manure differ in composition as follows:

1. Rotted manure is richer in plant-food constituents. This concentration of plant nutrients is due to shrinkage in dry weight, which would automatically raise the level of plant food. One ton of fresh manure may lose one-half its weight in the rotting process. The loss occurs principally in the organic nitrogen-free constituents which contain no mineral plant-food elements.
2. More of the nitrogen in fresh manure is soluble. The decrease in soluble nitrogen is brought about by microorganisms in the synthesis of their body tissues during active decomposition of the organic constitutents of the manure. There is considerable utilization of urinary nitrogen in the formation of complex proteins during the decomposition of manure. Unless rotting conditions are carefully controlled, there may be considerable loss in total nitrogen through volatilization.
3. The solubility of phosphorus is greater in decomposed manure. If no leaching occurs, there is no change in the total quantities of phosphorus or potassium.

That the increase in concentration of plant nutrients in rotted manure is obtained at the expense of large losses of organic matter and considerable losses of nitrogen, especially available nitrogen, has not been fully appreciated. In general, it can be said that benefits derived from the rotting process may be more than offset by losses. If manure can be applied to the field daily, there is little justification for allowing it to rot except for special uses.

## USE OF SOIL FOR ANIMAL WASTE DISPOSAL

The concentration of large numbers of animals on specialized livestock farms has created serious animal waste disposal problems. Odors from stored or rotting manure are commonly offensive. Burning animal wastes can pollute the air. Laws exist prohibiting the disposal of animal wastes into water courses. Chemical processing of animal wastes is expensive. As a consequence, animal waste disposal is a major problem for many livestock producers. The soil is a natural medium for waste disposal and much attention is now being directed toward using the soil for animal waste disposal.

### Magnitude of the Animal Waste-Disposal Problem

An interesting comparison between the production of human and animal wastes has been made for Minnesota.[1] In Minnesota there are about 14 million chickens and turkeys, 4 million cattle, 1½ million dairy cows, 2½ million hogs, and ¾ million sheep. These animals produce waste equivalent to over 60 million people. Since the population of Minnesota is about 4 million, the animal wastes produced are about 15 or more times greater than the human wastes.

Some feedlots in the United States carry as many as 10,000 head and produce waste equivalent to a city of 164,000 people (see Fig. 13–5). Some data on animal population and quantity of waste produced in the United States are given in Table 13–2. The data show that about 1½ billion tons of waste are produced annually and that about two-thirds is solid manure and about one-third is liquid. If the waste resulting from dead animal carcasses, paunch manure from abattoirs, and used bedding are included, the total amount of animal waste produced is about 2 billion tons per year.

### Capacity of Soil for Waste Disposal

Under normal conditions, the soil microorganisms exist in a near state of starvation. Readily available food or energy supplies are usually limited. The addition of decomposable wastes, such as manure, produce a rapid increase in the numbers and activity of

[1] "Soil as an Animal Waste Disposal Medium," W. P. Martin, *Journal of Soil and Water Conservation*, March–April 1970, pp. 43–45.

**Fig. 13–5** Each head of cattle produces body wastes equivalent to about 16 people. A farm with 1000 head of cattle has a body waste disposal problem equivalent to a city of 16,000 people. (USDA photo.)

**Table 13–2** Animal Population and Waste Production in the United States

| Livestock | 1965 Population, millions | Annual Waste Production, million tons | |
|---|---|---|---|
| | | Solid Waste | Liquid Waste |
| Cattle | 107 | 1004.0 | 390.0 |
| Horses | 3 | 17.5 | 4.4 |
| Hogs | 53 | 57.3 | 33.9 |
| Sheep | 26 | 11.8 | 7.1 |
| Chickens | 375 | 27.4 | – |
| Turkeys | 104 | 19.0 | – |
| Ducks | 11 | 1.6 | – |
| Total | – | 1138.6 | 435.4 |

From "Wastes in Relation to Agriculture and Forestry," C. H. Wadleigh, United States Department of Agriculture Misc. Pub. 1065, p. 41, 1968.

microorganisms. As a result, the soil can dispose of enormous quantities of animal waste. It seems reasonable to expect that a one-inch layer of manure could be incorporated into the soil every year without difficulty, assuming favorable soil moisture and temperature. Assuming also that the manure is largely water (see Table 13–1), a one-inch thick application of manure would be less than 100 tons per acre. If the manure contains about 10 pounds of nitrogen per ton, the nitrogen application would be about 700 to 900 pounds per acre. On the basis of the discussion in Chapter 5 of the capacity of the ecosystem to immobilize nitrogen without danger of nitrate pollution of groundwater, nitrate pollution of groundwater would likely result in humid regions from 100 ton per acre applications of manure. From the practical standpoint, it appears that considerations other than the decomposing capacity of the soil limit the rate of waste application. Researchers in Canada analyzed the manure on many kinds of farms and concluded that to prevent nitrate contamination of groundwater and not adversely affect corn yields, a minimum one-half acre of land was needed for 1000 broilers, 100 laying hens, 10 hogs (30 to 200 pounds), 2 feeder cattle (40 to 1100 pounds), or 1 dairy cow (1200 pounds).[2] On this basis a 10,000 head feedlot operation would require 2500 acres for the disposal of the manure.

[2] "Animal Wastes," L. R. Webber, Dept. of Soil Science Annual Progress Report, University of Guelph, Guelph, Ontario, pp. 45–49, 1967.

# 14

# Soil Erosion and Its Control

In the development of a new country little attention is given to conservation. Usually the natural resources that the country affords are present in such quantity that they appear inexhaustible. The limitation of supply is in manpower and in those essentials of life which the country does not produce. Later, when population density has become great, the necessity for conserving resources becomes apparent and frequently acute. Often, as in the United States, the need for conserving soil is one of the last to be recognized. In fact, many people in our country do not yet realize the urgency for conserving soil resources. Our surplus production of a few crops, like cotton, wheat, and corn, tends to obscure the waste in soil productivity that is being sustained. Although in a few areas, as in the southeastern and southwestern states, some consideration had been given to soil erosion control, the problem as a whole attracted no widespread attention until about 1933. Since that time public interest in the work has developed rapidly.

### SOIL EROSION DEFINED

The basic definition of the word, "erosion," is to wear away. Since the earth was first formed, there has been a continual wearing away of the surface. Many agents are responsible, but the discussion here will be limited to cultivated fields. It will be restricted to water and wind erosion.

### Geological Versus Soil Erosion

Erosion which takes place under natural conditions, that is, when the land surface and native vegetative cover have not been disturbed by man's activities, is called natural or *geological erosion*. On the other hand, when timberland is cleared or grassland is broken up, processes of erosion are accelerated and we have unnatural or soil erosion. Whenever erosion is speeded up as a result of man's activities so that it removes all or part of the topsoil, we call the process *soil erosion*.

Geological erosion is a relatively slow process under many conditions, and soil formation may keep pace with the removal of the surface soil. Soil erosion, on the contrary, is very rapid when environmental factors favor erosion. For example, at the Statesville North Carolina Experiment Station, land planted to cotton year after year suffered an annual loss of 31 tons of soil per acre. The same kind of land, however, under natural conditions lost only 0.002 ton of soil per acre annually by geological erosion. Nevertheless, over long periods of time geological erosion moves very large quantities of soil material to lower levels as pointed out in the following section.

### Geological Erosion a Natural Process

Geological erosion is a natural process which tends to bring the earth's surface to a uniform level. Whenever one part of the surface of the globe is elevated above the surrounding portions, erosion immediately begins the work of leveling off the high land. The leveling process may result in a very rough topography in the early stages by the cutting of gullies or of canyons in a mountainous region, but the ultimate result is a comparatively level surface. Evidence of this geological process is seen in peneplanes, mesas, valley fills, alluvial plains and deltas, extensive deposits of wind-laid material, and numerous other geological formations. Some concept of the extent of erosional activity may be gained from the fact that the Appalachian Mountains are said by geologists to be only about one-half of their original height.

## REASONS FOR EMPLOYING EROSION CONTROL PRACTICES

A relatively short trip through virtually any section of our country reveals evidences of soil mismanagement. In the midst of a productive agricultural area many farms will be observed which have been depleted in organic-matter content, as evidenced by a lighter color and lower productivity than surrounding soil. In areas of undulating to rolling topography, slopes may be seen which have been denuded of the surface or plow soil, leaving the lighter colored subsoil exposed. Again, many instances are seen of sandy soils which have been cleared of their forest cover, cropped for a few years until the virgin fertility was exhausted, and then abandoned to become covered with weeds and brush or to be blown about by the wind.

As pointed out in Chapter 17, many nations have inadequate food supplies, and a greatly increased food production will be needed to feed the expected world population in the future. This situation affords another reason for giving careful attention to soil conservation.

## Damage Done to Agricultural Land by Soil Erosion

The damage done to farm lands in the United States since they were occupied by the white man can only be estimated roughly. From data supplied through soil surveys, erosion surveys, and other inventory methods, however, it has been suggested that erosion has ruined for agricutural use or seriously damaged some 282,000,000 acres in the United States. A far larger area, possibly three-quarters of a billion acres, has suffered the loss of a greater or lesser portion of its topsoil.

Of land used for crop production, it is estimated that around 50,000,000 acres have been rendered useless for crop production and a like amount is approaching that condition. An additional 100,000,000 acres, although still being cultivated, have lost one-half or more of the surface soil, and on a similar acreage erosion is carrying on its insidious work of destruction. H. H. Bennett, former chief of the Soil Conservation Service, has expressed the opinion that erosion control "is the first and most essential step in the direction of correct land utilization on about 75 percent of the present and potential cultivated area of the nation."[1]

In considering the damage done by erosion one should keep in mind the fact that a large share of the soil lost by this process is the surface or plow soil. It is this soil layer which contains the highest percentage, in an available condition, of many of the essential plant-food elements. Furthermore, studies of the soil eroded from fields in many parts of the country have shown these to be made up largely of the finer soil particles (clay, silt, very fine sand, and humus). These particles contain a higher percentage of several of the plant nutrients than do the coarser particles.

[1] By permission from *Soil Conservation*, by H. H. Bennett, McGraw-Hill. Copyright 1939.

## Erosion Damage Not Confined to Soil Loss

Erosion by water opens the way for at least five types of loss or damage.

1. The loss of the water causing the erosion. It might have been useful in crop production had it entered the soil instead of running off over the surface.
2. The soil carried away by erosion frequently ceases to be of value in crop production, and, furthermore, the remaining soil, denuded of the surface or plow soil, is much decreased in productivity.
3. The soil carried away frequently causes much damage. Especially during gully formation, a layer of infertile subsoil may be deposited over an area of productive soil, thus greatly reducing the crop-producing power (Fig. 14–1).
4. Another damage resulting from gully formation is the cutting up of fields into irregular pieces. As these gullies get too deep to cross with farm implements, great inconvenience and loss of efficiency in cultivating the land and planting and harvesting crops result.
5. The soil removed through erosion may be deposited in streams, harbors, and reservoirs, thus increasing floods, impeding navigation, and reducing water-storage capacity.

**Fig. 14–1** Soil eroded from poor land may greatly damage crops growing on good soil. This cherry tree is being buried by coarse sand washed from adjacent slopes. (Courtesy of Soil Conservation Service.)

### Rapid Sedimentation of Reservoirs

The amount of sediment carried by various streams is enormous, as is shown by the estimates made by the U.S. Geological Survey given in Table 14–1. When water storage reservoirs are built on rivers

**Table 14–1** Tons of Sediment Carried by Several Rivers

| River | Tons Per Year | River | Tons Per Year |
|---|---|---|---|
| Hudson | 240,000 | Savannah | 1,000,000 |
| Susquehanna | 240,000 | Tennessee | 11,000,000 |
| Roanoke | 3,000,000 | Missouri | 176,000,000 |
| Alabama | 3,039,000 | | |

carrying large amounts of sediment, the reservoirs quickly fill and lose their water-storage capacity. An extreme example of the loss of storage capacity as a result of erosion is furnished by the Washington Mills Reservoir at Fries, Virginia. In the course of 33.5 years, 83 percent of the storage capacity had been lost.

A study of Lake Decatur, municipal-water-supply reservoir of Decatur, Illinois, showed that it had lost 1.0 percent of its storage capacity annually between the date of construction (1922) and 1936. Between 1936 and 1946 the annual rate of sedimentation had increased to 1.2 percent of the capacity. Also, in a land-utilization-project area near Pierre, South Dakota, the annual rate of silting of stock ponds was found to vary between 1.10 and 5.56 percent.

The Roosevelt Dam, which supplies water for electric power and for irrigation of the great Salt River Valley in Arizona, lost 7 percent of its storage capacity through sedimentation during the first 24 years of its existence. Likewise, the Elephant Butte Dam on the Rio Grande River in New Mexico decreased in storage capacity about 17 percent during the period between 1915 and 1947. Such losses of storage capacity are serious matters, especially when the supply of water is scarcely adequate to meet demands.

## CONTROLLING EROSION CAUSED BY WATER

In discussing erosion control it is well to remember that this phenomenon really involves *two* processes: (1) the loosening or detachment of soil particles or groups of particles from the main body of the soil, and (2) the transportation or removal of these particles or particle groups from their position. Four fundamental principles which guide many of the methods employed to reduce erosion by water are: (1) protect the soil surface from impact of raindrops; (2) prevent water from concentrating and moving down a slope in a narrow path; (3) impede the movement of water so that it moves down a slope slowly; and (4) encourage more water to enter the soil.

## Types of Water Erosion

Erosion by water may be divided into four categories, namely, splash, sheet, rill, and gully. Strictly speaking, sheet erosion refers to the quite uniform removal of soil from the surface of an area in thin layers. For sheet erosion alone to occur it is necessary that there be a smooth soil surface, which is seldom the case. Usually a soil surface designated smooth contains small depressions in which water will accumulate. Overflowing from these at the lowest point, the water cuts a tiny channel as it moves down the slope. Duplicated at innumerable points, this process presently creates a surface cut by a multitude of very shallow trenches which are called rills. None of these may grow to appreciable size or depth, and so the surface soil is rather uniformly removed from the field. Accordingly, sheet erosion and rill erosion work hand in hand and the combined process is usually called sheet erosion, as distinguished from gully formation.

A type of erosion which received little attention until recent years is the splashing or scattering of the smaller soil particles by the impact of raindrops. At first thought this action seems trivial, but, when consideration is given to the large number of raindrops that strike a square inch of soil surface during a one-hour rain and the force with which they strike, it is seen that the net effect in loosening and moving soil particles may be considerable.

**Sheet Erosion Insidious.** Rills are obliterated by the first cultivation of a field even by a harrow, and, in consequence, sheet erosion is not at first obvious. A farmer may be conscious that the surface soil is unusually deep at the foot of a slope, but it does not occur to him to worry about it. So sheet erosion goes on largely unnoticed until the yellow or brown subsoil begins to appear on the steepest part of the slope and crop growth at that point is thin and unthrifty. Even then many farmers take the occurrence as a matter of course and do not feel the need of putting into practice erosion control methods. A little extra fertilizer or manure is spread on that part of the field, and farming goes on as usual. The very fact that sheet erosion fails to attract attention accounts for the vast damage done by this process. Also, the occurrence of sheet erosion on slopes which are quite gentle leads one to overlook the loss of soil. Although far less noticeable, the total damage done by sheet erosion greatly exceeds that caused by gullying. On many farms sheet erosion is likely to occur on the best land.

**Gully Erosion Spectacular.** Although sheet erosion may pass unnoticed by the average observer, gullies attract immediate attention. They disfigure the landscape and give the impression of land neglect and soil destruction (Fig. 14–2). Not only do gullies result in soil loss but also, as previously mentioned, the eroded material is usually deposited over more fertile soil at the foot of the slope. Also, fields dissected by gullies offer many problems in farming operations.

Gullying proceeds by three processes: (1) waterfall erosion, (2) channel erosion, and (3) erosion caused by alternate freezing and thawing. Usually more than one process is active in a gully. Water falling over a soil bank undermines the edges of the bank, which then caves in, and the waterfall moves upstream. This process produces U-shaped gullies, particularly if the underlying soil material is soft and easily cut. Gullies which are V-shaped are produced by *channel erosion* through the cutting away of the soil by water concentrated in a drainageway. This type of gully usually forms when the underlying soil horizons are of finer texture and more resistant to erosion than the surface horizons. Soil loosened from sides of

**Fig. 14–2** This gully was formed in 25 years by an improper terrace emptying out into a road. (Courtesy of Soil Conservation Service.)

gullies by alternate freezing and thawing sloughs off and is then carried away by heavy rains. Gullies are sometimes classified as small if they are less than 3 feet deep, medium-sized if from 3 to 15 feet deep, and large if over 15 feet in depth.

## Concentration of Water to Be Avoided

As pointed out, gully formation results from water concentrating in a narrow channel to flow down a slope over unprotected soil. Careless and often seemingly trivial mismanagement practices in the handling of land are most frequently responsible for the starting of gullies. It is much simpler and less expensive to prevent the concentration of water before it flows down a slope than it is to stop gully formation when the process is well established. Some of the procedures which have been found effective in preventing water from accumulating in a channel are based on two principles: (1) cause the absorption of excess water in the soil of the area which feeds the channel, and (2) divert runoff water so that it runs around rather than into the drainageway. Both in curing gullies and in preventing them, attention should be given first to absorption of the excess water. When these measures are not sufficient to prevent water concentration, use should be made of the second principle.

Some of the practices used in increasing absorption of rainfall are: (1) a good crop rotation, including ample use of sod crops and cover crops, (2) permanent pasture when necessary, (3) strip cropping, (4) contour cultivation alone or in combination with terracing, and (5) minimum tillage. Varying types of contour furrowing and listing make long trenches which hold large quantities of runoff water until it can be absorbed. Terraces as well as diversion ditches carry excess water around existing or prospective gully heads. The safe disposal of such diverted water is a most important problem.

## Damage Due to Impact of Raindrops

Soil particles must be dislodged from their resting place before they can be moved. This is one part of the erosion process. Except in very sandy soils, dislodgment usually involves the breaking down of small clods or soil granules. The impact of raindrops on unprotected soil is one of the main forces of particle dislodgment. The kinetic energy of a raindrop is proportional to the product of its mass and the square of its velocity. There are data which indicate that the velocity of falling rain is of the magnitude of 20 miles per hour, and that a 2-inch rain on an acre would have 6,000,000 foot-pounds of kinetic energy. This is sufficient energy to raise a 7-inch layer of soil 3 feet.

There are, of course, great variations in the drop impacts of different storms. Studies by W. D. Ellison, of the Soil Conservation Service, show that these variations may result in a difference in quantity of soil moved per acre of from less than 1 ton to more than 100 tons. Differences in the characteristics of different soils also cause vast variations in the quantity of soil splash. Measurements have shown that soil particles may be splashed to a height of over 2 feet and may be moved more than 5 feet horizontally on level surfaces. Soil particles thrown into the air will naturally fall downhill on steep slopes, and hence splash from raindrops may result in considerable soil movement. Probably, however, the dislodgment of soil particles so that they are carried by water moving on the soil surface is a more detrimental effect of falling raindrops (Fig. 14–3). Another type of soil damage caused by the impact of raindrops is discussed next.

**Disintegration of Soil Granules by Raindrops.** Rain soaks rapidly into soil composed mainly of large granules or of small clods. As long as these structural units remain intact, the spaces between them are sufficiently large to offer little resistance to the entrance of water. Water soon softens the binding materials which hold the particles together however, and then the granules and clods disintegrate. The impacts of the raindrops hasten this disintegration and also bring about a separation of the finer particles by causing them to splash into the air. This process results in an accumulation of the very fine particles on the soil surface. Evidence of this phenomenon is seen in the thin layer of fine sediment which tends to crack and curl up on the soil surface as it dries after a heavy rain. These minute particles fill any spaces between larger particles and result in a "skin" or layer over the soil which permits only very slow entrance of water. In consequence, a high percentage of the rainfall runs off over the soil surface during hard beating rains on sloping land. As would be expected, this "sealing" of the soil surface is more pronounced on soils containing a high percentage of colloidal clay, but it also occurs to a surprising extent on many sandy loams.

If the granules have a high stability, they do not disintegrate so readily and hence the soil may not seal over so badly in storms of moderate intensity. Soil granules produced by growing sod crops seem to have a greater stability than those produced by freezing and thawing, or by wetting and drying. Likewise, a high organic content in soils appears to keep them in a better condition for the intake of moisture. Needless to say, any soil covering, such as crop residues and mulches, which protects the soil from the impact of raindrops, will maintain it in the very best water-receptive condition.

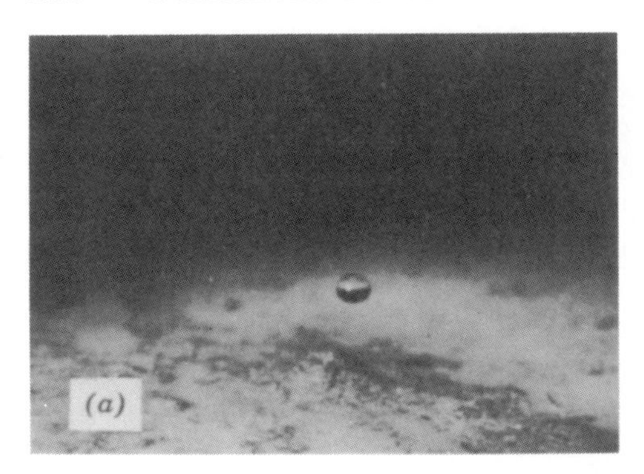

**Fig. 14–3** How raindrops cause soil erosion by "splash." (*a*) A drop just before it strikes the ground, (*b*) Just after the drop struck the soil surface. (*c*) What happens when the drop "explodes" like a small bomb, scattering fragments of soil in all directions. (Courtesy of Soil Conservation Service.)

## Torrential Storms Most Detrimental

Strange as it may seem, water erosion frequently causes more damage in regions of low rainrall than in parts of the country having a moderate annual precipitation. This results from the fact that, although rain is less frequent in semiarid and arid climates, the storms are more violent and a large amount of water falls in a short time. Under these conditions no soil is able to absorb the rain as rapidly as it falls, and hence a high percentage of it runs off even on soils of a sandy texture. The so-called dry washes, arroyos, and gulches of the Great Plains and the West bear witness to the erosive power of these torrential rains falling on lands without protective cover.

Regions with a large annual rainfall are also subject to severe erosion. Such areas usually have mild climates, and the soil is frozen only a small part of the year if at all. Under such conditions there is much more opportunity for soil washing than there is where soils are frozen for several months each year. High rainfall in areas of rolling topography offers ideal conditions for erosion. The four most intense storms during a 10-year period caused 40 percent of the soil loss on corn plots that were tilled up and down the slope at the experiment station at Clarinda, Iowa. In contrast, these four storms accounted for only 3 percent of the rainfall.

## Degree and Length of Slope Pertinent Factors

Other factors being equal, the steeper the slope the more rapidly water will run off over it. A fact to remember in this connection, however, is that the carrying capacity of water increases very rapidly with an increase in velocity. For example, if the rate of flow is doubled, the water can carry particles sixty-four times the size of those carried previously. The quantity of soil carried by water running rapidly is also much greater than that carried by slowly moving water. In this connection one must remember that soil cannot be carried away until it is dislodged. Clear water running over a soil surface at even a relatively high velocity was shown by Ellison to have little erosive power because it dislodged only a small amount of soil. When the water contained abrasive particles, however, erosion resulted because these particles cut loose the soil particles so that they could be transported by the water. It is readily seen then that steep slopes are more subject to erosion than gentle ones (Fig. 14–4).

Long slopes of moderate steepness may suffer more erosion than steep but very short slopes. In fact, control of erosion on long moderate slopes is a greater problem than it is on short, steep slopes because it is more difficult to find means for safely conducting a large volume

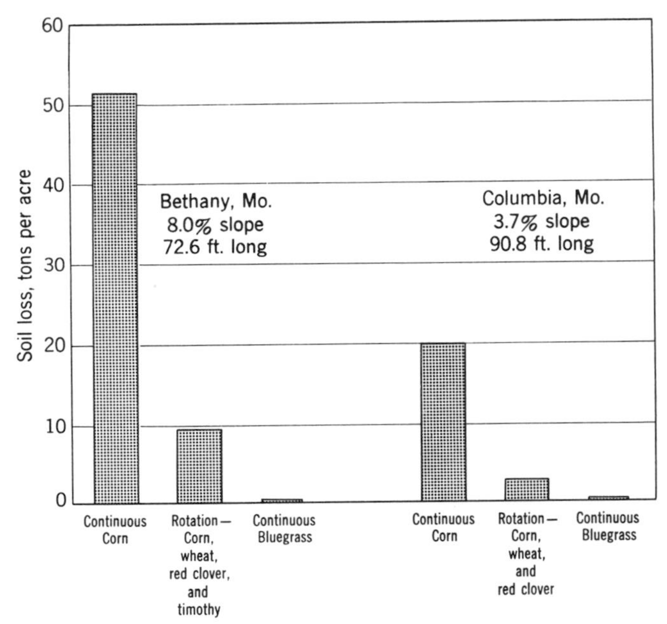


**Fig. 14–4** Effect of steepness of slope and cropping system on loss of soil by erosion. Although the two slopes studied are not equal in length and the two fields are not at the same location, it is believed that conditions are sufficiently similar so that the data are comparable. (*Missouri Agr. Exp. Sta. Bull.*, **518**.)

of water over a long distance than it is to control the movement of a small volume a short distance even though the flow is more rapid (Fig. 14–5).

### Time an Important Factor in the Absorption of Water by Soil

As has been pointed out, a highly granular soil absorbs water much more readily than one which is in a single-grained or puddled structural condition. Granulation and stability of granules can be increased by growth of sod crops, good cultural practices, mulch covers, and increased organic-matter content. Duley has shown that from two to four times as much water may run off bare soil during a heavy rain as from soil protected by crop residues. Under most conditions the movement of water downward into a soil is a relatively slow process. As a result, many storms supply water at a faster rate than most soils can absorb it. It becomes necessary, therefore, to keep the soil permeable and to devise methods of slowing the movement of water over the soil surface, that is, *to reduce the rate of runoff* in order to provide time for a more complete absorption of it. This is one of the fundamental principles underlying erosion control practices.

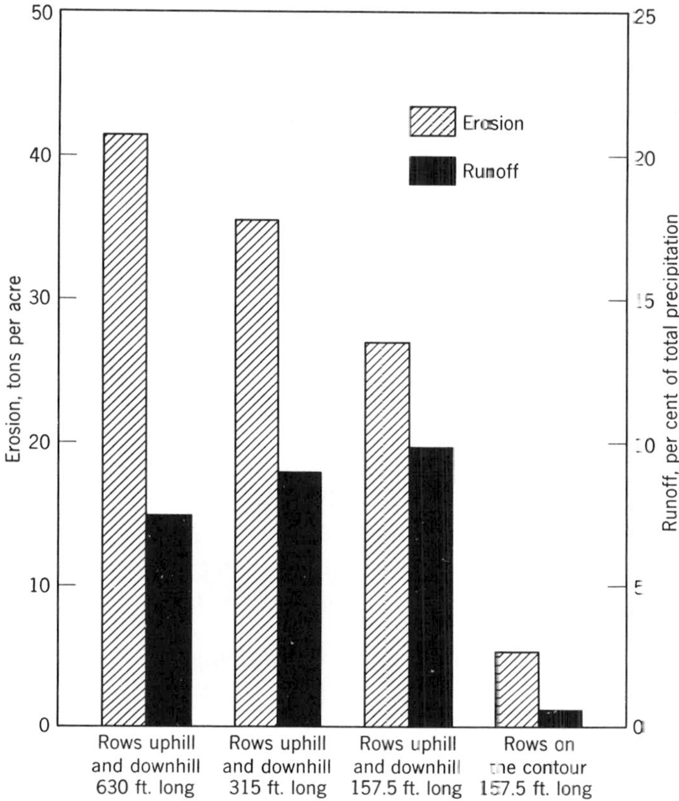


**Fig. 14–5** Effect of length of slope on percentage of runoff water and loss of soil by erosion. Also a comparison of losses when corn is planted on the contour versus up and down the slope. Soil was Marshall silt loam near Clarinda, Iowa. The annual precipitation was 28.33 inches. The cropping system consisted of corn annually planted in lister furrows. Although these data show a direct relationship between length of slope and soil and water losses, Professor Krusekopf of Missouri found that this relationship did not hold when a rotation of corn, wheat, and clover was followed. (Figure from *USDA Tech. Bull.*, **959**.)

### Use of Thick-Growing and Sod Crops to Decrease Erosion

Crops which cover the ground surface and fill the surface soil with fibrous roots tend to hold it in place and reduce erosion. Dense sods which are produced by many grass species are outstanding in this respect. They are followed closely by several legumes such as lespedeza, crimson and various other clovers, and alfalfa, which develop a sod-like root formation in the surface soil. Other crops, such as kudzu, which rapidly cover the soil surface with a dense growth are also effective in reducing erosion. A vegetative cover prevents the direct impact of raindrops on the soil.

The small grains, oats, wheat, and barley, are much less efficient than sod-producing plants in protecting the soil surface from rain and controlling soil washing, but in turn they protect the soil to a far greater extent than do crops grown in cultivated rows.

Aside from the effect of the root system and the interception of falling rain, "close-growing" crops or crops which have many stems to the square foot decrease erosion by slowing down the movement of water over the soil surface. Each stem offers an obstruction to water movement, and when the stems are close together the rate of runoff is greatly reduced. This action not only reduces the erosive power of the water but also provides time for more of it to soak into the soil.

In planning a rotation for land which is subject to erosion, liberal use should be made of hay, pasture, and small-grain crops. The greater the tendency toward erosion, the smaller is the proportion of time that cultivated crops should occupy the land. The data in Table 14–2 show the effect of different crops on the quantity of soil eroded.

## The Capability Factor in Land Management

Cropping systems may be devised for soils on the basis of any one of the following characteristics: slope, erodibility, need for drainage, moisture-retaining capacity, depth, fertility. Any such plan is faulty because it takes into consideration only one soil characteristic, whereas in reality the performance of land when being cropped is the resultant of the interaction of all the soil properties and the environmental factors. It is more logical, therefore, to classify or group soils on the basis of the result which may be expected from a crop or cropping system. Under such a scheme the result from a given crop will be approximately the same on all soils in the same class, although the soil characteristics and environmental factors which bring about the result may vary considerably.

The Soil Conservation Service usually divides land into two general groups: (1) that suitable for cultivation, and (2) that not suitable for cultivation but suitable for permanent vegetative cover. Each of these groups is then divided into four classes based on the degree that natural features of each soil limit its use or cause risk of damage if used for crops, grazing, woodland, or wildlife. Classes I, II, and III include soils which are suitable for regular cultivation. Soils in class I have the widest range of use and least risk of damage. Soils in class II have a medium range of suitability and will require some

**Table 14–2** Tons of Soil Lost Through Erosion per Acre per Year and Percentage of Precipitation Which Ran Off Under Different Cultural or Cropping Systems. (Average of 14 years. Slope 3.68%. Average precipitation 40.37 inches)

| | Plowed 8 Inches Uncropped | Blue-grass Sod | Wheat Each Year | Rotation: Corn, Wheat, and Clover | Corn Each Year |
|---|---|---|---|---|---|
| Tons of soil | 41.08 | 0.34 | 10.10 | 2.78 | 19.72 |
| Number of years to erode 7 inches of soil | 24 | 3043 | 100 | 368 | 50 |
| Percentage of runoff | 30.3 | 12.0 | 23.3 | 13.8 | 29.4 |
| Pounds of soil lost for each inch of runoff | 6734 | 142 | 2149 | 993 | 3314 |

[a] Prepared from "The Influence of Systems of Cropping and Methods of Culture on Surface Runoff and Soil Erosion," M. F. Miller and H. H. Krusekopf, *Missouri Exp. Sta. Res. Bull.* **177**, p. 22, 1932.

special management practices and protection. Soils in class III have a narrow range of use and require careful management, possibly terracing or the disposition of water on low, wet areas. Soils in class IV must be cultivated with great caution. It is usually better to leave this land in hay or pasture.

Classes V, VI, and VII include land not suited for cultivation. It should be used for grazing or forestry. In fact, land in class VI will need special care even for such uses because of steep slopes, thinness of soil, and similar characteristics. Land in class VII requires still more protective measures if it is to be used successfully for grazing or forestry. Class VIII includes land which is suited only for wildlife, recreation, and watershed protection. Its limitations may be extreme wetness, stoniness, steepness, erodibility.

Soils within a class may be further divided on the basis of the kind of hazards involved in their management. For example, one area may be subject to water erosion, while another area may be limited in use by a low moisture capacity. These subclasses may be divided again into land-capability units. This subdivision places in one unit those areas which require similar treatment.

The outline of the classification scheme is shown in Table 14–3, and a land-capability map is presented in Fig. 14–6.

**Table 14–3** Outline of the Land-Capability Classification

| Major Land-Use Suitability (Broad grouping of limitations) | Land-Capability Class (Degree of limitations) | Land-Capability Subclass (Kind of limitations) (Groupings of land-capability units) (Examples of possible subclasses in class III) | Land-Capability Unit (Distinctive physical characteristics) (Land-management groups based on permanent physical factors) (Examples) |
|---|---|---|---|
| Suited for cultivation | I. Few limitations; wide latitude for each use; very good land from every standpoint | | |
| | II. Moderate limitations or risks of damage; good land from all-round standpoint | | |
| | III. Severe limitations or risks of damage; regular cultivation possible if limitations are observed | 1. Limited by hazard of water erosion; moderately sloping land | 13C2. Moderately sloping, slightly acid soils on limestone<br>9C2. Moderately sloping, highly acid soils on sandstone or shale |
| | | 2. Limited by excess water; needs drainage for cultivation | |
| | | 3. Limited by low moisture capacity; sandy land | |
| | | 4. Limited by climate | |

| | | | |
|---|---|---|---|
| | IV. Very severe limitations; suited for occasional cultivation or for some kind of limited cultivation | | |
| Not suited for cultivation; suited for permanent vegetation | V. Not suited for cultivation because of wetness, stones, overflows, etc.; few land limitations affect grazing or forestry use | Groupings of range, pasture, or forest sites | Land-management groups based on permanent physical factors such as range sites or forest sites |
| | VI. Too steep, stony, arid, wet, etc., for cultivation; moderate limitations for grazing or forestry | | |
| | VII. Very steep, rough, arid, wet, etc.; severe limitations for grazing or forestry | | |
| | VIII. Extremely rough, arid, swampy, etc.; not suited for cultivation, grazing, or forestry; suited for wildlife, watersheds, or recreation | | |

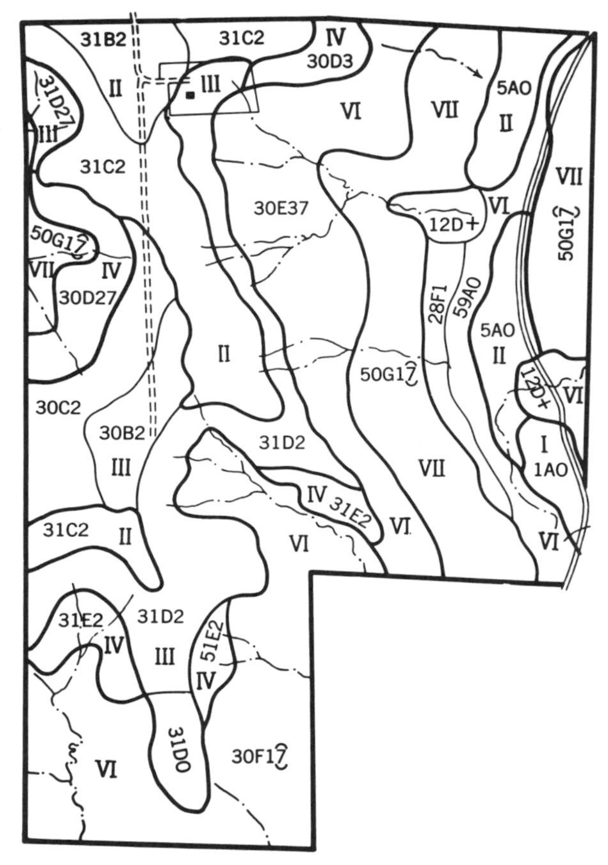


**Fig. 14–6** Land-capability map of a 200-acre dairy farm in Wisconsin. The Roman numerals designate the land-capability classes. The other numbers and letters indicate the land characteristics. For example: in 30E37, the 30 designates the type of soil, the E indicates the steepness of slope, and the 37 stipulates that over 75 percent of the topsoil has been lost and that there are occasional gullies. For explanation of land-capability classes see Table 14–3. (Courtesy of Soil Conservation Service.)

## Cultivation on the Contour

Effective water erosion control can seldom be obtained from contour cultivation alone. The capacity of furrows normally made by contour tillage is small. They can easily become filled with water and overflow during storms, resulting in considerable loss of soil. The practice is effective when the rainfall is gentle and slopes are short and only moderately sloping. Best results are obtained when contouring is used either with strip cropping or terracing.

**Fig. 14–7** Lister furrows on the contour not only prevent erosion but also hold much water until it can be absorbed into the soil, as is taking place on this Oklahoma field. (Courtesy of Soil Conservation Service.)

Lister furrows on the contour not only prevent erosion but also hold much water until it can percolate into the soil (Fig. 14–7).

### Strip Cropping, a Practical Means of Reducing Erosion

In view of the variation in the effect of different crops on erosion, it is logical to plant crops with different growth habits in strips across sloping fields. By this system, erosion, which may start in a strip of cultivated crop such as corn, cotton, or beans, can be arrested in a strip of hay growing next down the slope. A strip of small grain can then follow, then another strip of hay, and then a cultivated crop (Fig. 14–8). The width of the strips of each crop should be determined by the steepness of the slope, the amount and nature of the precipitation, and the erodibility of the soil. The strips should be planted on the contour as nearly as possible.

At first thought the planting of crops in strips across a slope seems inconvenient and inefficient to many farmers. Experience, however, has proved it otherwise. Many times fences must be removed and a new field arrangement worked out. Strip cropping frequently results in longer rows and, hence, fewer turns with planting, tillage, and

**Fig. 14–8** Strip cropping, alone or in conjunction with terracing, makes possible the utilization of much land which would otherwise be severely damaged by erosion. (Courtesy of Soil Conservation Service.)

harvesting machinery. Also, since the strips follow the contour of the land, work is done approximately on the level in place of up and downhill. This arrangement requires less power.

Strip cropping is adapted to land with long and fairly regular slopes. With a "choppy" topography involving short and very irregular slopes, such as is found in some glaciated areas, the system has less general application.

### Terraces Remove Runoff Safely

Many agriculturists in the northern states are inclined to look upon terracing as "something new" in soil management in America, but terracing is not new. Various types of ridges and depressions or channels across slopes were used in Colonial times to control runoff water. The term *terrace* was applied to these structures as early as 1847. From the southeastern states the practice of terracing spread westward as farmers migrated in that direction. Terraces were used in Texas in 1903 and in Oklahoma by 1907.

There are several types of terraces. Three types will be mentioned, but a detailed description of them and their uses cannot be given here. The bench terrace is the oldest type, having been used by ancient peoples in South America, Europe, and Asia. It is used primarily on steep slopes and where land for crop production is limited. The narrow, quite level benches are supported by very steep and sometimes practically vertical risers, sometimes covered with vegetation and sometimes consisting of retaining walls of rock. In the United States such terraces have been used to some extent in parts of the southeastern states but are no longer recommended.

During the period immediately before and somewhat after 1800, farmers in southeastern United States began making shallow ditches across the slopes in their fields. These were known as "hillside ditches" and have developed into what is now known as *graded-channel* or *interception-and-diversion* terraces.

The most widely used type of terrace was developed around 1885 by a farmer named Priestly Mangum, near Wake Forest, North Carolina. This structure is known as the *broad-base* or Mangum terrace. It consists of a low, broad mound crossing the slope approximately on the contour. The shallow depression on the uphill side has sufficient fall to carry excess water around the hill to an outlet channel which carries it safely down the slope. The ridges may be constructed by repeated plowing of a strip of land, the soil always being thrown toward the center, or by plowing a strip and working the soil toward the center with a grader, whirlwind terracer, or one-way disk plow. No special equipment is needed for the construction of terraces. The ridges are of such low relief that farm implements may work over them with comparative ease. The entire field, including the ridges and depressions above them, may be planted to the crop to be grown. Crops should be planted on the contour.

### Sod Waterways Effective and Not Inconvenient

Because no system of soil management will result in the complete absorption of all water from rain and melting snow, some provision must be made for conducting water down slopes with a minimum of gully erosion. Depressions or shallow channels into which water naturally drains as it moves down the slope may be kept in sod and so reduce gully erosion to a minimum (Fig. 14–9). At times some grading may be necessary to bring the water from various parts of a field into such a channel, and the drainageway itself may need to be broadened or the sides graded down to improve its usefulness. Some sodding is frequently necessary after alterations in the drainageway to prevent soil washing while a grassed waterway is being established by seeding. A straw mulch held down by wire netting or brush is often useful in controlling erosion while a stand of grass is developing.

**Fig. 14–9** This grassed waterway is furnishing considerable hay in addition to permitting water to flow down the slope without damage to the soil. (Courtesy of Soil Conservation Service.)

On larger areas it is more practical to apply a straw mulch on a well-prepared seedbed and then work it in with a rotary treader so that part of the straws stand on end and protrude above the surface. When working in a field with sod waterways, the farmer lifts his tillage implements or straightens his disks in crossing the channel so as not to damage the sod. Sod waterways are a simple and effective device for preventing gully formation and for stopping the enlargement of small gullies. A large field may have several of them, some quite short and others of considerable length.

## CONTROLLING EROSION CAUSED BY WIND

Wind erosion occurs on soil unprotected or only partially covered by vegetation. Soil blowing then may be expected during periods when land is being fitted for planting, before a young crop has reached sufficient size to protect the soil, and when land is being fallowed. Much wind erosion also occurs on range land which has been overgrazed.

Wind erosion reaches its greatest proportions in semiarid and arid regions. Nevertheless, much damage is caused to both crops and soils in humid areas by soil blowing, although the phenomenon is less spectacular and attracts comparatively little attention in these regions. In the United States more attention has been given to the control of wind erosion in the Great Plains than in other sections of the country. In North Dakota damage from this cause was reported as early as 1888, and in Oklahoma four years after breaking of the sod. Soil blowing was one of the hazards confronting the early settlers, and control of wind erosion was one of the first problems studied by agricultural experiment stations in the Plains states.

### Blowing Not Confined to Small Particles

Sand grains and clusters or granules of small particles are not picked up by wind so easily nor carried such great distances as are silt and clay particles. Nevertheless, these larger particles are moved over the soil surface by rolling and by short skips. In this way they may do much damage to a young crop and may be piled up into small ridges and great dunes. Smaller particles may be carried along somewhat above the soil surface, or they may be swirled high in the air and carried long distances. Rain containing considerable fine soil, which was gathered from the air, was reported in some of the eastern states during the great dust storms in the Plains states in the early 1930s.

### Blowing Not Confined to Sandy Soil

When the prairie sod in the "short grass" or western Great Plains is plowed, the organic matter appears to decay rapidly during the first years of cultivation. One of the binding materials which holds the soil particles in clusters is thus destroyed. Furthermore, during periods of continued drought soil granules disintegrate, leaving the soil surface covered with a powdery layer which blows with extreme ease. It is seen, then, that, although sandy, incoherent soils offer less resistance to blowing when wind velocity is sufficiently high, loam and clay soils also blow when conditions induce disintegration of the soil granules and small clods.

Soils composed largely of the partially decayed remains of plants, variously called muck and peat in different states, have very low specific gravities. As a result, these soils blow readily during periods of dry weather when the surface is not protected by a vegetative cover. The fact that such soils constitute level areas adds to the likelihood of wind erosion.

### Crops Damaged by Blowing Soil

If blowing occurs while crops are small, the soil may be blown away from the root system, and the plants actually "blown out" over considerable areas. This is one type of damage greatly feared by onion growers on muck soil during dry springs. In dry falls, winter wheat may be blown out over large areas in the hard winter-wheat region.

Young crops may also suffer severe damage through being bruised by blowing soil particles. There may not be sufficient blowing to raise large clouds of dust, but the soil particles moving over or just above the surface cut and bruise the tender young plants. This type of damage is not uncommon on level sandy soils or on muck soils.

Another source of danger to crops is the possibility of being covered with blowing soil. A crop in a well-managed field may be covered by blowing soil from a poorly managed field near-by. Likewise, soil from higher land may be deposited on the crop growing in lower, more sheltered areas.

### Damage to Fences, Buildings, and Highways

When blowing starts on a neglected or abandoned area of land, the soil will pile up around buildings and fences in an unbelievably short time.

Sand dunes around large lakes and in desert areas may cause much damage by moving over highways and encroaching on resort property. Summer cottages along lakes sometimes have to be protected from blowing sand.

### Three Types of Soil Blowing

Soil particles move in three ways during wind erosion. Fine soil particles, those from 0.1 to 0.5 mm in diameter, are rolled over the surface by direct wind pressure and then suddenly jump up almost vertically from a short distance to a foot or more. Once in the air, the particles gain velocity and then descend in an almost straight line, not varying more than 6 to 12 degrees from the horizontal. The horizontal distance traveled by a particle is from 4 to 5 times the height of its jump. Upon striking the surface, the particles may rebound into the air or knock other particles into the air and come to rest themselves. The major part of the soil carried by wind moves by this process, which is called *saltation*. It is interesting to note that around 93 percent of the total soil movement by wind takes place below a height of 1 foot, and probably 50 percent or more occurs between 0 and 2 inches.

Very fine dust particles are protected from wind action because they are too small to protrude above a minute viscous layer of air which clings to the soil surface. As a result, a soil composed entirely of extremely fine particles is very resistant to wind erosion. These dust particles are thrown into the air chiefly by the impact of particles moving in saltation, but once in the air their movement is governed by wind action. They may be carried very high and over long distances.

Relatively large particles (between 0.5 and 1.0 mm in diameter) are too heavy to be lifted by wind action but are rolled or pushed along the surface by the impact of particles in saltation. This process is called *surface creep.*

From the above facts it is evident that wind erosion is due principally to the effect of wind on particles of a suitable size to move in saltation. Accordingly, wind erosion can be controlled (1) if the soil particles can be built up into clusters or granules of too large a size to move in saltation; (2) if the wind velocity near the soil surface can be reduced by ridging the land, by vegetable cover, or even by developing a cloddy surface; and (3) by providing strips of stubble or other vegetative cover sufficient to catch and hold the particles moving in saltation. Some management practices designed to provide these conditions are discussed in the following paragraphs.

## Soil Areas To Be Left in Sod

Studies have shown that extensive wind erosion originates mainly in land where crops frequently fail because of inadequate moisture supply. Sandy soils and shallow soils suffer the most, particularly in the areas receiving 14 to 17 inches of annual rainfall. Accordingly, land in regions of low rainfall, where production of crop is dependent on a "favorable" season, is more safely left in native grasses. Unfortunately, during periods of high demand for food with correspondingly high prices, large areas of such land have been plowed and planted. Serious soil blowing is certain to result, and very quickly when a series of normal or of unusually dry years follows. Furthermore, it is very difficult to reestablish the sod on such land when wind erosion has once gotten well under way.

## Summer Fallow Dangerous

Any cropping system which leaves the land unprotected by a growing crop for an appreciable length of time opens the opportunity for soil blowing. In regions where annual precipitation or precipitation during the growing period is insufficient to produce a crop, either some system of moisture storage in the soil or irrigation must be resorted to. Keeping the soil in a fallow condition with enough cultivation to prevent weed growth is a practice frequently followed under such conditions. Unfortunately, there has been considerable trouble with soil blowing during the fallow year in parts of the Great Plains. Alternating strips of a summer crop such as sorghum with strips of fallow has proved an effective way of reducing soil blowing when fallowing is necessary. The use of subsurface tillage, as discussed in the section on stubble-mulch culture below, also offers a satisfactory means of handling fallow ground.

## A Smooth Soil Surface Undesirable

Furrows made by the lister or duckfoot cultivator at right angles to the wind are helpful in reducing soil blowing (Fig. 14–10). When these furrows become filled with the soil, new ones should be made.

Another method of keeping the soil surface rough is to plant crops in furrows instead of on the level, as is done in humid regions. Row crops such as corn and members of the sorghum family are planted in lister furrows. Drills which make shallow furrows have also been developed for planting small grains. Naturally all plantings should be made at right angles to the prevailing winds.

## Rotations Helpful

Rotations which involve a minimum of soil tillage are helpful in preventing wind erosion. A typical rotation consists of one year of

**Fig. 14–10** Listing on the contour not only protects the listed field from wind action but also may collect much soil blowing from an adjoining field, as illustrated in this Oklahoma field. The velocity of the wind is indicated by the posture of the man in the foreground. (Courtesy of Soil Conservation Service.)

cultivated crop, then a spring-seeded crop, then a year of fallow, followed by winter wheat. Alfalfa makes a wind-resistant sod but is objectionable in areas of low rainfall because it dries out the soil to a considerable depth. A rotation suggested by the Wyoming Agricultural Experiment Station consists of a cultivated crop followed by a spring-drilled crop, then rye, a fallow year, and winter wheat. Some wheat farmers object to rye because it becomes mixed with the wheat and lowers the market grade. Land in grass is maintained for a long time because of the difficulty of establishing a sod. Other rotations or sequences of crops suitable for growth in a given locality and arranged in an order to meet the hazards of wind and water erosion on given farms or fields can be obtained from state colleges and local technicians of the Soil Conservation Service.

### Strip Cropping Effective

The planting of crops in strips at right angles to the prevailing winds is proving valuable in reducing soil blowing. The width of the strips is determined by the nature of the soil, exposure to wind, and similar factors. On sandy soils the strips may be only 8 to 10 rods wide, but on fine-textured soils they may be from 15 to 20 rods in width.

### Stubble-Mulch Culture Helpful

Crop residues are an effective soil protection against wind erosion. Small-grain stubble reduces wind velocity and also catches soil particles moving in saltation. Strips of stubble left at frequent intervals across a field being fallowed or fitted for a spring crop form effective barriers. The height of the stubble as well as wind velocity influence the width of strip needed, but a 10-foot strip of short stubble should trap 90 to 98 percent of the moving soil.

Another effective cultural practice is to work crop residues partially into the soil but leave enough on the surface to obstruct the movement of soil particles and reduce the force of the wind. The one-way disk and various types of subsurface tillers are useful in this type of soil culture (Fig. 14–11).

In planting crops in the mulch-culture system of farming, drills and planters must be used that will work through the crop residues without covering them or leaving them in a position to interfere with seed germination or growth of young plants. The semideep-furrow drill or the deep-furrow drill accomplishes these purposes in seeding small grains and puts the seed down in moist soil. The low-down, press-wheel drill also is satisfactory. Row crops may be planted with a standard planter having furror openers mounted on stub runners which pass through the residues with little trouble.

**Fig. 14–11** By working crop residues partially into the soil and leaving a considerable portion on top, this Montana soil is protected from wind erosion. The soil is also in an excellent condition to absorb rainfall. The work is being done with a one-way plow. (Courtesy of Soil Conservation Service.)

Crops should be cultivated in such a way as to leave the residues on the surface; in other words, cultivate beneath them. This result can be accomplished by use of sweep-type cultivators. The stubble-mulch system of soil management has many possibilities and should be carefully investigated in areas where wind or water erosion is a serious hazard.

### Stabilization of Blowing Sand with Trees and Coarse Grasses

Sand dunes and hilly sandy areas which are blowing badly in humid areas are effectively stabilized by use of coarse grass, such as one of the beach grasses, and by reforestation. The grasses are transplanted in little bunches. Originally these were placed in a double row with plants about 18 inches apart. The pairs of rows were spaced from 12 to 40 feet according to conditions. Later, planting the pairs of rows in a serpentine pattern rather than straight was considered more effective. A more recent system is to plant the rows in a radiating pattern, like the spokes of a wheel, or in the form of a cross. This method of planting requires far less material and labor than the older systems.

The grass spreads by means of rhizomes and will establish a thick enough sod to hold the sand. Sometimes it is necessary to protect the new plantings by windbreaks of fencing or brush. Trees or shrubs are often planted to supplement the effect of the grass. With the older and thicker systems of setting the grass, trees may be planted immediately, but with the newer systems several years should elapse before the trees are planted.

The establishment of a forest cover on blowing sand plains and hills is desirable but is sometimes beset with difficulties. The sand may be blown out from around the young trees, or they may be severely cut by the blowing sand or covered with it unless some protective measures are adopted. Temporary windbreaks afford protection, and in special small areas the soil may be covered with a mulch of straw which is held in place by brush. This practice, however, is expensive.

In humid sections various species of conifers have proved satisfactory for planting on sandy soils subject to blowing, and in some areas the black locust has been very successful. Trees suitable for use under the climatic conditions which prevail in any given area should be selected.

In drier areas a coarse-growing plant such as sorghum or sudan grass may be grown and the residues worked into the soil to control blowing while a grass is being established.

## Trees for Windbreaks on Organic Soils

Organic soil areas of appreciable size are frequently protected from wind damage by planting windbreaks of trees around them. In addition, rows of trees are often planted across the area at right angles to the prevailing wind. Willows have been found satisfactory for this purpose, especially since they make a very rapid growth in organic soil. A number of species of conifers have also been used. Austrian and Scotch pine make a rapid growth and give promise of making satisfactory windbreaks in the north central states. White pine grows more slowly but otherwise appears suitable for such use. The use of trees as windbreaks must be limited because of the large amount of soil they take out of crop production on account of their extensive root system. A number of shrubs also make good windbreaks. Spirea is frequently used.

## Additional Protection from Wind Needed

Because of the limitation in use of trees, supplementary windbreaks are needed on muck areas of appreciable size. Picket fencing, such as is used for keeping the snow off highways in the northern states, makes very satisfactory temporary windbreaks. Fastening burlap bags to the lower part of the fence increases its effectiveness. Bags also may be hung from wires stretched between posts driven into the muck.

Occasionally rows of corn have been planted across the field at frequent intervals. A more common method is to drill five or six rows of rye across the field at intervals of 50 or more feet. All such plantings must be at right angles to the wind. Planting adapted grasses on ditch banks has been found a satisfactory method of protecting these exposed areas in Michigan and preventing filling of the ditch with wind-carried muck.

Moist muck does not blow appreciably. Accordingly, use of an overhead irrigation system is helpful during windy periods before the crop cover is sufficient to prevent soil blowing. Rolling the muck with a very heavy roller induces moisture to rise more rapidly by capillarity and so dampens the surface layer. This practice is not effective unless the soil layers below the surface are quite moist.

# 15

# Soils and Agriculture of Arid Regions

No generally accepted definition of arid region exists. A definition based solely on total annual precipitation is faulty because it does not take into consideration factors which influence the efficiency of a given amount of precipitation in the production of crops. Some of these factors are (1) distribution of the rainfall through the year, (2) temperature, (3) wind, (4) air humidity, and (5) amount of sunshine. A comparatively low rainfall, falling mostly during the growing season, may permit relatively high crop yields, while some areas, receiving as much as 70 to 80 inches of rain, with little of it falling during the growing season, require irrigation for crop production. Frequent fogs, moderate temperature, and little wind lead to low rates of evaporation of water from the soil and reduced transpiration of plants and hence reduce the moisture required for crop production.

For the purpose of this discussion the arid region is considered to include those areas with predominantly clear weather and annual precipitations as high as 15 inches, and also areas subject to fog, but with less precipitation, in which the soils have the same characteristics as those developed in the clear areas. Without irrigation, these lands are used principally for grazing, with a very small production of small grains in areas where conditions are especially favorable and under a system of one year of crop and one of fallow for moisture accumulation.

## CHARACTERISTICS AND UTILIZATION OF SOIL IN ARID REGIONS

Many misconceptions regarding soils of the arid regions have prevailed in the minds of laymen. These soils are not merely accumulations of more or less decomposed rock fragments with no profile characteristics, nor are they all of a coarse texture or sandy nature. They exhibit much diversity in properties. There are fine-textured soils, medium-textured soils, and sandy soils; young soils and soils with a pronounced profile development; saline soils and those without injurious salt accumulations in the arid regions.

The agricultural utilization of arid soils is also varied. With an abundance of water for irrigation, intensive agriculture may be carried on, as described in the following chapter. Without irrigation, forage for grazing ranges from the very scanty supply of the true desert regions to the thin grass cover of the arid prairies.

### Locations of Arid Regions

Arid regions of western United States occupy, in the main, broad valleys and basins between mountain ranges and extend up the slopes until increased precipitation or lower temperatures give rise to a different type of vegetation and to soils of different characteristics. In some places, however, arid conditions prevail over the entire surface of low mountain ranges, which are either bare of vegetation or are occupied by a thin cover of brush. The mountains serve as barriers to clouds and to cooling winds and produce areas of low rainfall and relatively high mean annual temperatures. With few exceptions, the relative air humidity is very low. The precipitation map in Fig. 15–1 shows the influence of the mountain ranges or elevation on precipitation in the arid region of southwestern United States.

### Vegetation of Arid or Desert Regions

Some desert regions, in Peru and Chile for example, are virtually rainless. The deserts of Peru and Chile are near the Pacific Coast where on shore winds are cold and dry, having travelled over the cold waters of the Humboldt current. In many arid regions, however, there may be as much as 10 or more inches of precipitation annually, which may support a considerable amount of vegetation. Desert shrubs dominate the most arid regions with shrubs giving way to desert (bunch) grasses with increasing moisture. Plants are widely spaced and use whatever soil moisture there is quite effectively. Perhaps one of the surprising things for a person who has lived all his life in the humid region and then takes a trip to the desert is the great diversity of plants and the considerable amount of vegetation (see Fig. 15–3).

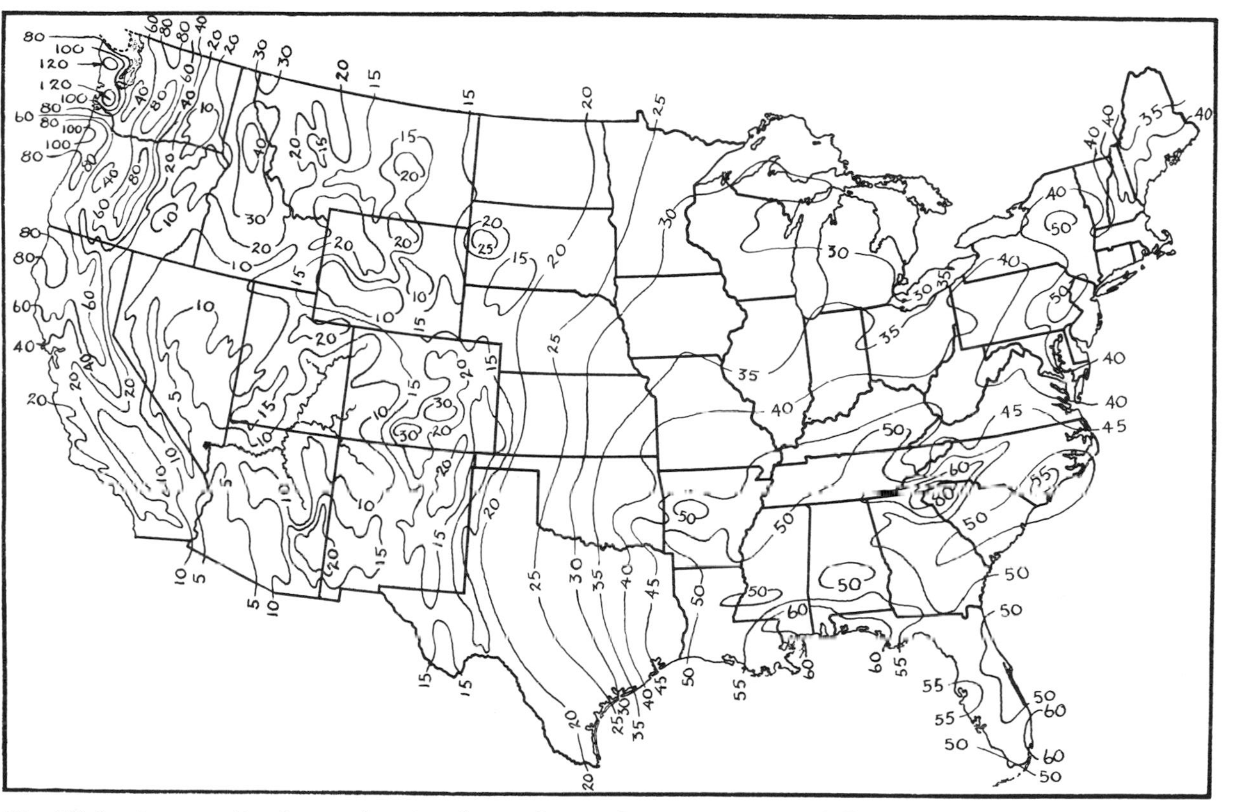

**Fig. 15–1** A generalized map showing the total annual precipitation in different parts of the United States. Precipitation includes rain and the water equivalent of snow, sleet, and hail. Some of the smaller areas, such as those receiving 100 or 120 inches, are enlarged on the map. (Drawn from a map prepared by the Weather Bureau and presented in Atlas of American Agriculture.)

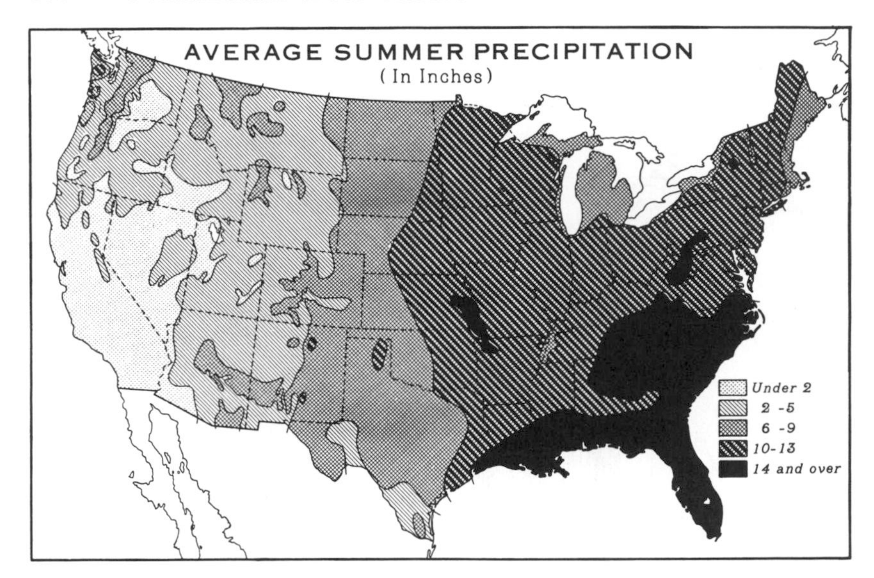


**Fig. 15–2** Total annual precipitation does not always indicate whether or not a crop can be grown without additional water. The amount of precipitation during the growing season is a better criterion. (From USDA Misc. Pub. 260, by O. E. Baker. Drawn from a map prepared by the Weather Bureau and published in Atlas of American Agriculture.)

**Fig. 15–3** Vegetation on the Sonoran Desert of southwestern United States where the annual precipitation is about 10 inches. Note the variety and abundance of shrubs and cacti.

Many desert plants grow and function during the wetter seasons of the year and go dormant during the driest seasons. On the more humid or eastern edge of the arid region of western United States, the desert bunch grasses give way to taller and more vigorous grasses and the Aridisols merge with the Mollisols.

## Aridisols—The Dominant Soils of the Arid Regions

The dominant soils of the arid regions belong to the Aridisol order. In Chapter 10 it was pointed out that Aridisols are the most abundant soils in the world and occupy about one-fifth of the land. In arid regions the soil-forming processes are similar to those of humid regions, but the rate of soil formation is much slower in arid regions. The lesser amount of plant growth and the potential for organic-matter decomposition produce soils with low organic-matter contents. Winds play a major role in the development of Aridisols. Winds move dust about and occasional rains wash soluble nutrients from the dust on its transient journey across the desert. A more obvious role of the wind is the blowing away of fine soil particles resulting in the formation of a concentration of gravel or the formation of desert pavement.

Water is less effective in leaching soluble salts and translocating colloidal material in arid regions because of the low amount of precipitation. Another factor is the torrential nature of much of the rainfall, which results in considerable runoff. A striking feature of most Aridisols is a zone at varying distance below the surface where calcium carbonate has been deposited by percolating water (calcic horizon). Many Aridisols have well developed argillic (Bt) horizons, which is evidence of considerable clay movement. The widespread occurrence of argillic horizons in many Aridisols suggests that many years ago a more humid climate existed than exists today. The Mohave is a common Aridisol from southwestern United States and some data of a Mohave is given in Table 15–1 to illustrate features or

**Table 15–1** Some Properties of the Mohave Sandy Clay Loam—an Aridisol

| Horizon | Depth, Inches | Clay, Percentage | Organic Matter, Percentage | C/N | CEC, me per 100 gms. | Exch. Na, Percentage | pH | $CaCO_3$, Percentage |
|---|---|---|---|---|---|---|---|---|
| A1 | 0–4 | 11 | 0.25 | 6 | 8 | 1.2 | 7.8 | – |
| B1 | 4–10 | 14 | 0.19 | 6 | 15 | 2.0 | 7.4 | – |
| B2t | 10–27 | 25 | 0.24 | 7 | 22 | 2.5 | 8.5 | – |
| B3ca | 27–37 | 21 | 0.25 | 8 | 17 | 4.1 | 8.9 | 10 |
| IICca | 37–54 | 17 | 0.08 | – | 6 | 12.7 | 9.2 | 22 |

Data adapted from "Soil Classification, A Comprehensive System," USDA, 1960.

properties commonly found in Aridisols. Note the presence of an argillic horizon (Bt), low content of organic matter and low carbon-nitrogen ratio of the organic matter, the presence of significant exchangeable sodium, high pH values, and accumulation of calcium carbonate (Ca) in the lower part of the profile (calcic horizon).

Aridisols are placed in suborders on the basis of the presence or absence of argillic horizons. The suborder Orthids includes Aridisols without argillic horizons and, by contrast, the suborder Argids includes Aridisols with argillic horizons. As we have already noted, the Mohave has an argillic horizon, therefore, the Mohave is an Argid. The general distribution of Orthids and Argids in the United States is shown in Fig. 15-4.

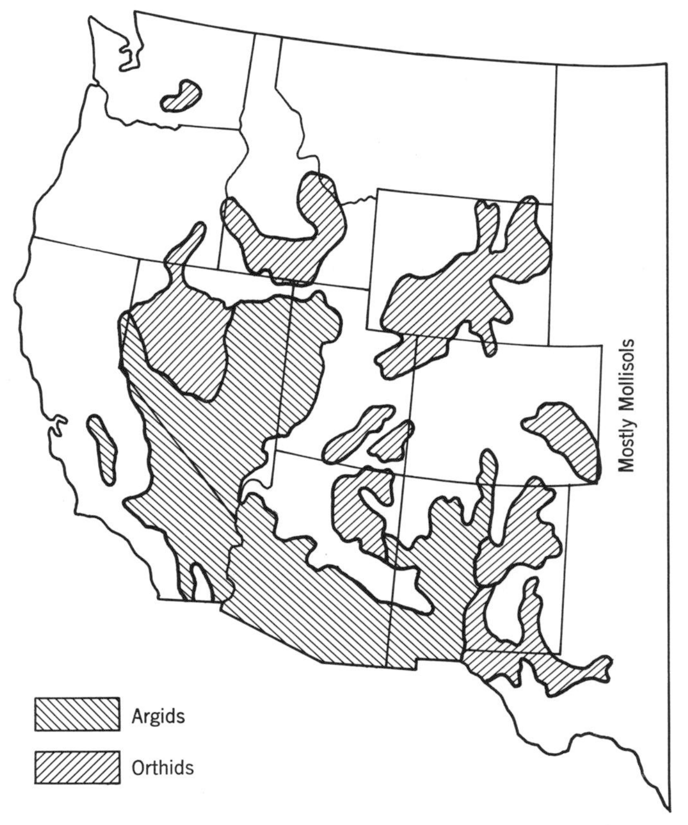


**Fig. 15-4** Generalized distribution of Argids and Orthids in the United States. (Drawn from map in Fig. 10-20.)

## Relationship of Land Surfaces to Age of Aridisols

It is believed that Orthids are the "younger" Aridisols and that the Argids are the "older" Aridisols. There is evidence that Orthids in United States have largely developed within the past 25,000 years in an arid climate. Orthids are mostly located where recent alluvium has been deposited. Argids are common on the older land surfaces in any landscape where there has been more time for the development of argillic horizons and greater likelihood that the soil has been influenced by a more humid climate more than 25,000 years ago. On the most recent sediments or land surfaces the Entisols are abundant. (See Fig. 15–5 for a photograph of a desert landscape showing land surfaces of greatly different ages.)

## Land Use in Arid Regions

The Aridisols of western United States occur almost entirely with a region called the "western range and irrigated region." As the name implies, grazing of sheep and cattle and the production of crops by irrigation are the two major uses of land. The use of land for grazing is closely related to the precipitation, which largely determines the amount of forage produced. Some areas are too dry for grazing, while other areas are more favorable and may also be able to take advantage of summer grazing on mountain meadows. As much as 75 acres or more in the drier areas is required per head of cattle, thus making large farms or ranches a necessity. Most ranchers supplement the range forage by producing some crops on a small acreage favorably located for irrigation (Fig. 15–5). The major hazard in the use of grazing lands is overgrazing, which results in the invasion of less desirable plant species and increased soil erosion.

Only about 1 or 2 percent of the land is irrigated because the production of crops by irrigation depends on a water supply. Most of the irrigated land is located on alluvial soils or Entisols along streams and rivers where the land is nearly level and the irrigation water can be distributed over the field by gravity (Fig. 15–6). In addition, the nearby rivers serve as a source of water from natural river flow and carry irrigation water released from water storage reservoirs. The alkaline nature of Aridisols may cause deficiencies of various micronutrients on certain crops. Major crops include alfalfa, cotton, citrus fruits, vegetables, and grain crops. In Arizona crop production only 2 percent of the land that is irrigated accounts for 60 percent of the total farm income. Grazing, by contrast, utilizes 80 percent of the land and accounts for only 40 percent of the total farm income.

**Fig. 15–5** Grazing lands dominated by Aridisols. Soils in the area are closely related to age of land surface with Aridisols on older surfaces and Entisols on the youngest surfaces. Note small irrigation water reservoir and irrigated cropland near ranch headquarters. (USDA photo by Bluford W. Muir.)

## Erosion on the Desert

Storms on the desert are frequently of torrential nature, which circumstance, coupled with the formation of crusts and layers of soil which resists infiltration of water, leads to severe erosion. Stream beds and dry washes, through most of the year entirely dry, carry torrents of water during storms and may wash out highways or cover them with streams of such depth and velocity that they are impassable for many hours. Gully formation is common, and the washes are often of great depth. Sheet erosion removes the surface soil from the gentler slopes. Water erosion is much more pronounced on the sandy loams and on soils of finer texture than on the very sandy soils, which are more receptive to rainfall.

The effects of wind erosion are evident everywhere. Each bush or clump of vegetation occupies a little mound, the soil of intervening spaces having been carried away to a greater or lesser extent. The

**Fig. 15–6** Lettuce being grown on Entisols (alluvial soils). The irrigation water from the river in the background (located where the line of trees are growing) is distributed over the field by gravity flow in the furrows between the rows. (Photo courtesy of USDA.)

formation of the desert pavement is an evidence of soil blowing as is the accumulation of wind-drifted sediment on the windward slopes of the hills and mountains. The giant sand dunes northwest of Yuma and extending some 50 to 60 miles in a northwesterly–southeasterly direction through southern California and northern Mexico are further proof of the erosive power of desert winds.

## DEVELOPMENT AND MANAGEMENT OF SALINE AND SODIC SOILS

In arid regions some soils develop under conditions of poor drainage in spots and there is more water evaporating than there is water coming into the area as precipitation. Under these conditions soluble salts and exchangeable sodium may accumulate in sufficient amounts to impair plant growth and to alter soil properties. The addition of irrigation water, which contains varying amounts of soluble salts, for crop production in arid regions also creates the potential for the accumulation of soluble salts and exchangeable sodium. *Saline* soils contain sufficient soluble salt to impair plant growth and *sodic* soils contain sufficient exchangeable sodium to impair plant growth and alter soil properties. The quantity, proportion, and nature of salts present may vary in saline and sodic soils. This gives rise to three kinds of soils, namely, saline, saline-sodic, and sodic soils. A consideration of the development, properties, and management of these soils follows.

## Saline Soils

These soils contain a relatively high concentration of soluble salts, made up largely of chlorides, sulfates, and sometimes nitrates. Small quantities of bicarbonates may occur, but soluble carbonates are usually absent. Frequently relatively insoluble salts, such as calcium sulfate, and calcium and magnesium carbonates are also present. The chief cations present are calcium, magnesium, and sodium, but sodium seldom makes up more than one-half of the soluble cations and is not adsorbed to an appreciable extent on the colloidal fraction of the soil.

The pH value of these soils is 8.5 or less, and the exchangeable sodium percentage is less than 15. White crusts frequently accumulate on the soil surface, and streaks of salt are sometimes found within the soil. Saline soils have a favorable structure because the colloids are highly flocculated. Soils in this group are similar to those designated as "white alkali" soils in older publications.

## Saline-Sodic (Alkali) Soils

Soils in this group are also characterized by a high concentration of soluble salts, but they differ from saline soils in that the exchange sodium percentage is greater than 15. As long as the large quantity of soluble salts remains in the soil, the high sodium content in the colloids may not cause trouble and the soil pH seldom exceeds 8.5. If, however, the soluble salts are temporarily leached downward, the pH goes above 8.5, the sodium causes the colloids to disperse, and a structure unfavorable for tillage, the entry of water, and root development develops. The movement of the soluble salts upward into the surface soil may lower the pH and restore the colloids to a flocculated condition. The management of this group of soils is a problem until the excess soluble salts and the exchangeable sodium are removed from the zone of root growth. Unless calcium sulfate or another source

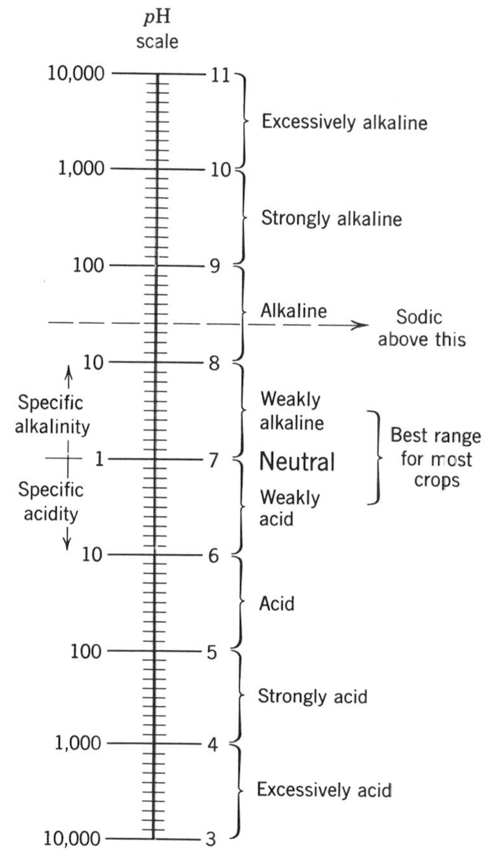


**Fig. 15–7** A chart showing the relationship between soil reaction (pH) and alkalinity. (Supplied by W. T. McGeorge.)

of soluble calcium is present, the drainage and leaching of these soils convert them into nonsaline sodic soils.

### Sodic (Alkali) Soils

These soils do not have so high a concentration of soluble salts as do those in the two preceding groups, but the exchangeable sodium percentage exceeds 15. The pH values usually range between 8.5 and 10. The sodium dissociates from the colloids, and small amounts of sodium carbonate may form. Organic matter in the soil is highly dispersed and is distributed over the surface of the particles, giving a dark color; hence the term "black alkali" which was formerly used to designate such soils. These soils frequently occur in small irregular areas in regions of low rainfall and are referred to as "slick spots."

Sodic soils may develop as a result of irrigation. Because of the dispersed state of the colloids, the soils are difficult to till and are slowly permeable to water. After a long period of time the dispersed clay may migrate downward, forming a very dense layer with a prismatic or columnar structure. When this phenomenon occurs, a few inches of relatively coarse-textured soil may be left on the surface.

The soil solution of sodic soils contains only small amounts of calcium and magnesium, but larger quantities of sodium. The anions include sulfate, chloride, bicarbonate, and usually small quantities of normal carbonate. In some areas an appreciable amount of potassium salts is also present.

## Formation of Saline and Sodic Soils

The groundwater of arid regions usually contains considerable quantities of soluble salts. If the water table is high, large amounts of water move to the surface by capillarity and are evaporated, leaving an ever-increasing accumulation of soluble salts. Through this process the upper parts of the soil may become so highly impregnated with salts that only salt-tolerant plants will grow. In locations where the water table is too deep for appreciable capillary movement to the surface, harmful accumulations of soluble salts will not develop. The rate of salt accumulation will be determined by (1) the rate of capillary movement of water to the surface, (2) the salt content of the ground water, and (3) the rapidity of evaporation. The nearer the water table is to the surface, the more rapid will be the salt accumulation, other factors being the same.

Under irrigation, saline soils may develop due to several different processes: (1) Excessive applications of water have raised the ground water level sufficiently to permit concentration of salts through evaporation. (2) Seepage from leaky canals and lateral ditches has resulted in a high water table. (3) The use of irrigation water with a high salt content has resulted in salt accumulation when (*a*) drainage is poor so that the salts cannot be leached out, (*b*) the application of water is so limited that the salts are left in the root zone in place of being washed out. If the water contains such quantities of sodium salts that the colloids become impregnated with them and the favorable structure of the soil is destroyed, sodic soils may develop.

## Sources of Soluble Salts

Soluble salts are derived primarily from the soil and rocks through which percolate the waters that supply the water table and the irrigation streams. The salt supply of the rocks and soil may be the result of accumulation when the area was a portion of the ocean bed or of the bottom of a saline lake, as in sedimentary deposits. The chemical decomposition of minerals has also contributed some salts. In igneous rocks, chemical weathering has supplied the soluble material.

## Detrimental Effects of Saline and Sodic Soils on Plants

High concentrations of neutral salts such as sodium chloride and sodium sulfate may interfere with the absorption of water by plants through the development of a higher osmotic pressure in the soil solution than exists in the root cells. Furthermore, the wilting coefficient of soils is raised by salt accumulations, and hence the quantity of water a soil will supply to plants may be reduced through the presence of salts. Detriment to plants may result also from soluble salts when the concentration is not sufficient to influence absorption of water. The entrance of nutrient ions into root hairs is influenced by the nature and concentration of other ions present. The salts may therefore result in nutritional difficulties in crops because of their inability to absorb needed nutrients from the soil.

The highly alkaline reaction due to the presence of sodium carbonate and the large quantity of adsorbed sodium represses the availability of several nutrients, especially iron, manganese, zinc, and phosphorus. Also, the alkaline soil solution has a corrosive action on the bark of roots and stems.

The exchangeable sodium in sodic soils results in a deflocculation of the colloids and hence in a breaking down of the soil structural units. This puddled condition renders the soil more or less impervious and retards entrance of irrigation and rain water and impedes drainage. In fine-textured soils the penetration of roots may be restricted by the density of the deflocculated zone. Aeration is also much reduced, setting up anaerobic conditions and resulting in the formation of reduced compounds which are toxic to plants.

The presence of certain "white alkali" salts, largely those of calcium, tends to counteract to some extent the detrimental effects of the adsorbed sodium and the sodium carbonate. The general action of the neutral salts is to flocculate the colloids and hence preserve the normal soil structure.

## Tolerance of Plants for Soluble Salts

Plants differ markedly in their tolerance of soluble salt, and likewise the salts differ greatly in their detrimental effects on plants. Hilgard set the limit of tolerance for the principal salts found in soils as follows:

Sodium carbonate, 0.10 to 0.25 percent
Sodium chloride, 0.20 to 0.50 percent
Sodium sulfate, 0.5 to 1.00 percent

Unquestionably, associated conditions such as temperature, moisture supply, organic-matter content, soil texture, and supply of nutrients, influence the tolerance of plants for these salts.

Shantz[1] lists some of the most salt-tolerant native plants, and the percentage of salts in soils in which they are found growing, as follows:

Greasewood, more than 0.5 percent
Seepweed, 2.5 percent or more
Pickleweed, 1.0 to 1.5 percent
Salt grass, somewhat less than 1.0 percent
Alkali sacaton, about 0.5 percent

The creosote bush of the southern deserts and sagebrush of the northern deserts indicate soils containing harmless quantities of salts. Dense stands of desert salt bush occur on fine-textured soils containing some salt but an amount insufficient to damage crops under irrigation unless further salt accumulation is permitted.

The relative tolerance of a number of fruit, field, truck, and forage crops for soluble salts is indicated in Table 15–2. The crops listed first in each group are considered most tolerant, and those named last, most sensitive, to salinity.

## Reclamation of Saline Soils

Good drainage is necessary for the reclamation of saline soils. It is essential in the reclamation process to remove the excess salts from the root zone, and this can only be done by the application of sufficient water to wash them into the lower soil depths. Unless there is ample drainage, the addition of so much water will raise the water table and hence lead to increased accumulations of salt in the surface soil rather than to a correction of the saline condition. Sufficient drainage should be provided to reduce the ground-water level well below the zone of root penetration. Kelley states that "preferably the ground water should never be less than 8 to 10 feet below the surface and every reasonable effort should be made to prevent its rising nearer than 5 to 6 feet from the surface even for brief periods."

[1] "Plants as Soil Indicators," H. L. Shantz, *USDA Yearbook* 1938, p. 852.

**Table 15–2** The Relative Tolerance of Several Crops for Salinity[a]

| Good Salt Tolerance | Moderate Salt Tolerance | | Poor Salt Tolerance | |
|---|---|---|---|---|
| | *Fruit Crops* | | | |
| Date palm | Pomegranate | | Pear | Apricot |
| | Fig | | Apple | Peach |
| | Olive | | Grapefruit | Orange |
| | Grape | | Prune | Lemon |
| | | | Almond | Avocado |
| | *Field and Truck Crops* | | | |
| Barley (grain) | Rye (grain) | Broccoli | Radish | |
| Sugar beets | Wheat (grain) | Tomato | Celery | |
| Milo | Oats (grain) | Cabbage | Beans | |
| Rape | Rice | Cauliflower | | |
| Cotton | Alfalfa | Lettuce | | |
| Beets (garden) | Sorghum (grain) | Sweet corn | | |
| Kale | Corn | Carrot | | |
| Asparagus | Foxtail Millet | Onion | | |
| | *Forage Crops* | | | |
| Alkali sacaton | White sweet clover | Wheat (hay) | White Dutch clover | |
| Salt grass | Yellow sweet clover | Oats (hay) | Meadow foxtail | |
| Nuttall alkali grass | Perennial rye grass | Orchard grass[b] | Alsike clover | |
| Bermuda grass | Mountain brome | Blue grama | Red clover | |
| Rhodes grass | Barley (hay) | Meadow fescue | Ladino clover | |
| Fescue grass | Birdsfoot trefoil | Reed canary | Burnet | |
| Canada wild rye | Strawberry clover | Big trefoil | | |
| Beardless wild rye | Dallis grass | Smooth brome | | |
| Western wheat grass | Sudan grass | Tall meadow oat grass | | |
| | Huban clover | Cicer milk vetch | | |
| | Alfalfa (California common) | Sour clover | | |
| | Tall fescue | Sickle milk vetch | | |
| | Rye (hay) | | | |

[a] From "Diagnosis and Improvement of Saline and Alkaline Soils," U.S. Regional Salinity Laboratory, Riverside, California, 1947.

[b] Salt tolerance of remaining crops is "fair" rather than "moderate."

With ample drainage provided, one may proceed to the leaching out of the salts. In fine-textured soils the reclamation process will be slow, and doubly so if the soil is underlain by a dense clay subsoil. In fact, the presence of a dense clay layer makes difficult the removal of salts from even medium- or coarse-textured soils. It is questionable if reclamation of soils with very deep clay subsoils is feasible from an economic standpoint.

Experiments have shown that leaching is all that is needed to reclaim saline soils that have adequate internal drainage. The addition of chemicals or plowing under of manure or green-manuring crops is unnecessary. No specific directions can be given regarding the frequency of irrigation or the quantity of water to apply at each irrigation. The main points to observe are (1) that the soil be kept moist so that the soil solution will not become sufficiently concentrated to damage the growing crop, (2) that sufficient water be applied at each irrigation to result in some leaching of salts into the drainage water, and (3) that the soil of each irrigation check be carefully leveled so that the water will enter the soil uniformly.

### Reclamation of Sodic (Nonsaline Alkali) Soils

All that has been said concerning the need for drainage and the application of sufficient irrigation water to cause leaching is of as much, if not more, importance in the reclamation of sodic soils as in the treatment of saline soil. Although it has been shown that the application of ample irrigation water, coupled with good farming practices, will ultimately result in the removal of exchangeable sodium as well as soluble salt, the reclamation process may be materially hastened through the application of various chemicals. The basis of the treatments is the replacement of exchangeable sodium in the colloidal fraction by calcium and the conversion of the replaced sodium and any occurring as the carbonate into neutral sodium sulfate (see Fig. 15–8).

The desired changes may be brought about by applications of considerable quantities of finely ground calcium sulfate (gypsum). Ground sulfur, however, will accomplish the same results somewhat more slowly. The sulfur must first be oxidized in the soil and then combines with water to make sulfuric acid. Other soluble sulfates such as iron or aluminum have also proved effective. A supply of soluble calcium is needed to complete the reactions. Acid resulting from sulfur additions dissolves $CaCO_3$ that might be present in the soil to supply soluble calcium. Harmful exchangeable sodium can than be replaced by calcium with a resulting improvement in the soil physical condition.

The growth and plowing under of sweet clover or the applying of manure is a good way to start the reclamation process whether or not chemicals are to be applied. The growing of Bermuda grass, a

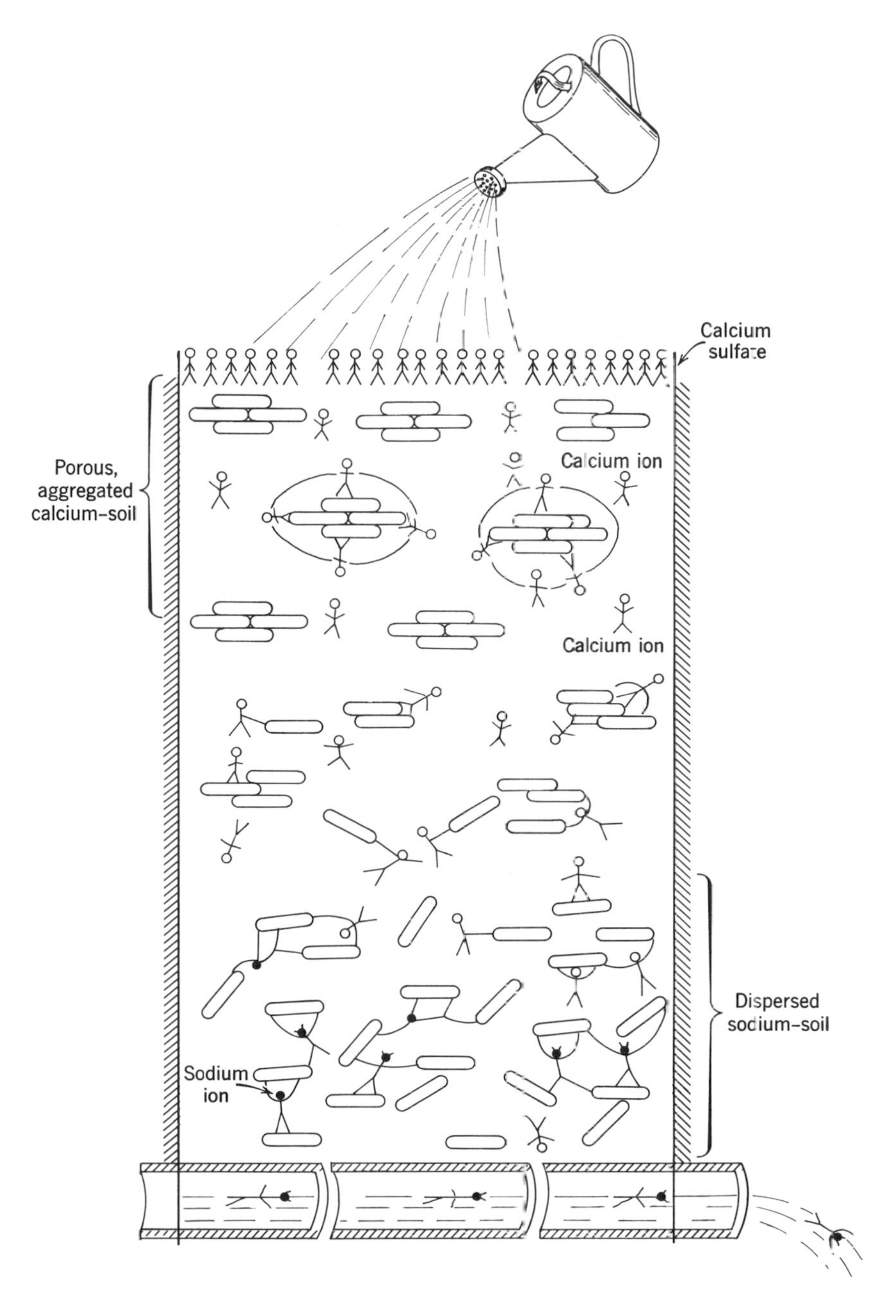


**Fig. 15–8** Illustration of the removal of exchangeable sodium by the addition of calcium sulfate. When irrigation water and calcium sulfate are applied to a sodic soil having dispersed particles and small pores, the calcium ions from the calcium sulfate replace the exchangeable sodium ions which form sodium sulfate and leach out of the soil. Soil pH is lowered, colloids flocculate, larger pores develop, and soil permeability is increased. (Adapted from "Chemical Amendments for Improving Sodium Soils," *Agr. Info. Bull.*, **195,** USDA, 1959.)

highly salt-tolerant plant, also has proved an advantageous method of starting alkali-removal practices. The data in Table 15–3 show the

**Table 15–3** Effects of Various Soil Treatments on the Yields of Crops Grown on Alkali Soil.[a] (Yields in pounds per acre.)

| Date | Crop | Sulfur 2 Tons, Gypsum 2½ Tons | Sulfur 2 Tons, Ground Limestone 2 Tons | No Treatment | Sulfur 2 Tons |
|---|---|---|---|---|---|
| 1925 | (Sweet clover plowed under as green manure on all plots) | | | | |
| 1926 | Alfalfa | 10.544 | 11,129 | 80 | 11,129 |
| 1927 | Alfalfa | 16,984 | 18,291 | 596 | 18,243 |
| 1928 | Alfalfa | 18,953 | 20,323 | 1693 | 19,679 |
| 1929 | Alfalfa | 10.484 | 14,113 | 5161 | 15,080 |
| 1930 | Alfalfa | 5000 | 6450 | 2095 | 5565 |
| 1931 | Alfalfa | 12,661 | 15,242 | 9596 | 14,233 |
| 1932 | Alfalfa | 14,033 | 15,161 | 14,516 | 18,106 |
| 1933 | Alfalfa | 12,422 | 16,814 | 16,885 | 18,549 |
| 1934 | Alfalfa | 12,097 | 16,289 | 17,823 | 18,385 |
| 1935 | Oat hay | 2742 | 3790 | 3838 | 4597 |
| 1936 | Oat hay | 6129 | 5645 | 6613 | 6853 |

[a] "The Reclamation of Alkali Soils," W. P. Kelley, *California Agr. Exp. Sta. Bull.* 617, p. 13, 1937.

effects of various treatments in the reclamation of an alkali soil in California, as measured by the yields of crops. The treatments were applied in May, 1923, after which the soil was allowed to lie idle, with an occasional light irrigation and cultivation, until February, 1925. After the plot was flooded twice to leach out salts, white sweet clover was seeded and the crop plowed under as green manure in September. The next February alfalfa was seeded.

It will be noted that although the yields on the untreated plot were quite low at first, they were very satisfactory after 6 years. So successful have been experiments in the reclamation of alkali soils that both Kelley of California and McGeorge of Arizona, the two men who have probably done more work on this problem than any other investigators, maintain that virtually any such soil may be reclaimed if adequate drainage can be established and if the soil is not underlain by a very impermeable clay subsoil. They are of the opinion, furthermore, that any soil under irrigation may be maintained sufficiently free of alkali and in a satisfactory structural condition if recommended management practices are followed, particularly those concerning irrigation.

### Effect of Sodium Removal on Soil Structure

Alkali soils are of notoriously poor structure, becoming very hard upon drying and breaking up into clods, which are very difficult to crush, when plowed. When a considerable quantity of soluble salts accompanies the alkali condition, the structural condition is usually good and a satisfactory seedbed can be prepared The leaching out of the soluble salt without replacement of the exchangeable sodium tends to leave the soil in an even worse physical condition. Experiments have shown, however, that applications of calcium, iron, or aluminum sulfates, and also sulfur, followed by proper irrigation, tend to flocculate the colloids and cause the rapid development of a granular structure. Soils from which the sodium is leached without the application of any of the chemicals mentioned are much slower in developing a granular structure. The growing of a crop such as Bermuda grass or the plowing under of a green-manuring crop hastens the granulation process.

# 16

# Irrigation

Irrigation is an ancient agricultural practice that was used 7000 years ago in Mesopotamia. Other ancient notable irrigation systems were located in Egypt, China, Mexico, and Peru. Today, about 11 percent of the world's cropland is irrigated. Some of the world's densest populations are supported by producing crops on irrigated land, as in the United Arab Republic (Egypt) where 100 percent of the cropland is irrigated. Other countries that have large percentages of irrigated cropland include Peru, 75 percent; Japan, 60 percent; Iraq, 45 percent; and Mexico, 41 percent. About 8 percent or 38 million acres of cropland are irrigated in the United States. "Nearly 2/3rds of the world's population lives in diet deficient countries having less than half of the world's arable land but with 3/4ths of the irrigated land."[1] Thus, we can see the great importance of irrigation in the world today and for the future.

## WATER SUPPLY AND LAND FOR IRRIGATION

Irrigation diminishes one of the greatest hazards in crop production, namely, inadequate water supply. In few instances is this factor entirely eliminated, however, as a shortage of water during some part of the growing season is not an uncommon occurrence on many irrigation projects because of improper irrigation scheduling.

[1] Quote and data from "The World Food Problem," Report of the Panel on the World Food Supply, The White House, May 1967.

The right to obtain water from streams and canals for irrigation purposes and the cost of the water are considered next.

## Obtaining Water from Streams and Canals

Water from streams is usually appropriated to individuals and canal companies according to priority rights, determined by the date the arrangements are made for water and the supply available. When two or more companies or individuals obtain water from a stream which is subject to wide variations in flow, the last to obtain water rights is the first to suffer when a scarcity arises. Although in some states the doctrine of riparian rights is recognized—that is, the owning of land along a stream entitles one to withdraw water from the stream for irrigation purposes—this right is modified by the doctrine of appropriation. In most of the western states the appropriation, diversion, and distribution of water from streams are under the direction of a state engineer or irrigation board.

When the government develops an irrigation project and supplies water to privately owned land, it operates through a water users' association. All the landowners become shareholders in the association through the purchase of water rights. The operation and management of the irrigation system, exclusive of storage reservoirs, usually passes into the hands of the landowners when payment of water rights for the major portion of the land has been made. The maintenance and operation of the system then becomes the responsibility of the landowners, and the cost is prorated among them.

Most of the irrigation water is surface water resulting from rain and melting snow. Many rivers in the world have their headwaters in mountains and flow through arid or semiarid regions. Examples include the Indus River that starts in the Himalaya Mountains and the numerous rivers on the western slope of the Andes Mountains that flow through the desert of Peru to the Pacific Ocean. Much of the water in these rivers comes from the melting of snow in the high mountains. In fact, the extent of the snow pack is measured to obtain information on streamflow and the amount of water that will be available the next season for irrigation (see Fig. 16–1).

Many farmers have their own source of irrigation water because they use well water. About 20 percent of the water currently used in the United States comes from irrigation wells.

## Selecting Land for Irrigation

In choosing land for irrigation, a careful examination should be made of the soil to determine (1) texture of soil to depth of several

**Fig. 16–1** Measuring the snow pack near Cooke City, Montana to obtain information for predicting streamflow and the quantity of water available for irrigation and other uses. (Photo USDA-SCS by P. E. Farnes.)

feet; (2) presence of an impermeable stratum or of gravel within a depth of 5 or 6 feet; (3) accumulation of soluble salts in injurious quantities; (4) slope and evenness of soil surface; and (5) behavior of the soil under irrigation. A desirable soil is readily permeable to water and yet is moisture retentive. Infiltration rates should be in the range of 0.1 to 3 inches per hour. It is well if the soil will absorb sufficient moisture in 24 hours to wet it to a depth of 2 or 3 feet. Some soils are so slowly permeable that they become wet to a depth of only a foot or less in 24 hours. Some soils are so coarse textured that water passes rapidly below the reach of plant roots, and little available moisture is retained. On the other hand, where fine- or medium-textured soil materials are underlain by coarser sands or gravels at a depth of several feet, the limited capacity of the coarser material to "pull" water downward by capillarity results in greater retention of water in the upper layer than if the soil was medium or fine textured throughout. The importance of soil properties in placing irrigated soils into land capability classes is illustrated in Table 16–1.

**Table 16–1** Guide For Placing Irrigated Soils Into Land Capability Classes (The Guide shows soil properties of importance in irrigation)

| Land Capability Class | Surface Texture | Coarse Fragments in Surface, Percentage | Available Water Holding Capacity: Surface Foot | Available Water Holding Capacity: Soil Profile to 60 Inches | Soil Permeability | Effective Depth | Salinity or Sodium Hazard |
|---|---|---|---|---|---|---|---|
| I | sandy loam to clay loam or silty clay loam | < 15% | > 1.5″ | > 7.5″ | Slow to Medium | > 40″ | None |
| II | loamy sand to silty clay loam or clay loam | 15–35% | 1.0–1.5″ | 5.0–7.5″ | Slow to Rapid | 30–40″ | Slight |
| III | sand to clay | > 35% | .75–1.0″ | 3.75–5.0″ | Slow to Rapid | 20–30″ | Moderate |
| IV | sand to clay | No Limits | < .75″ | < 3.75″ | Slow to Rapid | < 20″ | Severe |
| V, VI, VII, VIII | sand to clay | No Limits | No Limits | No Limits | Slow to Rapid | < 20″ | No Limits |

Adapted from Soil Conservation Service, USDA, Phoenix, Arizona State Guide by Donald Post.

Usually it is possible to judge the behavior of the soil under irrigation by observing similar soil in nearby irrigated fields. If this cannot be done, water may be applied to a small area for a trial. In general, if the soil is too clayey, it will absorb water slowly, be sticky and hard to cultivate when damp, and will crack and bake on drying, although some very fine-textured soils, particularly those with an adobe-like structure, absorb water readily and are not difficult to manage. Some sandy soils, on the other hand, form plow soles[2] very readily. Soil of too coarse a texture will absorb water very rapidly and show little coherence when dry.

The land surface should be comparatively smooth, for the cost of leveling land is high. A uniform slope of 10 to 20 feet to the mile is desirable, although much steeper slopes are in use. Land cut up by ravines, gullies, or buffalo wallows, or covered with sand dunes and hummocks should be avoided if possible.

## Providing Drainage

It is essential to have good drainage if irrigated land is to remain permanently productive. If the land is not naturally well drained, drainage ditches must be provided before the land is in use many years. It is difficult to apply sufficient irrigation water for maximum crop growth without getting some excess. In fact, some excess water frequently is desirable in order to wash out soluble salts. If the drainage water is not removed, it may raise the water level, thus increasing the danger of salt accumulation. Furthermore, the subsoil may become waterlogged with a resultant restriction of root development and crop damage.

## Irrigation Ditches

The main canals of an irrigation system are constructed and maintained by the company or association (Fig. 16–2). The ditches carrying the water to and over individual farms, however, are constructed and maintained by the farmers themselves.

The capacity, location, and design of farm ditches depend on several factors, among which the following deserve especial consideration: (1) depth and permeability of the soil; (2) whether the water is delivered in a continuous stream or for short periods with long intervals between; (3) the method to be used in applying the water to the land; (4) the acreage to be irrigated; and (5) the water requirements of the crops to be grown.

[2] A plow sole is a dense layer of soil at the bottom of the furrow resulting from compression of the tractor tire.

**Fig. 16–2** Concrete-lined canal bringing irrigation water to a farm in Utah. Although much of the water originates from the melting of the snow pack, it is collected in a reservoir and then discharged into canals for delivery to farms as needed (Photo USDA-SCS.)

If the soil is very permeable, there is much loss of water by seepage unless it contains sufficient silt and clay to form a seal on the sides and bottom of the ditch. In extreme cases pipes are used in place of ditches. These pipes usually are of concrete or plastic.

If water flows continuously, smaller ditches will suffice than if all the water needed must be obtained during a limited time at infrequent intervals. Likewise, large acreages and crops which require much water necessitate the construction of large ditches. In general, ditch capacities range from less than 1 to more than 10 cubic feet per second. Methods used to supply water to the land will be discussed in another section.

## IRRIGATION PRACTICES

Artificial watering adds certain complexities to farming practice. Judgment and experience are necessary to know when to apply water and how much to apply for different crops. Irrigation offers the opportunity for distribution of weed seed over the fields. Likewise, improper watering and irrigation without adequate drainage may result in salt accumulation. Water costs and the maintenance of ditches are also important considerations in the expense of crop production.

## Methods of Applying Irrigation Water

Choice of the various methods of applying irrigation water, of which there are about a dozen, is influenced by a consideration of (1) seasonal rainfall, (2) slope and general nature of the soil surface, (3) supply of water and how it is delivered, (4) crop rotation, and (5) permeability to water of the soil and subsoil. The methods of distributing water can be classified as flooding, furrow, sprinkler, or subirrigation. A brief discussion of each of these methods is given.

**Irrigation by Flooding.** The earliest irrigation method used was *flooding*. This method is applicable principally to rather smooth land with a regular and moderate slope and where there is an abundant supply of water. In Fig. 16–3 some large siphon tubes are being used to transfer water from a head ditch into a border area. Flood irrigation by the border method is suitable on land having a slope in one direction of 1 inch to 2 feet per 100 feet, although the system has been used on land with much steeper slopes. The field is laid out in strips varying in width from 6 to 60 feet and in length according to the head

**Fig. 16–3** A method of distributing water from a head ditch into a border area, using siphon tubes. (Courtesy of James C. Marr.)

of water available, the nature of the soil, and the degree of slope. If the slope of the field is too great, the strips may be laid out diagonally across it; otherwise they run down the slope. The strips must be of such width and length that they may be entirely covered with water without letting the water in at a rate that would cause erosion of the soil or without permitting too much or too little water to soak into the soil at one end before it reaches the other end.

A simple formula has been proposed for figuring the proper area of the strips. It is based on two factors: (1) the head of water in cubic feet per second, and (2) the nature of the soil. For sandy soils the ratio of irrigating head to acres of land is 20 to 1, and for clay soils 2 to 1. Thus, with a flow of water of 10 cubic feet per second, the size of the strips would be 0.5 acre on sandy soil and 5.0 acres on clay soil. The strips are set off by low, broad, flat-topped mounds of earth from 5 to 12 inches high, known as levees or borders. The borders are made in this shape so that farm implements may cross them readily and so that grazing stock will not destroy them when the field is in pasture. Needless to say, the soil surface in each strip must be leveled so that water will spread across it readily when let in through the two or more headgates (depending on the width of the strip) provided in the head ditch. It is not necessary that the strips in a field be at the same level.

In fields with irregular surface or with a considerable fall, ridges called checks are thrown up on the contour, thus dividing the land into plots. Irrigation is generally started with the plots highest up the slope. When one plot has been irrigated, water is let into the next lower one from the ditch. It is not considered good practice to let the water from one plot flow into the next except in irrigating rice. This is the contour-check method shown in Fig. 16–4.

Another modification of the flooding system of irrigation is known as the *basin method* (Fig. 16–5). The procedure consists in forming small basins by throwing up small earth dams or levees which retain the water while it soaks into the soil. A ditch is provided between each two rows of basins, and the water is run from the ditch into first one basin and then another as the irrigation proceeds. This system frequently is used in orchards and vineyards.

**Irrigation by the Furrow Method.** Crops commonly irrigated by furrow irrigation include row crops such as potatoes, sugar beets, corn, grain sorghum, cotton, vegetables, and fruit trees. Furrows are made across the field, leading down the slope. Water is let into the upper end of the furrow from a "head ditch" or pipe line running across the end of the field. Siphon tubes are commonly used to transfer the water from the head ditch into the furrows as shown in Fig. 16–6. Gated pipe is also used extensively.

**Fig. 16–4** The irrigation of a peach orchard by the contour-check method. (From *California Agr. Ext. Service Circ.*, **73**, by J. B. Brown.)

**Fig. 16–5** The basin method of irrigation. (From *USDA Farmers' Bull.*, **1518**, by Samuel Fortier.)

**Fig. 16–6** Use of siphon tubes to transfer water from head ditch into furrows. (Photo USDA.)

The *corrugation system* is a modification of the furrow method and is used most frequently in the northwestern states for the irrigation of small-grain and hay crops because of the uneven topography, use of small streams of water, and prevailing methods of planting and harvesting (Fig. 16–7). The irrigation furrows are placed at such distances apart and are of such length that will provide for the wetting of the soil between them by horizontal seepage. The permeability of the soil, slope of the land, and volume of water available are determining factors in spacing the furrows.

**Sprinkler Irrigation.** Everyone is familar with the sprinklers used to water or irrigate lawns. Sprinkler systems are versatile and have special advantages where high infiltration rates or topography prevents proper leveling of the land for surface distribution of water. The rate of application can also be carefully controlled. The portable nature of many sprinkler systems makes them ideally suited for use where irrigation water is used to supplement the natural rainfall.

**Fig. 16–7** Irrigation by the corrugation method. Four corrugations are being supplied from one outlet. (From *USDA Farmers' Bull.*, **1348,** by James C. Marr.)

When connected to a soil-moisture measuring apparatus, sprinkler systems can be made automatic.

Large self-propelled sprinkler systems have been developed in recent years. A self-propelled system that can irrigate most of the land in a quarter-section (160 acres) is shown in Fig. 16–8. If you have flown over the western part of the United States you have probably seen large green irrigated areas produced by self-propelled sprinklers.

**Subirrigation.** Subirrigation is irrigation by water movement upward from a free-water surface some distance below the soil surface. In arid regions where almost all of the water used to grow crops is from irrigation, subirrigation would cause serious salt accumulation problems in the upper part of the soil. Subirrigation works best where natural rainfall removes any salts that may accumulate. In many poorly drained areas, as in the Sand Hills of Nebraska, there are areas where subirrigation is a natural occurrence. Artificial subirrigation is practiced in the Netherlands where tile-drainage systems in polders are used in wet seasons for drainage and in periods of drought for subirrigation. Subirrigation, as compared to other systems, is inefficient in use of water and is adapted only for rather special situations. However, subirrigation distribution systems are expensive and not always reliable and the salinity hazard may be increased.

**Fig. 16–8** Circular patterns made by self-propelled sprinklers in Holt County, Nebraska. A unit is operating in the front-left quarter-section. (Photo USDA-SCS.)

## Quantity of Water to Apply

An ideal application of water would be a sufficient quantity to bring the soil, to the depth of the root zone of the crop, up to its field capacity. More water may result in the waterlogging of a portion of the subsoil or in the loss of water by drainage. If much water percolates below the root zone, it may accumulate under low areas, thus raising the water table unless suitable drainage facilities are provided. On the other hand, unless enough water is applied to result in appreciable drainage, it is difficult to remove excess salts from soils in which there is a tendency for salt to accumulate.

The quantity of water available to plants which may be stored in soils ranges from ½ to 2 acre-inches per foot of depth. If the water-holding capacity of the soil per foot in depth is known, the approximate quantity of water to apply may be obtained by multiplying this figure by the depth in feet of root penetration of the crop. The readiness with which soils absorb water influences the quantity and rate of application. Most dry soils absorb water rapidly. Little change in the rate of water intake occurs in very permeable soils as the depth of wetting increases. Some clay soils, however, absorb water very slowly after the first foot or so has become saturated. Because of this situation, it is advisable to hold the water for a long time on clay soils and for only a short time on coarse-textured soils.

## Quality of Water

Careful attention should be given to the nature of the water to be used before an irrigation system is constructed. Sometimes the available water carries so high a concentration of soluble salts that its use for irrigation is not advisable, particularly on land which already contains a considerable concentration of soluble salt. Sodium salts in the water are much more objectionable than are salts of calcium and magnesium because of the tendency for sodium to cause deflocculation of the colloidal fraction of the soil and to develop an undesirable structure. Some waters also contain sufficient boron to be toxic to plants with continued use. To make an estimate of the quality of irrigation water, the measurement of the following three characteristics is essential: (1) conductivity or total concentration of salts, (2) sodium-absorption ratio (SAR),[3] and (3) the concentration of boron. Water quality as a function of conductivity and sodium-absorption ratio is given in Fig. 16–9 and as a function of boron concentration in Table 16–2. The relative tolerance of plants to boron is given in Table 16–3.

In making an estimate of the quality of water the effect of the salts on both the soil and the plant must be taken into consideration. Various factors such as drainage, soil texture, and kind of clay minerals present influence the effects on the soil. The ultimate effect on the plant is the result of these two and other factors operating simultaneously. Consequently, there can be no method of interpretation which is absolutely accurate under all conditions. The schemes for interpretation are accordingly based chiefly on experience.

## Special Practices for Salinity Control

Irrigation water commonly contains one or two tons of dissolved salt per acre foot. The accumulation of salt in the soil, particularly where natural precipitation is not sufficient to cause leaching, must be guarded against. Crop failures result in some cases where salt accumulates and low salt-tolerant crops are grown. Sugar beets, cotton, and barley are crops that can often be grown profitably on "saline" soils. The salinity tolerance of many crops is given in Table 15–2.

Practices that allow for the uneven distribution and downward movement of water causes differences in the distribution of salt in the soil. In the case of furrow irrigation, leaching of the soil occurs below the furrow while water moves by capillarity to the top of the

[3] $SAR = \frac{Na^+}{\sqrt{\frac{(Ca^{2+} + Mg^{2+})}{2}}}$, where the concentrations of the cations are given in milliequivalents per liter.

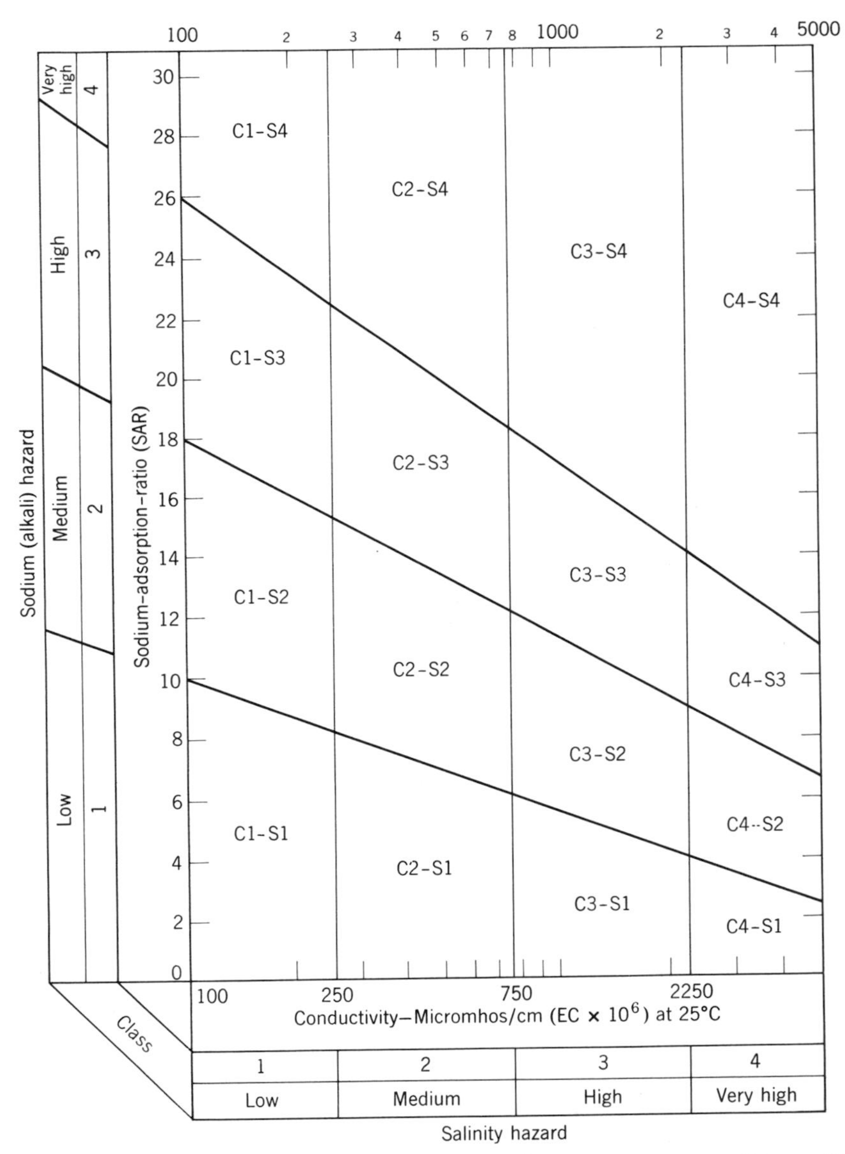


**Fig. 16–9** Diagram for the classification of irrigation waters. (From *USDA Agriculture Handbook*, **60**, 1954.)

**Table 16–2** Permissible Limits for Boron of Several Classes of Irrigation Water[a]

| Classes of Water | | Sensitive Crops, ppm | Semitolerant Crops, ppm | Tolerant Crops, ppm |
|---|---|---|---|---|
| Rating | Grade | | | |
| 1 | Excellent | < 0.33 | < 0.67 | < 1.00 |
| 2 | Good | 0.33–0.67 | 0.67–1.33 | 1.00–2.00 |
| 3 | Permissible | 0.67–1.00 | 1.33–2.00 | 2.00–3.00 |
| 4 | Doubtful | 1.00–1.25 | 2.00–2.50 | 3.00–3.75 |
| 5 | Unsuitable | > 1.25 | > 2.50 | > 3.75 |

[a] After C. S. Scofield.

**Table 16–3** Relative Tolerance of Plants to Boron (In each group, the plants first named are considered as being more tolerant and the last named more sensitive)

| Tolerant | Semitolerant | Sensitive |
|---|---|---|
| Athel (*Tamarix aphylla*) | Sunflower (native) | Pecan |
| Asparagus | Potato | Black walnut |
| Palm (*Phoenix canariensis*) | Acala cotton | Persian (English) walnut |
| Date palm (*P. dactylifera*) | Pima cotton | Jerusalem artichoke |
| Sugar beet | Tomato | Navy bean |
| Mangel | Sweetpea | American elm |
| Garden beet | Radish | Plum |
| Alfalfa | Field pea | Pear |
| Gladiolus | Ragged Robin rose | Apple |
| Broadbean | Olive | Grape (Sultanina and Malaga) |
| Onion | Barley | Kadota fig |
| Turnip | Wheat | Persimmon |
| Cabbage | Corn | Cherry |
| Lettuce | Milo | Peach |
| Carrot | Oat | Apricot |
| | Zinnia | Thornless blackberry |
| | Pumpkin | Orange |
| | Bell pepper | Avocado |
| | Sweet potato | Grapefruit |
| | Lima bean | Lemon |

From *Agriculture Handbook* 60, USDA, 1954.

ridges and evaporates to produce a salt accumulation on the ridge. Note the marked differences in salt distribution in a soil where cotton was grown and water of medium salinity was used as shown in Fig. 16–10. The distribution of salt had an effect on the absorption of water by the cotton plants. The regions of low salt were regions where the soil moisture has been depleted the most (Fig. 16–11). Control of

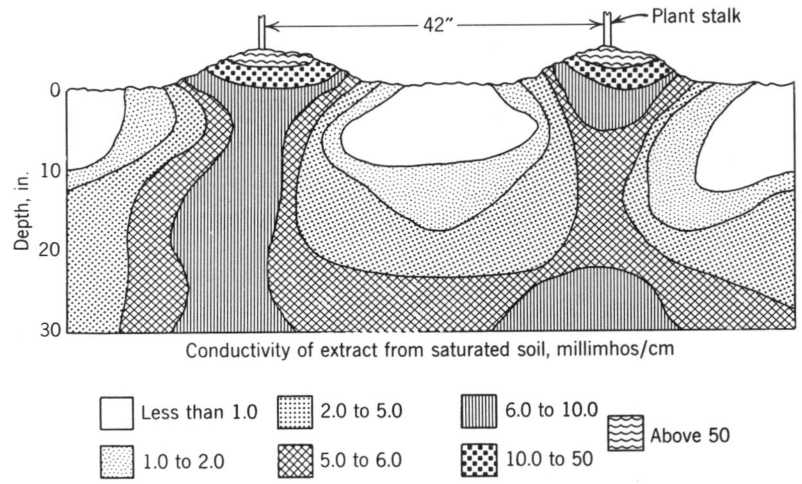


**Fig. 16–10** Salt distribution under furrow-irrigated cotton for soil initially salinized to 0.2 percent salt and irrigated with water of medium salinity. (From *USDA Agriculture Handbook*, **60**, 1954.)

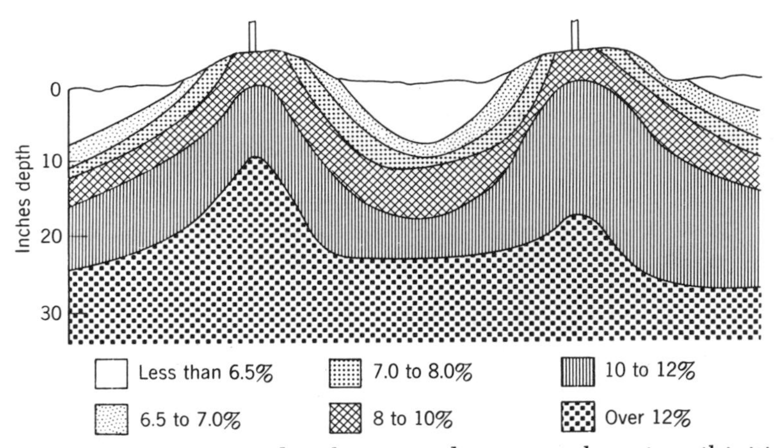


**Fig. 16–11** Moisture distribution under cotton plants in soil initially salinized to 0.2 per cent. (From "Salt Distribution Under Furrow and Basin Irrigated Cotton and Its Effect on Water Removal," C. H. Wadleigh and M. Fireman, *Soil Sci. Soc. Am. Proc.*, Vol. **13**, p. 530, 1948.)

salinity is facilitated if the land is level, the water is uniformly distributed, and leaching is uniform.

Seed germination is frequently a problem in salty soils. Irrigation water can be applied in such a way that the salt content of the seed zone is minimum, thus insuring maximum germination and plant stand. Maintaining a higher than normal water content in the soil at germination also minimizes the effects of salt.

## Efficiency of Water Use

There are many places where irrigation water can be lost from storage reservoir to plant root. These losses are summarized in Fig. 16–12. In western United States only 1.7 acre feet of water is used by crops for every 5.8 acre feet of water diverted for irrigation. The overall efficiency of water use is only about 30 percent.

Where irrigation water comes from rain and melting snow, the future supplies can be considered to remain unchanged since a change in supply would necessitate a climatic change. However, many farmers dependent on well water have been removing water from underground water reservoirs at rates that exceed natural recharge so that water tables are falling. For some areas of the high

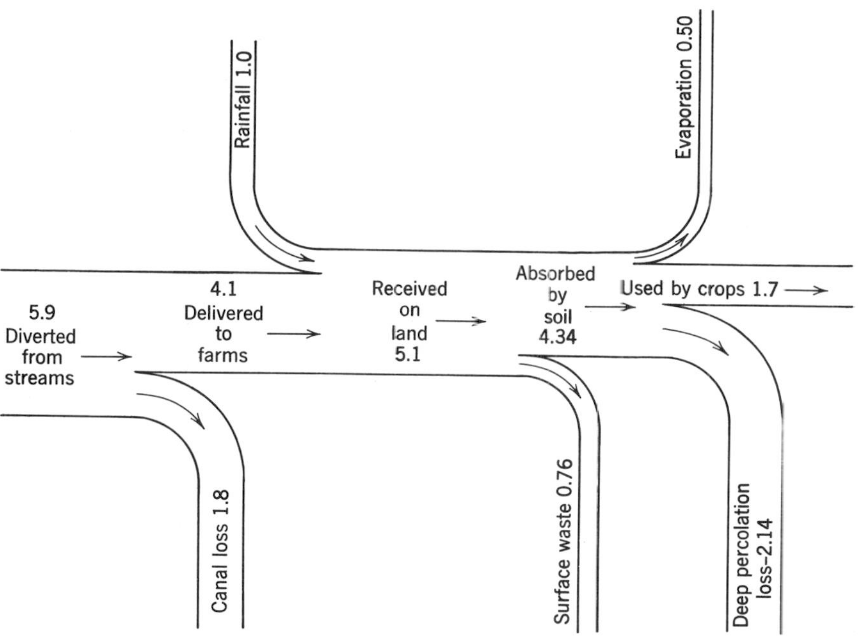


**Fig. 16–12** Disposal of water diverted for irrigation. (From "Our Unseen Waters," D. A. Williams, *Soil Conservation*, p. 52, October 1963.)

plains it is estimated that water is being pumped out at a rate that exceeds the normal rate of recharge by 100 to 200 times. Increased pumping depths result as water tables fall, which in turn forces farmers to grow higher value crops or abandon land. These concerns plus increased competition for water by growing cities and industries have created interest in more efficient use of water, and in particular, a new irrigation method called the *drip* or *trickle method.* The method originated recently in Israel and involves the slow application of water at particular points to the soil where the water immediately enters the soil. The water can be applied continuously at a very slow rate and sometimes the water is applied by seepage from underground pipes. The result is a constant soil moisture tension for the plants and a greatly reduced loss of water. Israel soil scientists have obtained 50 to 100 percent greater yields by using only half as much water as the conventional methods.[4]

[4] "Irrigating A Drop At A Time," A. W. Marsh and C. D. Gustafson, *Crops and Soils Magazine*, pp. 9–11, March 1971.

# 17

# Soil Resources and Population

As food, textiles, and housing materials are largely products of the soil, either directly or indirectly, an ample acreage of productive land is one of the most essential and most stabilizing resources a nation can have. A nation with soil resources so limited that it must depend mainly on imported food and fiber is always in a precarious economic position because not only war but also changes in trade conditions may curtail the supply of imports. Furthermore, manufacturers of many commodities must depend on agricultural workers to purchase a considerable portion of the manufactured products, even though the workers live in different countries. Unless the people who live on the land have reasonable incomes, their purchasing power is limited. It is extremely important, therefore, to the city producer that there be ample good land for farming purposes and that the land be maintained in a sufficiently productive state to afford a reasonable income to farmers and to provide an adequate food supply.

In considering the soil resources of the United States and other countries, the adequacy of these resources to supply the needs of the people concerned will be discussed.

## ACREAGES OF PRODUCING AND POTENTIAL CROPLAND IN THE UNITED STATES

An inventory of land resources in the United States was made by federal and state authorities. Within the United States, there are

Major Uses of Land, 48 Contiguous States, 1959
Total Area of 48 States, 1,902 Million Acres

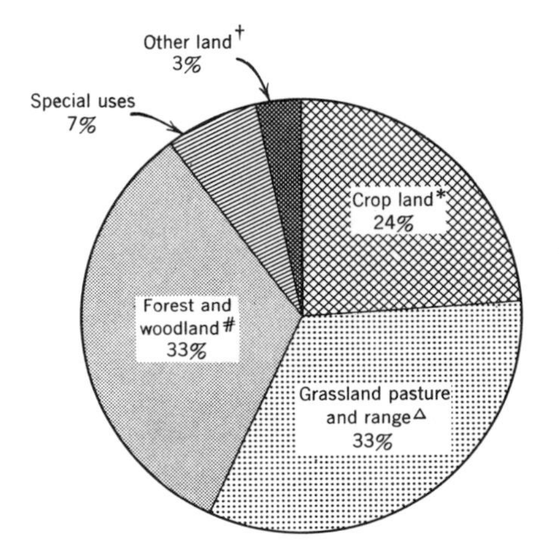


Major Uses of Land, 50 States, 1959
Total Area of 50 States, 2,271 Million Acres

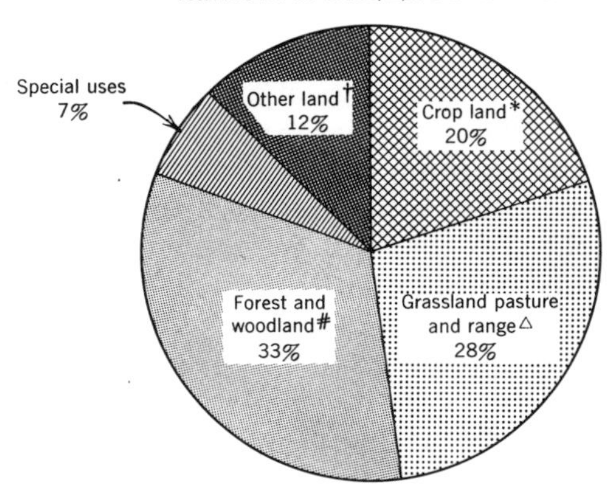


*All crop land, including crop land used only for pasture.
△Grassland pasture and range, private and public.
#Excludes forested areas in parks and other special uses.
°Urban and other built up areas, parks and special facilities etc.
†Desert, swamp, dunes etc.

**Fig. 17–1** Major uses of land in the United States.

approximately 2.3 billion acres of land. Three-fifths of this acreage is in farms or privately owned. About one-third of the land is in federal ownership and the state and local governments hold about 5 percent.

## Land Use in the Continental United States

Approximately one-third of the land in the continental United States is currently used for forest and woodland, one-third for grassland and range, and one-fourth is devoted mainly to crop land. Special and other uses account for 10 percent of the land use (Fig. 17–1).

From 1880 to 1950 the acreage of crop land more than doubled. This increase occurred primarily by conversion of available pasture and range land into crop land. Since 1920 the acreage of crop land has remained about the same and the acreage of pasture and range land has shown a small decrease. Since 1880 woodland and forest land have remained about the same. Important changes in land use have occurred since 1920, however, but some of these changes have, in large part, canceled each other. For instance, west of the Mississippi River the acreage of crop land has increased, and east of the Mississippi River it has decreased. For the country as a whole, the acreage has remained relatively stable since 1920 (Fig. 17–2).

At present, the development of new farm land nearly equals the conversion of farm land to other uses. From 1910 to 1959 the increase in land devoted to cities, towns, parks, airports, and other uses rose

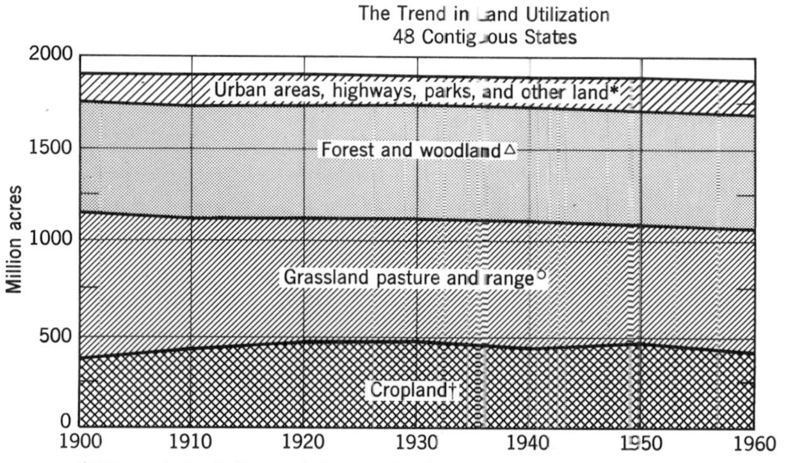


*Urban and other built areas, highways, railroads, airports, parks and other land.
△ Excludes forested areas reserved for parks and other special uses.
o Includes grassland pasture and range, private and public.
†Cropland planted, cropland in summer fallow, soil improvement crops, land being prepared for crops and idle.

**Fig. 17–2** Trends in land use in the United States 1900 to 1960.

from 53 to 113 million acres, or at a rate slightly over 1,000,000 acres per year. Cities continue to absorb about a million acres a year at the present time.

### Acreages of Land Suitable for Cultivation and Land Cultivated

The National Inventory of Soil and Water Conservation Needs taken in 1967 places land in one of eight classes. The classes range from I, land suited for crops or most other agricultural uses, to class VIII, land suited only for recreation, wildlife, or water supply, having only limited agricultural use. The amount of land suitable for cultivation and the amount used for crops of the various classes are given in Table 17–1. The data in Table 17–1 show that only about 60 percent of the best land is currently being used for crops in the United States

**Table 17–1** Land Suited for Cultivation and Land Used for Crops in the United States in 1967

| Land Class | Available Cultivatable Land, Acres | Land Used for Crops in 1967, Acres |
|---|---|---|
| Classes I, II, & III (no limitations for farming) | 630 million | 360 million |
| Class IV (land suited for only limited farming) | 180 million | 49 million |
| Classes V to VIII (best suited for grazing, forests, recreation, wildlife, and water supply) | 628 million | 25 million |

From "Fact Book of U.S. Agriculture," Mesc. Pub. 1063, USDA, 1970. Data is for entire United States including Alaska and Hawaii.

and that considerable other cultivatable land is available. For quite a few years to come, the supply of land for crops appears to be very adequate.

### Distribution of Land of Different Quality Among the States

A study has been made with the purpose of dividing the land into five groups on the basis of the land's capacity to produce grains, grasses, and legumes without irrigation, drainage, or fertilization. The group names used are excellent, good, fair, poor, and unfit for crop production. The distribution of the first three classes of land is also shown graphically in Fig. 17–3.

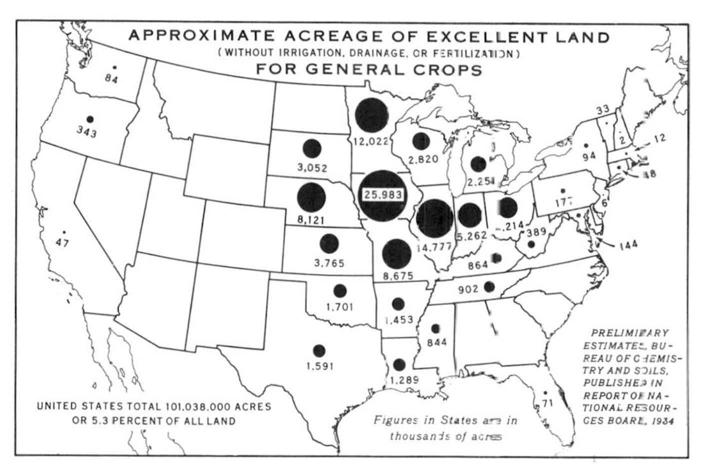


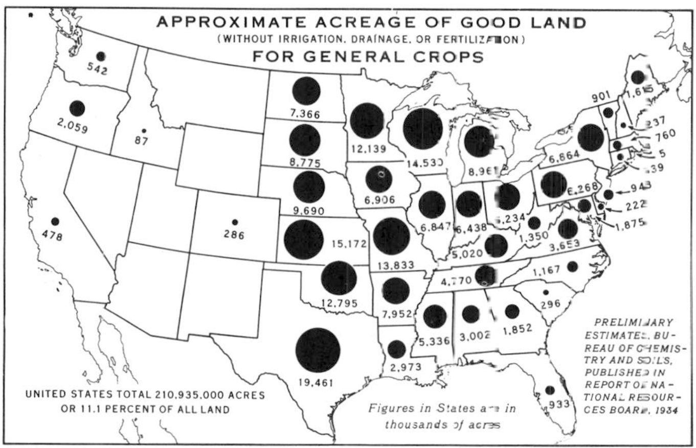


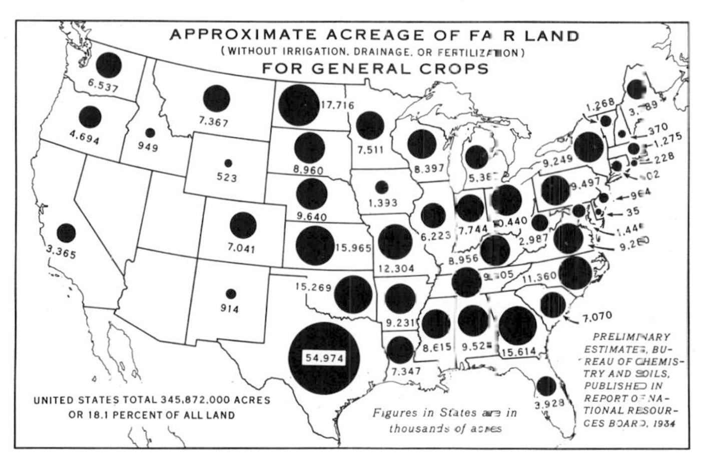


**Fig. 17–3** The distribution of excellent, good, and fair land in the various states. (From *USDA Misc. Pub.*, **260**.)

The high percentage of excellent and good land in the Corn Belt states and the states immediately adjoining them is worthy of consideration. The question may well be asked, what conditions gave rise to the development of so much land of superior quality in that location? It should be remembered, furthermore, that some of the land designated "fair" in this classification produces superior yields when fertilized. In fact, some of the states which have very little if any "excellent" land show higher average acre yields for several of the grain crops than do the Corn Belt states. Likewise, the production on large acreages of land classified "fair" but which is irrigated is much greater than on high-quality land without irrigation.

## ACREAGES OF ARABLE LAND AND LAND REQUIREMENTS

A knowledge of the acreage of farm land in different countries in relation to population leads to a better understanding of some of the problems confronting these nations. The interdependence of nations is also brought out by such a study. The more one learns about good farm land and the part it plays in the welfare of people, the more respect one has for the soil.

### Meaning of Arable

A broad and variable interpretation of the term arable is possible on the basis of experience and viewpoint. Strictly speaking, arable land is land which is suitable for plowing or cultivation. No mention is made, however, of whether such land may be cultivated profitably at any set price level for agricultural products or whether the land will deteriorate rapidly under cultivation unless special tillage practices and cropping systems are used. Probably there is a tendency to apply the term arable from a purely physical viewpoint and to consider any cleared land arable which is not too stony, too rough topographically, or too badly in need of drainage to grow crops, or which is not in an area of very limited rainfall during the growing season. It is evident that land which one person or group of persons would consider arable might not be so classed by another group of persons. For this reason it is difficult to compare accurately the acreages of good land in different countries. In the following discussion the comparisons are based on the best data available and arable land is considered the same as cultivatable land.

### About One-Tenth of the World's Land in Cultivation

In general, about one-half of the land in the world is considered entirely unsuitable for cultivation. This portion is made up of areas permanently covered with ice and snow (11 percent), the tundra (4 percent), high mountains (16 percent), and deserts and semideserts (17 percent). Although appreciable areas of desert may be irrigated, this acreage is not large enough to change the general picture. By no means all of the other one-half of the world's land can be easily placed under cultivation. Much of it is so rocky, so sandy, so hilly, so full of soluble salts, or so badly in need of drainage that it may never be cultivated. Estimates of the percentage of the world's land now in cultivation vary widely, some being greater and some less than 10 percent. Certainly not a large part of the land surface is being cultivated. How much this area can be practically increased depends on economic conditions and scientific developments, but any large increase in the immediate future appears doubtful.

## Acres of Cultivatable Land per Person in the United States

We shall consider the land data given in Table 17–1. Assuming a population of 210 million, there were about 1.7 acres of good cropland used per person and about 3 acres of good cropland available. Considering all the available cropland, there is about 6 acres per person. The data show that the United States yet has an abundance of cultivatable land.

## Land Required to Produce Food

There arises the question whether there is enough good land in continental United States to produce the foods needed to supply adequate diets for all our people. To obtain an answer to this question we must know how many acres are required to produce the foods needed by one person. This point has been given careful consideration. Table 17–2 is quoted from "Efficient Use of Food Resources in the United States," by R. P. Christensen, *U.S. Department of Agriculture Technical Bulletin* 963. In calculating the acres of cropland

**Table 17–2** Acres of Cropland Required To Produce Different-Cost Adequate Diets and the Percentage of Food Energy from Livestock Products in Each Diet, United States

| Diet Plan | Crop Land Required per Person, Acres | Food Energy in Diet from Livestock, Percentage |
|---|---|---|
| Low-cost | 2.12 | 30 |
| Moderate-cost | 2.57 | 36 |
| Liberal-cost | 3.15 | 44 |

required per person, the acreage of pasture was converted into cropland equivalent.

It is evident that the number of acres required to feed one adult varies considerably with the proportion of the diet supplied from livestock products. This is understandable when it is remembered that a given quantity of wheat supplies from 5 to 10 times as many calories when eaten directly as it does when it is fed to livestock and the meat, lard, butter, or milk is consumed.

The acreage needed to feed one person also varies with the efficiency of production. Harrison Brown estimates that in a hunting and food gathering situation there would need to be about two square miles of fertile land in the natural state to support one person, and that the world could support a population of ten million people under these conditions.[1] Values of acres required per person need to be interpreted with caution. Many countries or regions do not have an abundance of land per person for food production and these will be discussed in the next sections.

In some countries much of the power requirements are met with the use of draft animals. Some 55,000,000 acres of land formerly used to grow feed for horses and mules in the United States is now used to produce food products for human consumption. Substitution of tractors for draft animals is one of the fastest ways of increasing the available food in some countries and thus is also a factor affecting the number of acres needed to supply the food for one person.

## THE FOOD AND POPULATION PROBLEM OF THE WORLD

Political and social unrest are frequently related to population and food production problems. This provides a basis for concern about the possibilities for providing most of the world's population with an adequate diet.

### Food Production Currently Increasing Faster than Population

During the past decade the Food and Agriculture Organization of the United Nations reported that the world's population was increasing at an annual rate of 1.8 percent, whereas food production increased at an annual rate of 2.9 percent.[2] The major increase in food production, however, occurred primarily in those areas where food

[1] See "The Challenge of Man's Future," Harrison Brown, p. 14, Viking Press, 1954.

[2] "The State of Food and Agriculture 1962," Food and Agriculture Organization of the United Nations.

is already abundant or in surplus. Although the actual number of persons in the world who are receiving an adequate diet is increasing, the percentage of the population that is receiving an adequate diet is decreasing. In the pre-World War II era 38.6 percent of the world's population had an average diet under 2200 calories per day, and in a recent postwar period this had increased to 59.6 percent of the population.[3] The areas that are increasing food production at the slowest pace are also those expected to experience an accelerating rate of population growth, resulting from improved hygiene and medical care, in the decades immediately ahead.

## Oceans as a Source of Fish

The oceans play a very important role in providing food, as fish products are rich in protein and serve to balance the calorie intake of starchy diets for many of the world's people. One factor commonly overlooked in food problems is that a person cannot eat enough low-protein food in the form of rice, wheat, or corn to satisfy his daily protein requirement. Stated another way, if a person ate all the corn he could, the corn would not contain sufficient protein to meet his daily body needs. Consideration of increased food production from the oceans in the form of fish is, therefore, of more than ordinary importance because of the protein considerations.

A large increase in ocean-fish production can be expected with current technology. It is easy to be over optimistic, however, because the oceans occupy three-fourths of the earth's surface and photosynthesis can occur at great depths. Consequently, the total amount of photosynthesis carried out by the plants in the oceans is many times greater than that of the land plants. However, two factors are cited to point out the difficulties in achieving large and easy increases in world fish production. First, fish production is very inefficient in terms of the use of primary plant calories. The production of a pound of fish may require from 1000 to 100,000 pounds of primary plankton depending on the number of times one organism is devoured by another before it is caught for food. Secondly, much of the ocean water is nutrient poor. Fish production is limited in large areas because of insufficient phosphorus or nitrogen for plankton production.

## Potential for Increasing Arable Land

It has already been pointed out that only 60 percent of the "best" land in the United States is being used for crop production. What

[3] "Man and Hunger," World Food Problems 2, FAO, Rome, Italy, 1957.

about the situation on a world basis? The data clearly show that the world's supply of arable land could be doubled. Data in Table 17–3 indicate that 3.4 billion acres of land are now being cultivated and that 7.8 billion acres in the world could potentially be cultivated. However, it should be kept in mind that the best lands are being cultivated and that expanding the acreage of cultivated land will usually mean the use of poorer land and increased production costs.

**Table 17–3** Present Cultivated Land on Each Continent Compared with Potentially Arable Land

| Continent | Area, Billions of Acres | | | Acres of Cultivated Land per Person | Ratio of Cultivated to Potentially Arable Land, Percentage |
|---|---|---|---|---|---|
| | Total | Potentially Arable | Cultivated | | |
| Africa | 7.46 | 1.81 | 0.39 | 1.3 | 22 |
| Asia | 6.67 | 1.55 | 1.28 | .7 | 83 |
| Australia and New Zealand | 2.03 | 0.38 | 0.08 | 2.9 | 21 |
| Europe | 1.18 | 0.43 | 0.38 | .9 | 88 |
| North America | 5.21 | 1.15 | 0.59 | 2.3 | 51 |
| South America | 4.33 | 1.68 | 0.19 | 1.0 | 11 |
| U.S.S.R. | 5.52 | 0.88 | 0.56 | 2.4 | 64 |
| Total | 32.49 | 7.88 | 3.47 | 1.0 | 44 |

From "The World Food Problem," Vol. 2, A Report of the President's Science Advisory Committee, The White House, May 1967.

On a world basis there is a very unequal distribution of the potentially arable land. Two continents are essentially "filled" with people in the sense that nearly all the arable land is now in use. These continents are Asia, with 83 percent of the arable land now in use, and Europe, with 88 percent (see Table 17–3). Furthermore, the land used per person is low, being less than one acre per person.

Large acreages of potentially arable land exist on the African continent in the Congo Basin and in South America in the Amazon Basin. Here, great obstacles for expansion of cultivated land are found in the great infertility of the soils, lack of transportation, and so forth. The problem in Asia is that there is a very large population with little land per person and very little additional land that is potentially arable. Where living standards are good and considerable land is now being used per person, as in Australia, North America, and the Soviet Union, there is also considerable potential for future development of cultivated land. Thus, the uneven distribution of the cultivated and potentially cultivated land presents a difficult problem in terms of improving the food supply in those regions of the world where food is in shortest supply.

## Yields Can Be Increased

Crop yields appear to be a function of the overall economic development of a country. High acre yields generally depend on fertilizers, pesticides, and machinery, all which are the products of an industrial society. Wheat yields in Britain were nearly constant or decreased slightly from 1100 to 1350 A.D. under the feudal system (Fig. 17–4). Conversion of land into private ownership and the use of grass to rejuvenate exhausted land caused a significant increase in wheat yields from 1350 to 1550. Very modest increases in yield occurred between the years 1550 and 1900 from improvements in farming evolved by trial and error. Note, however, that in a recent 60-year period the yields increased more than during the preceding 800 years (Fig. 17–4). There is no reason to doubt that, with the current level of knowledge, large increases can be made on a large portion of the cultivated land in the world today. It is very reasonable to again suggest what C. E. Kellogg of the United Soil Conservation Service pointed out in 1948, that "there is sufficient soil resources to feed the world's population, providing the economic, social, and political problems can be solved."[4]

In the final sense, food production from the crop land is limited by the ultimate amount of carbon that is fixed in photosynthesis. Although the energy absorbed by the plant is used efficiently, as compared to the efficiency of man-made engines, only a slight percentage

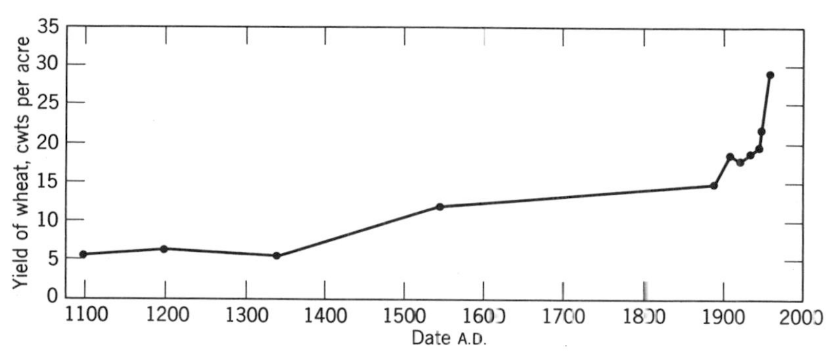


**Fig. 17–4** Wheat yields in Britain 1100 to 1960 (From "Man the Fertility Maker," G. V. Jacks, *J. Soil and Water Conservation*, 1962.)

[4] See *Am. Scientist*, **38**:535, 1948.

of the sun's energy is utilized when maximum yields are obtained. Some of the most important studies in agriculture today are those aimed at improving our knowledge of plant growth so that ultimately crop plants may be able to utilize more of the sun's energy through photosynthesis (Fig. 17–5).

## Future Outlook Not Very Bright Unless Population Can Be Controlled

It has been mentioned that current soil resources and technology could result in sufficient food production to feed the people in the world today. The food-population problem is not bright, however. First, the current population of slightly over 3 billion is expected to double by the year 2000. Although the highly industrialized nations may have no problem in meeting their food needs for the remainder of the century, it appears that increasingly less food per person for millions of people will be the case because populations are increasing most rapidly in areas where food production increases are the most difficult to achieve. One of the authors' objectives has been to bring to the attention of the reader, the fact that soil resources and tech-

**Fig. 17–5** An experimental setup at Cornell University to study light intensity, carbon dioxide concentration of the air, temperature, and water utilization in relation to the growth of corn. Experiments such as this one provide the knowledge that forms the basis for new practices that will enable us to increase the growth of plants.

nology are sufficient for the twentieth century and that a major goal of a nation such as the United States should be to bring about those economic, social, and political changes that will result in a better life for an ever increasing portion of the world's people. The task is enormous and cannot be ignored if there is to be world peace.

# Glossary

**A Horizon.** The surface horizon of a mineral soil having maximum biological activity, or eluviation (removal of materials dissolved or suspended in water), or both.
**ABC Soil.** A soil with a complete profile, including an A, a B, and a C horizon.
**AC Soil.** A soil with an incomplete profile, including an A and a C horizon, but no B horizon. Commonly such soils are young, like those developing from alluvium or on steep, rocky slopes.
**Active Acidity.** The activity of hydrogen ion in the aqueous phase of a soil. Its activity is measured and expressed as a pH value.
**Adsorption.** The adherence of one particle, ion, or molecule to the surface of another. Usually applied to the molecular or ionic state of division; for example, $NH_3$ gas concentrated on the surface of charcoal, potassium ions concentrated on the surface of soil colloids.
**Aeration, Soil.** The process by which air and other gases in the soil are renewed. The rate of soil aeration depends largely on the size and number of soil pores and on the amount of water clogging the pores. A soil with many large pores open to permit rapid aeration is said to

In the compilation of the Glossary extensive use was made of the glossary in USDA Yearbook of 1957, the terminology report in the Soil Science Society of America Proceedings, Vol. 29, May–June 1965, pp. 330–351, glossary in *Journal Soil And Water Conservation,* Vol. 25, January–February 1970, and USDA *Soil Classification, a Comprehensive System,* 1960.

be well aerated, whereas a poorly aerated soil either has few large pores or has most of those present blocked by water.

**Aerobic.** Living or active only in the presence of molecular oxygen. Pertaining to or induced by aerobic organisms, as aerobic decomposition.

**Aggregate (Soil).** A single mass or cluster of soil particles, such as a ped, crumb, or granule.

**Aggregation.** The cementation of particles into masses called aggregates. Essentially a process of flocculation and cementation.

**Albic Horizon.** A horizon from which clay and free iron oxides have been removed. Generally equivalent to A2 horizon.

**Alfisols.** Soils with gray to brown surface horizons, argillic horizons and medium to high supply of bases (over 35 percent in argillic horizons). These soils have formed mostly under forest or savannah vegetation in climates with slight to pronounced seasonal moisture deficit.

**Alkali Soil.** See sodic soil.

**Agric Horizon.** An illuvial horizon of clay and organic-matter accumulation just under the plow layer resulting from long-continued cultivation.

**Argillic Horizon.** An illuvial horizon of silicate clay accumulation.

**Alluvial Soils.** Soils developed from transported and relatively recently deposited material (by streams) characterized by a weak modification (or none) of the original material by soil-forming processes.

**Alluvium.** Fine material, such as sand, silt, clay, or other sediments, deposited on land by streams.

**Ammonification.** Production of ammonia as a result of the biological decomposition of organic nitrogen compounds.

**Anaerobic.** Living or active in the absence of molecular oxygen.

**Anion.** An ion carrying a negative charge of electricity.

**Anthropic Epipedon.** Molliclike horizon that has a very high phosphate content resulting from long-time cultivation and fertilization.

**Aridisols.** Soils of dry regions with pedogenic horizons, low in organic matter, that are never moist as long as three consecutive months.

**Autotrophic.** Capable of utilizing carbon dioxide as a source of carbon and of obtaining energy for the reduction of carbon dioxide and other life processes from the oxidation of inorganic elements or compounds, e.g., sulfur, hydrogen, ammonium, and nitrite salts, or from light. Contrast with heterotophic.

**Azonal Soils (obsolete term).** Soils without distinct genetic horizons. A soil order in 1949 system.

**B Horizon.** A soil horizon, usually beneath an A horizon, or surface soil, in which (1) clay, iron, or aluminum, with accessory organic matter, have accumulated by receiving suspended material from the A horizon above it or by clay development in place; (2) the soil has a blocky or prismatic structure; or (3) the soil has some combination of these features. In soils with distinct profiles, the B horizon is roughly equivalent to the general term "subsoil."

**Base Saturation.** The extent to which the cation-exchange complex of a soil is saturated with exchangeable cations other than hydrogen and aluminum. It is expressed as a percentage of the total cation-exchange capacity.

**BC Soil.** A soil with a B and a C horizon but with little or no A horizon. Most BC soils have lost their A horizons by erosion.

**Bog Soils.** An intrazonal group of soils with a muck or peaty surface soil underlain by a peat; developed under swamp or marsh types of vegetation, mostly in a humid or subhumid climate (1949 system).

**Brown Soils.** A zonal group of soils having a brown surface horizon which grades into lighter-colored soil and finally into a layer of carbonate accumulation; developed under short grasses, bunch grasses, and shrubs in a temperate to cool semiarid climate (1949 system).

**Brunizem (Prairie) Soils.** A zonal group of soils having a dark-colored, granular $A_1$ horizon 6 or more inches thick which rests on a brownish-colored subsoil, commonly having a blocky structure and usually a higher silicate clay content than the adjoining horizons. The organic-matter content of the surface horizon gradually decreases with depth. The exchange complex contains fewer exchangeable H than other cations. These soils are usually developed under grass vegetation in a humid to semihumid climate (1949 system).

**Buffering.** The resistance of a substance to an abrupt change in acidity or alkalinity.

**Bulk Density.** Mass per unit bulk volume of soil that has been dried to constant weight of 105° C. Commonly expressed in grams per cubic centimeter.

**C Horizon.** The unconsolidated rock material in the lower part of the soil profile.

**Calcareous Soil.** Soil containing sufficient calcium carbonate (often with magnesium carbonate) to effervesce visibly to the naked eye when treated with hydrochloric acid. Soil alkaline in reaction owing to the presence of free calcium carbonate.

**Calcic Horizon.** A horizon of secondary carbonate accumulation more than 6 inches in thickness, has a $CaCO_3$ equivalence of more than 15 percent and at least 5 percent more $CaCO_3$ than the C horizon.

**Cambic Horizon.** An altered horizon resulting from movement of soil particles by frost, roots, and animals to such an extent to destroy original rock structure.

**Caliche.** A more or less cemented deposit of calcium carbonate or of mixed calcium and magnesium carbonate, characteristic of soils of warm or hot desert and semiarid regions.

**Capillary Water.** Water held by adhesion and surface-tension forces as a film around particles and in the capillary spaces. Moves in any direction in which capillary tension is greatest.

**Carbon-Nitrogen Ratio.** The relative proportion, by weight, of organic carbon to nitrogen in soil or organic matter. The number obtained by dividing the percentage of organic carbon by the percentage of nitrogen.

**Carbonate Accumulation, Soil Horizon of.** A developed soil horizon, beneath the surface, containing more calcium (or magnesium and calcium) carbonate than the soil above it or the material below it.

**Catena.** A sequence of soils from similar parent material and of similar age in areas of similar climates but whose characteristics differ because of variations in relief and drainage.

**Cation.** An ion carrying a positive charge of electricity.

**Cation Exchange.** *See* Ion exchange.

**Cation-Exchange Capacity.** The sum of exchangeable cations adsorbed by a soil, expressed in milliequivalents per 100 grams of oven dry soil.

**Chernozem Soils.** A zonal group of soils having a deep, dark-colored to nearly black surface horizon, rich in organic matter, which grades below into lighter-colored soil and finally into a layer of lime accumulation; developed under tall and mixed grasses in a temperate to cool subhumid climate. From the Russian for black earth. Sometimes spelled Tschernosem (1949 system).

**Chestnut Soils.** A zonal group of soils having a dark brown surface horizon which grades below into lighter-colored soil and finally into a horizon of lime accumulation; developed under mixed tall and short grasses in a temperate to cool and subhumid to semiarid climate. They occur on the arid side of Chernozem soils, into which they grade (1949 system).

**Chlorosis.** A condition in plants resulting from the failure of chlorophyll to develop caused by a deficiency of an essential nutrient. Leaves of chlorotic plants range from light green through yellow to almost white.

**Class, Soil.** Classes of soil based on the relative proportion of soil separates.

**Clay.** The small mineral soil grains or particles, less than 0.002 mm in diameter.

**Clay Mineral.** Naturally occurring inorganic crystalline (or amorphous) material in soils or other earthy deposits of clay size—particles usually less than 0.002 mm in diameter.

**Claypan.** A dense and heavy soil horizon underlying the upper part of the soil; hard when dry and plastic or stiff when wet; presumably formed in part by the accumulation of clay brought in from the horizons above by percolating water.

**Colloid Soil.** The term colloid is used in reference to matter, both inorganic and organic, having very small particle size and a correspondingly high surface area per unit of mass. Many mineral colloids exhibit crystalline structure. Colloid comes from the Greek words for glue and like.

**Colluvium.** Heterogeneous deposits of rock fragments and soil material accumulated at the base of comparatively steep slopes through the influence of gravity.

**Concretions.** Hard grains, pellets, or nodules from concentrations of compounds in the soil that cement the soil grains together. The composition of some concretions is unlike that of the surrounding soil. Concretions can be of various sizes, shapes, and colors.

**Consistence.** The combination of properties of soil material that determine its resistance to crushing and its ability to be molded or changed in shape. Consistence depends mainly on the forces of attraction between soil particles. Consistence is described by such words as loose, friable, firm, soft, plastic, and sticky.

**Consumptive Use.** The water used by plants in transpiration and growth, plus water vapor loss from adjacent soil or snow, or from intercepted precipitation in any specified time. Usually expressed as equivalent depth of free water per unit of time.

**Contour.** An imaginary line connecting points of equal elevation on the surface of the soil. A contour terrace is laid out on a sloping soil at right angles to the direction of the slope and level throughout its course. In contour plowing, the plowman keeps to a level line at right angles to the direction of the slope, which usually results in a curving furrow.

**Cover Crop.** A crop used to cover the soil surface; to decrease erosion and leaching, shade the ground, and offer protection to the ground from excessive freezing and heaving.

**Creep.** Slow mass movement of soil and soil material down steep slopes, primarily by gravity but facilitated by saturation with water and alternate freezing and thawing.

**Deflocculate.** To separate or break down soil aggregates into their component particles. Usually refers to particles of colloidal dimensions.

**Denitrification.** The reduction of nitrates to nitrites, ammonia, and free nitrogen, as in soil by soil organisms, particularly certain anaerobic organisms (those living or active in the absence of air or free oxygen).
**Dispersion.** The destroying of soil structure (breaking up the granules) so that each individual soil particle behaves as a unit.
**Drift.** Material of any sort deposited in one place after having been moved from another. For example, glacial drift includes glacial deposits, unstratified (till) and stratified glacial outwash materials.
**Drumlin.** An oval hill of glacial drift, normally compact and unstratified, usually with its longer axis parallel to the movement of the ice responsible for its deposition.
**Duripan.** An indurated horizon cemented with such materials as silica, calcium carbonate, or aluminum silicate.
**Dust Mulch.** A loose, dry surface layer of a cultivated soil.
**Eluviation.** The movement of soil material from one place to another within the soil, in solution or in suspension, where there is an excess of rainfall over evaporation. Horizons that have lost material through eluviation are referred to as eluvial, and those that have received material as illuvial.
**Entisols.** Soils that have no diagnostic pedogenic horizons.
**Epipedon.** A diagnostic surface horizon used to classify soils in the comprehensive (1960) system.
**Erosion (Geological or Normal).** Erosion suffered by land in its natural state undisturbed by human activity. Erosion taking place on land used for crop production or on land whose natural condition has otherwise been modified through man's activities is sometimes referred to as "soil erosion."
**Evapotranspiration.** The loss of water from a soil by evaporation and plant transpiration.
**Exchangeable-Sodium Percentage.** The extent to which the adsorption complex of a soil is occupied by sodium. It is expressed as follows:

$$\text{ESP} = \frac{\text{Exchangeable sodium (meq/100 g soil)}}{\text{Cation-exchange capacity (meq/100 g soil)}} \times 100$$

**Fallow.** Crop land left idle in order to restore productivity, mainly through accumulation of water, nutrients, or both. Summer fallow is a common practice before cereal grain in regions of limited rainfall. The soil is tilled for at least one growing season to control weeds, to aid decomposition of plant residues, and to encourage the storage of moisture for the succeeding grain crop. Bush or forest fallow is a rest period under woody vegetation between crops.
**Family, Soil.** A category in soil classification between series and great

soil group; a taxonomic group of soils having similar profiles, composed of one or more distinct soil series (1949 system).

**Fertility (of Soil).** The quality that enables a soil to provide the proper compounds, in the proper amounts and in the proper balance, for the growth of specified plants when other factors, such as light, temperature, and the physical conditions of the soil, are favorable.

**Fertilizer.** A material supplying one or more of the plant nutrients in a condition suitable for use by plants.

**Field Capacity.** The amount of water held in soil after the gravitational water has drained away.

**Fixation.** The conversion of a soluble material, such as a plant nutrient like phosphorus, from a soluble or exchangeable form to a relatively insoluble form.

**Flocculate.** To aggregate individual particles into small groups or clusters. Usually refers to particles of colloidal dimensions.

**Flood Plain.** The nearly flat surface subject to overflow along streams.

**Fragipan.** Dense and brittle pans or layers in soils that owe their hardness mainly to extreme density or compactness rather than to high clay content or cementation. Removed fragments are friable, but the material in place is so dense that roots cannot penetrate it and water moves through it very slowly because of small pore size.

**Friable.** Easily crumbled in the fingers; nonplastic.

**Gilgai.** Microrelief of clays that have high coefficients of expansion and contraction with changes in moisture; usually a succession of microbasins, and microknolls in nearly level areas or of microvalleys and microridges that run with the slope. Common feature of Vertisols.

**Gley.** A soil horizon in which the material is usually bluish gray or olive gray, more or less sticky, compact, frequently without definite structure; developed under the influence of excessive moisture.

**Gravitational Water.** The water that moves under the force of gravity; it is not retained by the soil or is converted to capillary water when it drains into unsaturated soil layers.

**Gray-Brown Podzolic Soils.** A zonal group of soils having a comparatively thin organic covering and organic-mineral layers over a grayish brown leached layer which rests upon an illuvial brown horizon; developed under deciduous forest in a temperate moist climate (1949 system).

**Gypsic Horizon.** A horizon of secondary calcium sulfate enrichment that is more than 6 inches thick, has at least 5 percent more gypsum than the C horizon or underlying stratum and in which the product of the thickness in inches and the percent gypsum is equal to or greater than 60 percent-inches.

**Hardpan.** A hardened or cemented soil horizon. The soil may have any texture and is compacted or cemented by iron oxide, organic material, silica, calcium carbonate, or other substances.

**Heterotrophic.** Capable of deriving energy for life processes only from the utilization of organic carbon compounds and incapable of using carbon dioxide as the sole carbon source for cell synthesis. Contrast with autotrophic.

**Histic Epipedon.** A surface horizon very high in organic matter and saturated with water at some time during the year unless artificially drained.

**Histisols.** Soils formed from organic material, formerly called organic soils.

**Horizon, Soil.** A layer of soil approximately parallel to the land surface with more or less well-defined characteristics that have been produced through the operation of soil-building processes. Each layer differs from the one above or below in some characteristic.

**Humic Gley Soil.** Includes Wiesenboden and those soils formerly grouped with Half-Bog soils having a thin muck or peat A0 horizon and an A1 horizon. Developed in wet meadows and forested swamps (1949 system).

**Humus.** The well-decomposed, more or less stable part of the organic matter of the soil.

**Hydrolysis.** A double decomposition reaction in which water is a reactant.

**Hygroscopic Coefficient.** The maximum amount of water (percentage based on weight of dry soil) adsorbed on the surface of soil particles from an atmosphere slightly below 100 percent relative humidity.

**Hygroscopic Water.** Water which is adsorbed from atmospheric water vapor and held on the surface of particles by forces of adhesion.

**Igneous Rock.** A rock produced through the cooling of melted mineral material.

**Illuvial Horizon.** A soil horizon that has received material from other portions of the soil profile through the process of eluviation.

**Immature Soil.** A young soil. A soil lacking a well-developed profile.

**Immobilization (of Plant Nutrients).** The conversion of an available plant nutrient in the soil from an inorganic to an organic form in living tissue. Thus the addition of fresh straw or sawdust to the soil may greatly increase the number of bacteria. These remove available nitrogen and phosphorus from the soil and immobilize those nutrients within their cells.

**Inceptisols.** Soils that are usually moist with pedogenic horizons of alteration of parent material but not of illuviation.

**Infiltration.** The process by which water enters the soil through the surface. The rate at which water soaks into the soil is called the "infiltration capacity."

**Intrazonal Soil (obsolete term).** Any of the great groups of soils with more or less well-developed soil characteristics that reflect the dominating influence of some local factor of relief, parent material, or age over the normal effect of the climate and vegetation (1949 system).

**Ion.** An electrically charged element or group of elements in an electrolyte. An electrically charged particle.

**Ion Exchange.** The replacement in a colloidal system of an ion by another with a charge of the same sign. Cation exchange refers to the interchange of positively charged ions (cations), and anion exchange refers to the replacement of negatively charged ions (anions) by other anions.

**Irrigation Efficiency.** The ratio of water consumed by crops of an irrigated farm or project to the water diverted from a river or other natural source into the farm or project canals.

**Kame.** An irregular ridge or hill of stratified glacial drift.

**Lacustrine Deposits.** Materials deposited by lake waters.

**Land, Arable.** Land which, in its present condition, is physically capable, without further substantial improvement, of producing crops requiring tillage.

**Landscape (As Used in Soil Geography).** The sum total of the characteristics that distinguish a certain area on the earth's surface from other areas. These characteristics are the result not only of natural forces but of human occupancy and use of the land.

**Latosol (Laterite) Soils.** It has been recommended that the use of the term "laterite" be discontinued as defined in "Soils and Men." The term "latosol" has been suggested as a replacement to refer to the more friable and iron-rich soils of the hot, moist, or wet-dry tropics referred to as "red-loams" in European, Asiatic, and Australian literature (1949 system).

**Lattice Structure.** Orderly arrangement of atoms in crystalline material.

**Leaching.** Removal of materials in solution.

**Lime, Agricultural.** Any compound of calcium or magnesium or both used to correct the harmful effects of acid soils on plant growth.

**Limestone.** A general name for rocks composed essentially of calcium carbonate. There are several kinds of limestone, and they vary in physical and chemical composition.

**Lithosols.** An azonal group of soils having no clearly expressed soil morphology and consisting of a mass of rock fragments from consolidated rocks which are imperfectly weathered. They are found primarily on steeply sloping land (1949 system).

**Loess.** A fine-grained aeolian deposit dominantly of silt-sized particles.

**Luxury Consumption.** The intake by a plant of an essential nutrient in amounts exceeding what it needs. Thus if potassium is abundant in the soil, alfalfa may take in more than is required.

**Marine Material.** Material deposited in the waters of oceans and seas and exposed by elevation of the land or the lowering of the water level.

**Marl.** A soft, earthy deposit consisting chiefly of calcium carbonate mixed with sand, clay, organic matter, and other impurities in varying proportions. Frequently used as a liming material.

**Mature Soil.** A soil with well-developed characteristics, produced by the natural processes of soil formation, and in equilibrium with its environment.

**Mechanical Analysis.** The separation by mechanical means of the different size groups (separates) and the determination of the percentage of each group in a given soil sample (or a particle-size analysis).

**Metamorphic Rock.** A rock the constitution of which has undergone pronounced alteration. Such changes are generally effected by the combined action of pressure, heat, and water, frequently resulting in a more compact and more highly crystalline condition of the rock. Gneiss, schist, and marble are common examples.

**Mineral.** A naturally occurring combination of inorganic elements and ions in the form of salts or other compounds either in crystalline or amorphous condition.

**Mineral Soil.** A general term used in reference to any soil composed chiefly of mineral matter. The mineral material is dominant over the organic matter in determining the characteristics of the soil.

**Mineralization.** The conversion of an element that is in organic combination to the available form as a result of microbial decomposition.

**Mollic Epipedon.** A surface horizon that is thick, dark-colored, high base saturation and strong structure so that the soil is "soft" instead of hard and massive when dry.

**Mollisols.** Soils with mollic epipedons and have 50 percent or greater base saturation. Includes soils formerly classified as Brunizem, Chernozem, and Chestnut (also some Brown, etc.).

**Morphology, Soil.** The physical constitution of the soil, including the texture, structure, porosity, consistence, and color of the various soil horizons, their thickness, and their arrangement in the soil profile.

**Mottled (Mottling).** Irregularly marked with spots of different colors.

**Muck Soil.** An organic soil composed of highly decomposed organic material.

**Natric Horizon.** Illuvial horizon of silicate clay accumulation, over 15 percent exchangeable sodium and columnar or prismatic structure.
**Necrosis.** Death associated with discoloration and dehydration of all or parts of plant organs, such as leaves.
**Nitrification.** Formation of nitrates from ammonia as in soils by soil organisms.
**Nitrogen Fixation.** In soils, the assimilation of free nitrogen from the air by soil organisms, eventually making the nitrogen available to plants.
**O Horizon.** Organic horizon of mineral soils.
**Ochric Horizon.** Thin, light-colored horizon low in organic matter.
**Order.** The highest category in soil classification.
**Organic Soils.** Soils containing organic matter in sufficient quantities to dominate the soil characteristics. Frequently all soils containing 20 per cent or more organic matter by weight are arbitrarily designated as organic soils. Many now classified as Histosols.
**Ortstein.** The B horizon of Podzols that are cemented by the accumulated sesquioxides and/or organic matter.
**Osmotic.** A type of pressure exerted in living bodies as a result of unequal concentration of salts on both sides of a cell wall or membrane. Water will move from the area having the least salt concentration through the membrane into the area having the highest salt concentration and, therefore, exert additional pressure on one side of the membrane.
**Oxic Horizon.** An altered subsurface horizon consisting of a mixture of hydrated oxides of iron or aluminum and 1:1 clays and/or quartz.
**Oxisols.** Soils with oxic horizons. Many formerly classified as Latosols.
**Pans.** A layer or soil horizon within a soil that is firmly compacted or is very rich in clay. Examples include hardpans, fragipans, claypans, and traffic pans.
**Parent Material.** The slightly altered or unweathered material from which the soil was formed.
**Parent Rock.** The rock from which parent materials of soil are formed.
**Particle Density.** The average density of the soil particles not including the pore space. Usually expressed in grams per cubic centimeter.
**Peat.** Unconsolidated soil material consisting primarily of undecomposed or slightly decomposed organic matter accumulated under conditions of excessive moisture.
**Ped.** An individual natural soil aggregate such as a crumb, prism, or block, in contrast to a clod, which is a mass of soil brought about by digging or other disturbance.
**Pedalfer (obsolete term).** A soil in which there is a zone of alumina and

iron oxide accumulation in the profile but with no horizon of carbonate accumulation.

**Pedocal (obsolete term).** A soil with a horizon of accumulated carbonates in the soil profile.

**Pedology.** The science of the soil in which different soils are considered natural units and attention is given to development, physical, chemical, and biological relationships, and to their dynamic nature.

**Percolation.** The process whereby water moves through the soil in response to the force of gravity and the downward pull of soil pores.

**Permafrost.** Permanently frozen material underlying the solum or a perennially frozen soil horizon.

**pH.** A notation to designate or indicate the degree of acidity or alkalinity of systems. Technically, the common logarithm of the reciprocal of the hydrogen-ion concentration (grams per liter) of a system.

**Phase, Soil.** That part of a soil type having minor variations in characteristics such as relief, stoniness, or erosion which distinguish it from characteristic's normal for the type of soil.

**Plaggen Epipedon.** A surface horizon that is very thick, over 20 inches, produced by long continued manuring.

**Planosol.** An intrazonal group of soils with leached surface horizons underlain by strongly contrasting B horizons which are more strongly illuviated, cemented, or compacted than those in normal soils. They are developed in humid or subhumid climates and under grass or forest cover.

**Plinthite.** A highly weathered mixture of sesquioxides of iron and aluminum with quartz and other diluents which occurs as red mottles and that changes irreversibly to hardpan on exposure to drying.

**Podzol Soils.** A zonal group of soils having an organic mat and a very thin organic-mineral layer above a gray leached layer which rests upon an illuvial dark brown horizon; developed under the coniferous or mixed forest or under heath vegetation in a temperate to cold moist climate. Iron oxide and alumina, and sometimes organic matter, have been removed from the A and deposited in the B horizon (1949 system).

**Porosity, Soil.** The degree to which the soil mass is permeated with pores or cavities. It is expressed as the percentage of the total volume of the soil.

**Prairie Soils.** See Brunizem soils.

**Primary Mineral.** A mineral which occurs, or originally occurred, in igneous rocks; examples are micas and feldspars.

**Productivity (of Soil).** The capability of a soil for producing a specified plant or sequence of plants under a specified system of management.

**Profile, Soil.** A vertical cross section of the soil from the surface into the underlying unweathered material.

**Puddle.** To deflocculate or to destroy the structure of the soil.

**Reduction.** The process by which an element loses valence, that is, gains electrons. A decrease in positive valence or an increase in negative valence.

**Regosols.** An azonal group of soils consisting of deep soft mineral deposits (unconsolidated rock) in which few if any clearly expressed soil characteristics have developed; largely confined to recent sand dunes, loess, and glacial drift of steeply sloping lands (1949 system).

**Relief.** The elevation of inequalities of a land surface considered collectively.

**Rhizosphere.** The bounding surface of plant roots. The soil space in the immediate vicinity of the plant roots in which abundance and composition of the microbial population are influenced by the presence of roots.

**Rock.** Any relatively homogeneous mass of mineral material that forms a portion of the earth's crust, including loose, incoherent masses as well as solid masses. The common concept that material must be hard, as granite or limestone, to be classed as rock is not correct from a geological point of view.

**Salic Horizon.** A horizon at least 6 inches thick with secondary enrichment of salts more soluble in cold water than gypsum.

**Saline Soil.** A soil containing sufficient soluble salts to impair its productivity.

**Salt.** The product, other than water, of the reaction of a base with an acid.

**Sand.** Small rock or mineral fragments having diameters ranging from 2 to 0.05 mm.

**Sedimentary Rock.** A rock composed of particles deposited from suspension in water.

**Separate, Soil.** One of several groups of soil particles having definite size limits.

**Series, Soil.** A group of soils having genetic horizons similar in differentiating characteristics and arrangement in the soil profile, except for the texture of the surface soil, and developed from a particular type of parent material. A series may include two or more soil types which differ from one another only in the texture of the surface soils.

**Silt.** Small mineral soil grains the particles of which range in diameter from 0.05 to 0.002 mm (or 0.02 to 0.002 mm in the international system).

**Sodic Soil.** A soil in which the exchangeable sodium percentage is 15 percent or more or that contains sufficient sodium to adversely affect soil physical properties and permeability.

**Sodium-Adsorption-Ratio** *(SAR)* =

$$\frac{Na^+}{\sqrt{\frac{Ca^{2+} + Mg^{2+}}{2}}}$$

where the concentrations of the cations are in milliequivalents per liter.

**Soil.** The natural medium for the growth of land plants on the surface of the earth. A natural body on the surface of the earth in which plants grow; composed of organic and mineral materials.

**Soil Monolith.** A vertical section taken out of a soil profile and mounted for display or study.

**Soil Morphology.** The constitution of the soil body as expressed in the kinds, thicknesses, and arrangement of the horizons in the profile, and in the texture, structure, consistence, porosity, and color of each horizon.

**Soil Survey.** The systematic examination, description, classification, and mapping of soils in an area. Soil surveys are classified according to the kind and intensity of field examinations.

**Solonchak.** An intrazonal group of soils having a high concentration of soluble salts; usually light colored; without characteristic structural form; developed under salt-loving grass or shrub vegetation mostly in an arid, semiarid, or subhumid climate (1949 system).

**Solonetz Soils.** An intrazonal group of soils having a variable surface horizon of friable soil underlain by dark, hard soil, ordinarily with columnar structure; usually highly alkaline; developed under grass or shrub vegetation, mostly in a subhumid or semiarid climate (1949 system).

**Soloth Soils.** An intrazonal group of soils having a thin surface layer of brown, friable soil over a gray leached horizon resting on a brown or dark brown horizon. They develop in a semiarid or subhumid climate under shrubs, grasses, or a mixture of grass and trees (1949 system).

**Solum.** The upper part of the soil profile, above the parent material, in which the processes of soil formation are taking place. In mature soils this includes the A and B horizons, and the character of the material may be and usually is greatly unlike that of the parent material beneath. Living roots and life processes are largely confined to the solum.

**Spodic Horizon.** Illuvial accumulation of free iron and aluminum oxides and organic matter.

**Spodosols.** Soils with spodic horizons. These soils are formed in acid, mainly coarse-textured materials in humid and mostly cool but also in tropical climates. Many soils formerly classified as Podzols.

**Strip Cropping.** Strip cropping is the practice of growing ordinary farm crops in long strips of variable widths, across the line of slope, approximately on the contour, in which dense-growing crops are seeded in alternate strips with clean-tilled crops.

**Structure, Soil.** The morphological aggregates in which the individual soil particles are arranged. Common types include plocky, platy, granular, and prismatic aggregates.

**Subsoil.** Roughly, that part of the solum below plow depth.

**Sulfur Oxidation.** The biochemical oxidation of sulfur and its compounds.

**Surface Soil.** That part of the upper soil of arable soils commonly stirred by tillage implements or an equivalent (5 to 8 inches) in nonarable soils.

**Terrace (for Control of Runoff, Soil Erosion, or Both).** A broad-surface channel or embankment constructed across the sloping lands, on or approximately on contour lines, at specific intervals. The terrace intercepts surplus runoff to retard it for infiltration or to direct the flow to an outlet at nonerosive velocity.

**Texture, Soil.** The relative proportion of the various size groups of individual soil grains. The coarseness or fineness of the soil.

**Till, Glacial.** A deposit of earth, sand, gravel, and boulders transported by glaciers. Till is unstratified.

**Till Plain.** A level or undulating land surface covered by glacial till.

**Tilth, Soil.** The physical condition of the soil in relation to plant growth. A term indicating the conditions of soil structure produced by tillage or cultivation.

**Top Soil.** A general term applied to the surface portion of the soil, including the average plow depth (surface soil) or the A horizon, where this is deeper than plow depth.

**Transpiration Ratio.** The pounds of water transpired by a plant per pound of dry matter produced above ground.

**Tundra Soils.** A zonal group of soils having dark brown, highly organic layers over grayish horizons which rest on an ever-frozen substratum; developed under shrubs and mosses in cold, semiarid to humid climates, that is, in Arctic regions (1949 system).

**Type, Soil.** A group of soils having genetic horizons similar in differentiating characteristics, including texture and arrangement in the soil profile, and developed from a particular type of parent material.

**Ultisols.** Soils that are low in bases (less than 35 percent base saturation) in argillic subsurface horizons and to considerable depths. Normally, bases are removed by leaching faster than bases are released by weathering.

**Umbric Horizon.** A thick dark-colored surface horizon (epipedon) similar to a mollic horizon but with low base saturation and may be highly H saturated.

**Varnish, Desert.** A glossy coating of dark-colored compounds, probably composed mainly of iron oxides, covering pebbles, stones, and large rock surfaces exposed in hot deserts.

**Vertisols.** Clayey soils with high shrink-swell potential that have wide, deep cracks when dry. Most of these soils have distinct wet and dry seasons. Formerly classified as Grumusols.

**Vesicular Structure.** Soil structure characterized by round or egg-shaped cavities or vesicles.

**Volume Weight (obsolete term).** *See* bulk density.

**Water Logged.** State of being saturated with water.

**Water Requirement (of Plants).** Generally, the amount of water required by plants for satisfactory growth during the season.

**Water Table.** The upper limit of the part of the soil or underlying material wholly saturated with water.

**Weathering.** The physical and chemical disintegration or decomposition of rocks and minerals under natural conditions.

**Wilt Point.** The percentage of water in the soil (based on dry weight of the soil) when permanent wilting of plants occurs. It refers to that moisture content at which soil cannot supply water at a rate sufficient to maintain the turgor of a plant and it permanently wilts.

**Zonal Soil (obsolete term).** Anyone of the great groups of soils having well-developed soil characteristics that reflect the influence of the active factors of soil genesis—climate and living organisms, chiefly vegetation (1949 system).

# Index


A horizon, defined, 5–7, 204
ABC soil, 6
Absorption processes, of nutrients, 15, 83
  of water, 70, 83
AC soil, 210
Acid soils, causes, 180
  factors limiting plant growth, 194
  relation to base saturation, 182
Acidity, effect, on earthworms, 187
    on microflora, 104
  increased, by fertilizers, 322, 323
    by sulfur, 200, 201
  neutralized by lime, 195
  role, of aluminum, 181
    of exchangeable cations, 180
Acre furrow slice, weight, 48
Actinomycetes, cause of potato famine, 114
  characteristics, 102
  distribution in profile, 105
  weight in soils, 102
Adhesion, water, 65, 74
Aeration, effect, on microflora, 104
    on nutrient uptake, 279
  need for in paddy soils, 109
  versus plant growth, 52
Aeration pore space, 50, 52
Aerial photograph, 234
Agric horizon, defined, 241
Aerobic bacteria, 100
Air, amount in soil, 2
Air movement, relation to pore size, 51
Alfalfa, optimum pH, 190
  soil-root monolith, 18
  water removal from soil, 82, 87
Alfisols, development related to Planosols,
    209
  general discussion, 258–259
  physical properties, 36
  world area, 247
Algae, characteristics, 102
  classification of, 98
  importance in soils, 102
Alluvial sediments, 219, 220
Alluvial soils, 244
Aluminum, solubility versus soil pH, 186
  as source of acidity, 181
  toxicity, 188, 194
Aluminum octahedron, structure of, 161, 164
Amazon Basin, potential of arable land, 422
Ammonia, formation in manure, 331
  liquor, 302
  loss from manure, 335
  manufacture, 300
Ammonification, 109

Ammonium, in manure, 331
production in soil, 106
Ammonium nitrate, 301, 302
Ammonium phosphate, 301–303
Ammonium sulfate, effect on soil pH, 322, 324
as fertilizer, 301, 302
Amoebae, 120
Anaerobic bacteria, 100
Anatase, weathering stage, 158
Anhydrous ammonia, manufacture, 300
properties, 302
Animal wastes, disposal in soil, 341–343
production of, 341, 342
Animals, kinds and abundance in soils, 119
Anion, defined, 15
Anion exchange, importance, 176
source, 176
Anions, 15, 16
Anthropic horizon, 241
Ants, activities in soils, 123
AR soil, 5, 244
Arable land, by continents, 422
meaning, 418
needed per person, 419
potentials, 421, 422
in world, 418, 419
Archeology, relation to earthworms, 122
Argentina pampa, Mollisols, 249
Argids, in western states, 380
Argillic horizons, in Aridisols, 379
defined, 35, 241
relation to Bt horizons, 35
Arid region, soils and agriculture, 375–393
use of land, 381–383
vegetation, 376
Aridisols, age of, 381
general discussion of, 247–248
properties, 379
in western United States, 379–381
world area, 247
Arthropods, 119
Autotrophic bacteria, 99
Availability of plant nutrients, 15
Available water, 79, 84
Azonal soil, 238
Azotobacter, 110

B horizon, defined, 6, 7, 204
Bacteria, classification of, 98
distribution in profile, 105
kinds in relation to oxygen supply, 100
pH preferences, 104
weight in soils, 99
Bacteroids, 111, 112
Basalt, as parent material, 226
Base saturation, of Alfisols, 259
defined, 173
of forest nursery soils, 174
of Mollisols, 248
relation to pH, 182
in Tama silt loam, 172
of Ultisols, 260
Basic slag, 304
Belgian Congo, potential arable land, 422
Bg horizon, 230
Biological transformation of nutrients, 106–109
Black alkali, 385
Blacklands, soil structure, 38
Blueberries, optimum pH, 190
Bluegrass region soils, 218
Bog soils, availability related to pH, 186; *see also* Organic soils
Boron, deficiency symptoms, 292
in plants, 288
Brown soils, *see* Aridisols; Mollisols
Bt horizon, defined, 35, 206
Buffering, 192
Bulk density, defined, 47
determination, 47
in horizons of Miami loam, 48
of organic soils, 48
of peats, 145
range in soils, 47

C horizon, defined, 6, 7, 204
Calcareous soils, defined, 182
management problems, 201
pH of, 183
Calcic horizon, in Aridisols, 379, 380
Calcimorphic soil, 239
Calcium, atomic radius, 154
availability versus soil pH, 186
deficiency in soils, 194
deficiency symptoms, 293
percentage of earth's crust, 151
in percolate, 81
in plants, 288

Calcium carbonate, accumulation in Aridisols, 247
effect on soil, 182
equivalent, 197
Caliche, 251
Cantaloupe, effected by mulch, 62
water removal from soil, 82
Capillarity in soils, 68, 69
Capillary pore space, 50
Capillary water, 79
Carbon dioxide, in soil air, 52
Carbonic acid, effect on soil pH, 184
Carbon-nitrogen ratio, effect on available nitrogen, 141
of humus, 130
of organic materials, 141
Carboxyl, source of exchange sites, 130
Catena, 231
Cation, definition, 15
Cation constancy principle, 283
Cation exchange, from carboxyl, 130
defined, 169
from exposed edges of clays, 166
from isomorphous substitution, 164
origin in clay minerals, 166
replaceability series, 170
Cation exchange capacity, of clays, 167, 171
determination, 177
due to pH dependent charge, 177
effect on altering pH, 192
of forest nursery soils, 174
of horizons of Tama silt loam, 172
of organic matter, 171
related, to climate, 214
to plant growth, 174
of soils, 171
Cations, effect on structure, 42
as plant nutrients, 15, 16
radii, 154, 170
replacing power, 170
Chernozem soils, *see* Mollisols
Chestnut soils, *see* Mollisols
China, land use, 10
loess deposits, 224
Chlorine, in plants, 288
from precipitation, 297
Classification of soil, 277
Clay, distribution in Planosols, 52
effect, on cation exchange capacity, 171
on organic matter, 128
in Hawaiian soils, 169
importance, 28
movement in soils, 7, 35
as a separate, 28, 29
size limits, 29
as a soil class, 32
Clay minerals, adsorption of enzymes, 128
effect on cation exchange capacity, 171
origin, 159
in Oxisols, 262
structure and properties, 160–167
type related to climate, 215
in Vertisols, 264
Clay-pan soil physical properties, 51, 52
related to Mollisols, 210
Clay-skins, 35
Climate, effect on organic matter, 131, 212
influence on water use by crops, 84–87
in relation to soil development, 212–215
relation to soil productivity, 24
Clostridium, 110
Cobalt, in animal feed, 279
Colloids, effect on structure, 42
Color, factors affecting, 59
influence on soil temperature, 61
measurement, 59
mottled, 60
relation to drainage classes, 96
Composting, 143
Consumptive water use, 85
Contour cultivation, 362
Copper, amounts in crops, 276
availability versus pH, 186
deficiency symptoms, 293
in plants, 288
Corn, dry matter accumulation, 286
nitrogen content of organs, 286
nutrient accumulation, 285
nutrient composition, 276
seasonal use of water, 87
world production, 254
yields related to texture, 33
Corrals, as source of nitrate, 115
Crop rotations, effect on structure, 44, 45
for water erosion control, 357, 358
for wind erosion control, 369–371
Cultivated land, in United States, 416
Cultivation, to conserve water, 88
frequency versus yield, 56, 57
weed control, 55

Cutans, 35

Damping-off disease, 187
DDT, degradation in soils, 116
Decomposition of organic matter, 106–109
Deficiency symptoms, indicators of
fertilizer need, 313
of plants, 290–294
Denitrification, 108, 109
Desert pavement, 248
Deserts, erosion on, 382
soils and agriculture, 375–393
vegetation, 376
Detergents, degraded in soils, 118
Dewater, *see* Drainage
Diagnostic horizons, definitions, 240
Diammonium phosphate, 303, 305
Disease, caused by nematodes, 121
by soil organisms, 113
Dokuchaev, 4
Drainage, classes, 95
depth and spacing of tile, 92
effects on plants, 89
effect on soil development, 230
on soil temperature, 61
loss of nutrients in, 81
related, to plants, 96
to rooting depth, 96
of sewage effluent in soil, 95
systems, 91
Drip irrigation, 412
Drumlins, 223
Dune sand, 223
Dust, in desert regions, 247

Earth's crust, chemical composition, 150
most abundant elements, 154
Earthworms, abundance, 121
affected by acidity, 187
archeological relations, 122
effect on structure, 122
Edina soil, physical properties, 52
location in landscape, 231
Eluviation, defined, 216
effect of vegetation, 217
England, wheat yields 1100–1960, 423
Entisols, in arid regions, 381
discussion of, 242–246
as a stage in soil development, 209
world area, 247

Environmental quality, importance of soil
organisms, 114–118
nitrate pollution, 115, 343
pesticide degradation, 116
relation to fertilizers, 325–328
waste disposal, 117, 118, 341–343
Epipedons, definitions, 241
Erosion, control by, contouring, 362
crops, 357
sod waterways, 367
strip cropping, 363
terracing, 364
damage, 347–349
definitions, 345, 346
in deserts, 382
effect of slope, 355–356
rain drop impact, 352
by water, control of, 349–366
types, 350
by wind, 366–374
Eskers, 222
Essential elements, characteristics, 274
forms absorbed by plants, 16
functions, 286–288
listed, 16, 274
Evapotranspiration, crop variation, 87
defined, 84
percent of precipitation, 63
relation to consumptive use, 85
Evolutionary nature of soil, 5
Exchangeable cations, kinds, and amounts
in soils, 172
related to climate, 214
as sources of H and OH, 180
in Tama silt loam, 172
weight per acre furrow slice, 172

Facultative aerobes, 100
Fallowing, erosion hazard, 369
under shifting cultivation, 146
for water conservation, 88
Family, 271
Farmers of forty centuries, 10
Feldspars, 151, 158, 160
Fertility of soil, affected by termites, 125
defined, 24
related to productivity, 25
Fertilizers, acidity or basicity, 324
application and placement, 317
effect on water use, 86

effects, on crops, 325
on environmental quality, 325–328
on soil pH, 322
evaluation of need, 313–317
formula calculations, 309
guaranteed analysis, 308
laws controlling sales, 313
liquid types, 311
micronutrient carriers, 307
mixed, 307–313
nitrate pollution, 327
nitrogen carriers, 300–302
nutrient interactions, 315
phosphate carriers, 302–304
phosphate pollution, 328
potassium carriers, 304–307
preparation, 308
salt effect, 319
time of application, 320
use, and application, 317–325
related to environmental quality, 325–328
Field capacity, 76
Field crops, optimum pH, 190
Fish, as source of food, 421
Flood plains, 219
Flooding, for irrigation, 401
Flowers, optimum pH, 190
Foliar analysis of plants, 315
Food and population, 420–425
Food production, rate of increase, 420
Forestry, fertilizer use, 324–325
nursery soil requirements, 174
optimum pH for tree species, 190
production related to texture, 33
windthrow, 14
Formica cinera, 124
Four components of soil, 2
Fragipan, symbol for, 206
Freezing and thawing, effect on structure, 43
Fruits, optimum pH, 190
Fungi, classification of, 98
as decomposers of lignin, 101
distribution in soils, 105
kinds and nature in soils, 100
numbers and weight in soils, 100
pH preference, 104
Fungus roots, 113

Gibbsite, formation, 159
protonation, 176
as source of anion exchange, 176
structure of, 176
weathering stage, 158
Gilgai, 265
Glacial parent material, 220
Gleyed layer, 230
Granite, as parent material, 226
Granular structure, 38
Gravel, influence on class name, 34
Gravitational water, 68, 75, 76, 78
Gray-speck disease, 108
Great groups, 270
Great soil groups, of 1949 system, 238–240
Green manuring, effects of, 144
Grumusols, *see* Vertisols
Gully erosion, 351
Gumbotil, 232
Gypsic horizon, in Aridisols, 247
Gypsum, use to reclaim sodic soild, 390

H layer, 207
Halite, weathering of, 154
weathering stage, 158
Halomorphic soils, 239
Harvester ants, 123
Hawaii, kind of clay, 169
soil nematodes, 121
Heat balance, of soil, 60
Heat capacity, of soil, 61
Heat of wetting, 65
Heterotrophic bacteria, 99
Hilgard, 4
Histic horizon, 241
Histosols, discussed, 267–268
world area, 247
Horizon, differentiation processes, 207
symbol designations, 206
Houston clay properties, 266
Humic Gley soils, classified in 1949 system, 239
in relation to Mollisols, 252, 255
Humus, composition, 128
defined, 127
formation, 128
properties, 129
Humus theory, 149
Hydrated cations, adsorbed on clays, 165–167
Hydration, of ions, 170
Hydrogen bonding, in kaolinite, 161

in water, 64, 65
Hydrogen saturation, defined, 173
Hydrogen sulfide, 109
Hydromorphic soils, 239
Hydrosol, 231
Hygroscopic coefficient, 84
Hygroscopic water, 79

Igneous rocks, composition, 151
Illite, formation, 159
potassium bridge in, 165
structure and properties, 165–167
Illuviation, defined, 206
Immobilization, of nutrients, 106, 109
Inceptisole, general discussion, 246
world area of, 247
Infiltration, effect on erosion, 356
importance in irrigation, 397
Inhibitory factors, of plant growth, 13, 17
Intrazonal soil, 238
Ireland potato famine, 114
Iron, availability versus pH, 186
in plants, 288
toxicity, 188
Iron deficiency, on pin oak, 188
symptoms, 293
varietal susceptibility, 201
Iron oxidation, 108
Iron oxide, effect on structure, 43
Iron oxides, color of, 59
Irrigation, ditches, 399
drip method, 412
land for, 396–400
methods, 401–405
quantity of water to apply, 406
salinity control, 407–411
water quality, 407
water resources, 395–396
water use efficiency, 411
Isomorphous substitution, in clay minerals, 164
defined, 164
Israel, drip irrigation, 412

Kainite, as potassium carrier, 305
Kames, 222
Kansas till, 232
Kaolinite, formation, 159
structure and properties, 161, 167
King, F. H., 10

L layer, 207
Lacustrine plains, 222
Lake Agassiz, 222
Lake Bonneville, 222
Land, cultivated acres per person, 422
irrigated, 395
quality in United States, 417
Land capability classes, defined, 358–362
for irrigation, 398
Land use, in an agricultural society, 8
in China, 10
for effluent disposal, 117
in a food gathering society, 7, 420
in Netherlands, 11
in a technological society, 10
in United States, 413–416
Leopold, Aldo, 12
Levees, natural, 219
Liebig, 127, 149
Lime, application rates, 195
forms, 195
methods and time of application, 198
neutralizing value, 198
Lime requirement, determination, 193
Limestone, particle size versus use, 198
weathering rate, 211
Lithosequence, 229
Lithosol, 245
Loess, 223, 224, 229, 230–232
Losses, from soil, 208
Lumbricus terrestris, 122
Luxury consumption, 283
Lysimeter data, 81

Macronutrients, 15
Macropore space, 50
Magnesium, amount in crops, 276
atomic radius, 254
availability versus pH, 186
deficiency symptoms, 293
exchangeable cation, 172, 174
isomorphous substitution, 164
in percolate, 81
in plants, 288
Manganese, amount in crops, 276
availability versus pH, 186
in corn tissue, 323
deficiency on oats, 108
deficiency symptoms, 294
oxidation of, 108
in percolate, 81

in plants, 288
toxicity in plants, 188, 194, 323
Mangum terrace, 365
Manufactured soil, 3
Manure, application, 338
components, 330, 331
composition, 332
versus animal feed, 333
fresh versus rotted, 339
losses in handling, 334–336
management of, 336–341
use of superphosphate for conservation, 338
Marine parent material, 220
Massive soil, 40
Master horizons, 204, 205
Mechanical analysis, 28, 30
Methane, formation, 109
use in fertilizer manufacture, 300, 301
Methemoglobemia, 115
Miami loam, physical properties, 36
Miami soil, as catena member, 231
Micas, in igneous rock, 151
weathering of, 160
weathering stage, 158
Microbial activity, effect on structure, 43
Microfauna, role in decomposition, 120
Microflora, distribution in soil profile, 105
growth in soil, 103
relation to environmental quality, 114
role in pesticide degradation, 116
roles in soils, 105
Micronutrient, fertilizer carriers, 307
Micronutreints, 16
Micropore space, 50
Milliequvalent, defined, 171
Mineralization, 106, 109
Mineral matter, amount in soil, 2
Mineralogical composition, of rocks, 151
versus weathering stages, 157
Minimum tillage, 56
Mohave soil properties, 379
Mollic horizon, defined, 241
Mollisols, development of, 210
properties, location and use, 248–255
relation to other soils, 243
world area, 247
Mollusks, 120
Molybdenum, availability versus pH, 185
content in cauliflower leaves, 187
deficiency symptoms, 294
in plants, 283
Monoammonium phosphate, 303, 305
Montmorillonite, affect on soils, 33
formation, 159
structure and properties, 162, 167
in Vertisols, 264
weathering stage, 158
Moraines, 221
Mottled soil colors, 60, 230
Mottling, relation to drainage classes, 96
Muck soil, defined, 35
effect on roads, 2
Mulches, effect, on soil temperature, 61
on weeds, 62
Mushrooms, 101
Mycelium, of fungi, 100
Mycorrhizae, importance, 113
types, 113

Natric horizon, defined, 241
Nematodes, 120, 121
Netherlands, land use, 11
subirrigation, 405
Nipe soil, 8, 9, 263
Nitrate, accumulation in soil, 115
form absorbed by plants, 16
versus ground water pollution, 327
permissible in drinking water, 115
pollution, by corrals, 115
by fertilizers, 115, 326–328
by manure, 343
production in soils, 108
in soil percolate, 81, 326
Nitrification, effected by pH, 186
process, 109
Nitrobacter, 108
Nitrogen, amount, in air, 109
in crops, 276
in humus, 130
in manure, 331, 332
in peats, 145
availability versus pH, 186
deficiency symptoms, 290–294
effected by green manuring, 144
effects on plants, 289
fertilizer carriers, 300–302
as nitrate in percolate, 81
role in plants, 287
supplying power of soil, 295
Nitrogen carbon ratio, 130, 140–143

Nitrogen competition, between plants and microflora, 141
Nitrogen cycle, 109
Nitrogen factor, 143
Nitrogen fixation, amounts, 110, 111–113
  nonsymbiotic, 110
  as part of nitrogen cycle, 109
  symbiotic, 111
Nitrosomonas, 108
Nutrients, amount, in crops, 275, 276
    in forest vegetation, 108, 146
  availability affected by pH, 184
  biological transformation, 106–109
  losses in percolate, 80
  micro versus macro, 16
  requirements of microflora, 103
Nutrient recycling, defined, 106
  differences in, 216
  role of animals, 120
Nutrient uptake, factors affecting, 277
  process, 275

O horizon, defined, 204
Oceans, as source of food, 421
Ochric horizon, defined, 241
Octahedral sheet, 161, 164
Olivine, composition, 155
  weathering of, 155
  weathering stage, 158
Onions, pH preference, 190
  water use from soil, 82
Organic gardening, myth, 147
Organic materials, as parent material, 225
Organic matter, affect on soil color, 58
  amount, and distribution in soils,2,131, 216
  content, in grassland and forest soils, 131
    related to climate, 131
    versus texture, 138
  cultivation effects, 134
  decomposition in soil, 106
  equilibrium concept, 137
  in forest and grassland ecosystems, 133
  moisture tension as factor in decomposition, 104
Organic soils, as Histosols, 267
  bulk density, 48
  defined, 34, 35
  development, 139
  preservation of Tollund man, 139
  rate of accumulation, 139
Organisms, affected by soil, 103
  classification, of soil animals, 120
    of soil microflora, 98
  in soils described by Peter Farb, 97
Orthids, in western United States, 380
Outwash plains, 222
Oven dry state, 83
Oxic horizon, defined, 241
Oxide clays, formation, 159
  influence on soil properties, 168
  kinds, 167
Oxisols, general discussion, 262–264
  source of anion exchange in, 176
  world acreage, 247
Oxygen, atomic size, 154
  deficiency, 53, 89, 90
  diffusion rate in water, 89
  importance in soils, 13
  requirement of plants, 16
  in soil air, 52
Oxygen diffusion, importance, 52
Oxygen-silicon tetrahedra, arrangement in minerals, 156
  structure, 155

Paddy rice soils, need for aeration, 90, 109
Paper mill sludge, 142
Parent materials, of soils, 218–229
  role in soil development, 226–229
  stratification of, 228
Particle density, 49
Particle size analysis, 28, 30
Particle size classes, 29
Paulding clay, 44
Peats, properties and use, 145
Peat soil, defined, 35
  formation related to climate, 140
Pedalfers, 240
Pedocals, 240
Pedon, 271
Pennsylvania State University, sewage disposal project, 118
Percolation rate, classes, 51
  defined, 50
  relation to drain field operation, 95
Permeability, classes, 51
  importance, 51
Pesticide degradation, in soils, 116
pH, affect by tree species, 217
  alteration, 189

versus base and H saturation, 182
of calcareous soils, 182
causes in soil, 179
defined, 180
determination with dyes, 196
effect, of carbonic acid, 184
of fertilizers, 322
of nitrification, 186
of sulfur, 184
of sodium, 183
of forestry nursery soils, 174
of peats, 145
preferences of plants, 189
of saline soils, 184
significance, 184
range in soils, 179
pH dependent charge, 177
Phosphorus, amounts in crops, 276
availability versus pH, 186
fertilizer carriers, 303, 304
fixation, 295
deficiency symptoms, 292
effect on plants, 290
role in plants, 287
Phosphorus fixation, by anion exchange, 176
Phosphorus pollution, 328
Physical properties, of soils, 27
Plaggen horizon, defined, 241
Planosols, aeration pore space, 52
clay distribution, 52
permeability, 51, 52
Plant composition, related, to age, 284
to species, 282
effect of fertilizer, 325
of various plant parts, 285
Plant nutrients, effects on growth, 288–290
mobility in plants, 290
Plant requirements, diversity of, 22
Plant-soil water relations, 76
Plant tissue tests, 315
Plasticity, 168
Plow, early history, 53
Podzol, *see* Spodosols
Pollution, from animal wastes, 343
from nitrate, 327
from phosphate, 328
Pore space, calculation of, 49
determination, 49
distribution in profile, 50
importance, 13, 45, 50
types, 50
Potassium, amounts in crops, 276
availability versus pH, 186
deficiency symptoms, 292
effect on plants, 289
equilibrium in soils, 296
fertilizer carriers, 304–307
in percolate, 81
role in plants, 287
Potassium bridge, in illite, 165
Potassium chloride, fertilizer, 307
PPM, calculation of, 327
Profile, diagram of, 205
Protonation, 176
Protozoa, 113
Puddled soil, 40
Puerto Rico, Nipe soil, 9

Quartz, in sand and silt, 33
weathering stage, 158

R horizon, defined, 205
in soils, 5, 7
Raindrop impact, effect on erosion, 352
Regosols, 210, 244
Rendzina, 249, 266
Reservoir silting, 349
Rhizobium, 111
Rhizopus, 100
Rhizosphere, 103
Rockiness, 34
Roots, depth in relation, to drainage, 96
to rainfall, 82
distribution, of alfalfa, 18, 19
of oats, 14, 41
of soybeans, 19, 20
extent of soil contact, 19
extensiveness in soil, 19
Roses, optimum pH, 190
Rotations, for erosion control, 370
structure effects of, 45
Ruffin, Edmund, 262

Saline soils, development of, 383, 384
effects on plants, 387
pH of, 184
properties, 384
reclamation, 388–390
Salinity control, in irrigation, 407

Salt accumulation, in soil, 410
Salt effect, of fertilizers, 319
Sand, importance, 28
 size limits, 29
Sand Hills, subirrigation in, 405
Sandstone, mineralogical composition of, 151
 soils formed from, 5, 245
Sawdust, carbon-nitrogen ratio, 141
Septic tanks, use related to percolation rate, 51
Series, 271
Sewage effluent, disposal in soil, 95, 117
Shale, mineralogical composition, 96
Sheet erosion, 350
Shifting cultivation, land needed per person, 10, 146
 method, 146
 nutrient accumulation, 145
 shrink-swell potential, 33
Shrubs, optimum pH, 190
Silica sheet, 161, 164
Silicate clay, kinds, 159
Silicon, atomic size, 154
 in plants, 282
Silicon-oxygen tetrahedron, 154
Silt, importance, 28
 size limits, 29
Single grained soil, 40
Siphon tubes, use of, 401, 402
Slickensides, 265
Slope, effect, on soil development, 230
 on soil erosion, 355–356
 related to temperature, 62
Snow, source of irrigation water, 396
Sod waterways, 367
Sodic soils, development, 384–386
 effects on plants, 387
 reclamation, 390–393
Sodium, absorption ratio, 407
 effect, on pH, 183
 on structure, 393
 on water quality, 408
Sodium nitrate, equivalent basicity, 324
 as fertilizer carrier, 302
Soil, as four components, 2
 boundaries, 4
 definitions, 1
 evolutionary nature, 5
 as interface, 12
 as living filter, 117
 as medium for plant growth, 13
 natural body, 3
 natural resource, 8
Soil body, 3, 7
Soil classification, 237–243
Soil development, influence of drainage, 230
 in relation, to climate, 212
 to parent material, 218
 to time, 208
 to topography, 229
 to vegetation, 215
 rate of, 211
 stages of, 208
 time required, 209
Soil fertility, compared to productivity, 24
 defined, 24
Soil forming factors, listed, 203
Soil mapping, 232
Soil maps, purpose of, 234
 use in wildlife management, 235
Soil moisture tension, effect on organic matter decomposition, 104
 on plant growth, 82
 expression, 72
 measurement, 73
 versus air in soil, 72
 versus water content, 75
 versus air movement, 78
Soil orders, area in world, 247
 discussion of, 242–268
Soil organisms, *see* Organisms
Soil particles, number and surface area, 29
Soil productivity, concept, 22
 defined, 23
 relation to land use, 25
 versus fertility, 24
Soil profile, 7
Soil resources and population, 413–425
Soil series, number in United States, 203
Soil surveys, types, 233
Soil tests, for fertilizer recommendations, 314
Soil utilization, by plants, 21
 in society, 8–10
Soluble salts, accumulation in soils, 410
 effect on pH, 184
Solum, diagram of, 205
 thickness related to parent material, 227
Soviet Union, Mollisols, 249

Spheroidal weathering, 153
Spodic horizon, defined, 241
Spodosols, development in Alaska, 211
discussion of, 255–258
world area, 247
Springtails, 120
Sprinkler irrigation, 404
Squanto, 127
Stratified soil, water movement in, 93
Strip cropping, for erosion control, 363, 371
Structure, defined, 38
formation, 40
importance, 38
in tropical soils, 45
types, 38
versus crop yields, 44
Stubble mulch culture, 371
Subgroups, 270
Suborders, 1949 system, 238, 239
1960 system, 268, 269
Subirrigation, in Netherlands, 405
in Sand Hills, 405
Sulfur, to acidulate soil, 200
amount in crops, 276
deficiency symptoms, 294
effect on pH, 184
in percolate, 81
in plants, 288
in rainfall, 296
Sulfur oxidation, 108
Sulfur transformations, 108
Summer fallow, 369
Superphosphate, composition, 305
manufacture, 303
use to reduce losses in manure, 338
Support, for plants, 13
Surface area, importance in soils, 29
of soil separates, 29

Tama silt loam, chemical properties, 172
Temperature, control of, 61
effect of mulches, 61
on harvest date, 62
importance, 60
relation to location, 62
to microflora, 103
Tensiometer, 73
Termites, 125
Terraces, for erosion control, 364
stream, 220
Tetrahedral sheet, 156, 160
Texture, alteration of, 37
classes, 30, 31
defined, 27
engineering properties, 33
of Hawaiian soils, 169
importance, 27, 29
influenced by ants, 125
separates, 28, 29
triangle, 32
versus organic matter content, 138
versus crop yields, 33
Texture profile, defined, 35
influence on plants, 35
relation to wind erosion control, 37
Thiobacillus, 108
Tile drains, 91
Till, 222
Till plain, 221
Tillage, for crop residue management, 55
to conserve water, 88
effect on structure, 55
frequency of, 57
of the future, 57
minimum, 56
purposes, 54
for weed control, 54
Tillage and soil properties, 53
Time, in relation to soil development, 208
Topography, in relation to soil development, 229
Toposequence, 231
Trees, effect, on pH, 217
on soil development, 215
effected by mycorrhizae, 113
growth related to soil moisture tension, 83
optimum pH, 190
response to lime, 23
texture influence on growth, 33
Transpiration, by crops, 85
defined, 84
yearly variation, 85
Tropical soils, developmental stages, 209
structure of, 45
Tull, Jethro, 149
Tundra soils, 238, 246

Ultisols, brief definition, 157
general discussion, 260–262
world area, 247

Umbric horizon, defined, 241
Union of Soviet Socialist Republics,
Mollisols, 249
Unloading, 153
Urea, equivalent acidity, 324
manufacture, 301
as nitrogen carrier, 302
Urine, in manure, 330–338

Vegetable crops, optimum pH, 190
Vegetation, effect, on soil organic matter,
131
on soil development, 215
Vermiculite, formation of, 166
Vertebrates, 120
Vertisols, discussed, 264–267
world area, 247
Volatilization losses, from manure, 335–336

Water, absorption by roots, 170
adhesion, 65
adsorbed on soil particles, 167, 170
amount in soil, 2
available to plants, 83
classification of, 79
cohesion, 65
conservation by fallowing, 88
consumptive use, 85
content of peats, 145
content related to texture, 84
content versus tension, 75
drainage, 88, 91
effect on microflora, 104
energy and pressure relationships, 66
energy concept, 64
field capacity, 76
film thickness in soil, 66
gravitational, 68, 76
heat capacity of, 61
hydration of cations, 165–167
hygroscopic coefficient, 84
irrigation efficiency, 411
loss by percolation, 81
transpiration, 84
volatilization, 88
movement, as vapor, 88
in stratified soil, 93
in unsaturated soil, 69
of gravitational water, 76
related to pore size, 50
related to conservation of, 88
nitrate contamination of, 115, 326–328
needs of plants, 15
oven dry state, 83
percolation loss, 80
physical state in soils, 64
removal from soils by plants, 80, 82
requirement of crops, 84
role, nutrient absorption, 83
in structure formation, 42
sewage effluent disposal in soil, 95
structure of, 64
use affected by, cultivation, 88
fallowing, 88
fertilizers, 86
use efficiency, 86
use in United States, 63
wilt point, 80
world annual precipitation, 63
Water table, versus rooting depth, 90
Weathering, affected, by climate, 212
by crystal structure, 153
of bedrock, 5
defined, 152
stages of, 158
Weeds, control by tillage, 54
optimum pH, 190
Weight per acre furrow slice, 48
Wheat, production in the world, 253
water removal from soil, 82
yields in England, 423
White alkali, 384
Wilt point, 80
Wilting, due to oxygen deficiency, 17
Windbreaks, 373
Wind erosion, *see* Erosion
Windthrow, 14

Yeasts, 98

Zinc, amount in crops, 276
availability versus pH, 186
deficiency symptoms, 294
in plants, 288
Zonal soils, 238